Global
Environmental
Change

PETER D. MOORE BSc PhD
Division of Life Sciences
King's College London
University of London

BILL CHALONER FRS
Department of Geology
Royal Holloway
University of London

PHILIP STOTT BA
Department of Geography
School of Oriental and African Studies
University of London

b

Blackwell
Science

© 1996 by
Blackwell Science Ltd
Editorial Offices:
Osney Mead, Oxford OX2 0EL
25 John Street, London WC1N 2BL
23 Ainslie Place, Edinburgh EH3 6AJ
238 Main Street, Cambridge
   Massachusetts 02142, USA
54 University Street, Carlton
   Victoria 3053, Australia

Other Editorial Offices:
Arnette Blackwell SA
   224, Boulevard Saint Germain
   75007 Paris, France

Blackwell Wissenschafts-Verlag GmbH
   Kurfürstendamm 57
   10707 Berlin, Germany

   Zehetnergasse 6
   A-1140 Wien
   Austria

First published 1996

Set by Excel Typesetters, Hong Kong
Printed and bound in Great Britain at the University Press,
Cambridge

The Blackwell Science logo is a
trade mark of Blackwell Science Ltd,
registered at the United Kingdom
Trade Marks Registry

DISTRIBUTORS

Marston Book Services Ltd
PO Box 269
Abingdon
Oxon OX14 4YN
(*Orders*: Tel: 01235 465500
            Fax: 01235 465555)

USA
Blackwell Science, Inc.
238 Main Street
Cambridge, MA 02142
(*Orders*: Tel: 800 215-1000
            617 876-7000
            Fax: 617 492-5263)

Canada
Copp Clark, Ltd
2775 Matheson Blvd East
Mississauga, Ontario
Canada, L4W 4P7
(*Orders*: Tel: 800 263-4374
            905 238-6074)

Australia
Blackwell Science Pty Ltd
54 University Street
Carlton, Victoria 3053
(*Orders*: Tel: 03 9347 0300
            Fax: 03 9349 3016)

A catalogue record for this title
is available from the British Library

ISBN 0-632-03638-9

Library of Congress
Cataloging-in-publication Data

Moore, Peter D.
   Global environmental change/
   Peter D. Moore, Bill Chaloner,
   Philip Stott.
          p.   cm.
   Includes bibliographical references
   and index.
   ISBN 0-632-03638-9
   1. Global environmental change.
I. Chaloner, W. G. (William Gilbert)
II. Stott, Philip Anthony. III. Title.
GE149.M66   1996
   550—dc20

                                          96–6168
                                          CIP

Contents

Preface

There can be little doubt that oversimplification is the mother of misconception, and nowhere is this more true than in the field of global environmental change. Every day we read new stories of the horrors that lie in wait for us around the next corner and the continued use, or partial use and abuse of scientific data leaves many people confused. A state of confusion is less dangerous and is always scientifically preferable to one of misinformed conviction, since it leaves the mind with the humility and openness to new ideas that should characterize the mental condition of a true scientist. Firmly-held misconceptions, on the other hand, are increasingly common and have often been generated by oversimplification through the media of very complex and often incompletely understood issues.

Since the condensation and simplification of scientific thought and opinion can be so dangerous, it is with trepidation that we seek to set out here a very basic, simple, and sometimes personal, account of the current ecological state of the globe and its possible future. One of our basic guidelines in selecting and arranging material is that we can only appreciate the significance of an ecological change if we step back and examine it in its global spatial and temporal context. Ours is, therefore, both a broad and a historical approach.

A clear message that emerges from this approach is that we must avoid becoming caught up in irrational and ill-considered responses to every observed change in the ecology and environment of our planet. The earth is a dynamic system that has been subject to change for the last 4.5 billion years and its life-forms have evolved and perished in response to these changes. The last 2 million years have been particularly unstable in climatic terms and we have no scientific basis for expecting anything other than an unpredictable immediate future, quite apart from any involvement by ourselves in its development. We should not, as a consequence, resort to complacency and a *laissez-faire* attitude to our future environment, but neither can we assume that we can recreate some idyllic (and probably non-existent) past, or even hold on to the present in its current state, avoiding all future change. We are in a position to affect the processes that contribute to future developments, but we need to use this responsibility and this power in a carefully considered manner. Our expectations of life are currently dictated by the patterns of thought that underlie modern, industrialized western society, and we may well have to learn to modify these expectations as we continue to live in a rapidly changing world. Such is the message of this book. Many complex issues and processes are considered here and are simplified, perhaps, we admit, sometimes oversimplified. We hope, however, that those who have the patience to work through these pages will have their minds opened rather than closed by what they read.

Many people have contributed much to the gestation of this book, which has arisen mainly from an undergraduate Intercollegiate Course run for a number of years in the University of London via an interactive closed-circuit television network. Classes in the different University of London Colleges were able to benefit from the specialist expertise of many authorities in different aspects of global change and the need to provide a more permanent record of the approach that emerged from the course became apparent. We who have undertaken the assembly of the book wish to record our gratitude for the inspiration and support provided by contributors to that course, including Rick Battarbee, Nigel Bell, Dickie Clymo, Mick Crawley, David Hall, John Lawton and Lewis Owen.

Many other scientists and friends from around the world have also contributed ideas, criticisms, suggestions, illustrations and examples that have added to the diversity of this book, and we are most grateful to them all. In particular, we would thank David Foster of the Havard Forest for his help with some illustrations. We are also grateful for the (sometimes visually-impaired) faith and the relentless encouragement of Blackwell Science, physically embodied in Susan Sternberg and Ian Sherman. They have periodically chivvied us, constructively criticized us, and even occasionally bribed us with offers of food, all of which have assisted in the development of this work.

It is conventional, finally, for authors to acknowledge the patience of their long-suffering wives and families, but in our case our wives and families have not only suffered long, but deeply, as this book has grown like a carbuncle upon our lives. Hopefully, the excision being complete, we can now enjoy a period of recuperation, but meanwhile we wish to record our heartfelt gratitude for their support and tolerance.

Peter Moore,
Bill Chaloner, Philip Stott
London

Chapter 1

Global Change: Science, Scale and Society

The earth is not a static place—it has always been characterized by change and it almost certainly will always be so. Yet the current changes in the environment are generating a great deal of interest and concern. In this book we are going to ask a number of questions relating to current changes in order to establish whether these concerns are justified and whether remedial action needs to be taken and, if so, what kind of remedial action is appropriate and is liable to meet with any degree of success.

A special problem surrounds the way in which evidence for global environmental change is judged by both the scientific world and the general public. Some very worrying claims have been made about the likely consequences of global warming, and the effects of increased ultraviolet (UV) radiation ensuing from the thinning of the ozone screen around the earth. Much scientific research during the last decade has been funded by governments in response to genuine concerns about the impact of global environmental change in ways that could affect us all. As a result we have learnt a great deal about many matters which previously would not have been thought particularly important, such as the way in which the environment controls the distribution of animals and plants and the effect of low-dosage UV radiation on organisms.

Now, it is a matter of simple observation that bad news gets a bigger audience than good news, and certainly more than no news at all! Fears or predictions of major disasters always make better headlines than either an admission of uncertainty, or new evidence that the anticipated horror may not be as severe as had originally been supposed. As a result, scientists engaged in this field often tend to emphasize the bad news rather than the level of probability (or improbability) attached to their conclusions. The worse the perceived threat, the greater the prospect of further funding

for their research! This makes it all the more important that the evidence for all environmental prediction is viewed dispassionately and critically, and in a spirit of genuine scientific endeavour. In this book we attempt to assess the evidence for environmental change and its impact without apology for any deviation from political correctness in our views.

It would be an oversimplification of the situation, however, if we were to assume that science provides the only approach needed to the global problems we shall discuss. It is claimed that people are often responsible for global changes. It is entirely possible that the concerted efforts of people can cure many of the problems. But the central position of people in the whole concern means that considerations of sociology, politics and economics must play a central role in any future policies. Even scientific research is controlled and limited by the attitudes and behaviour of people.

Many of the most important of the questions relating to environmental change demand a historical perspective before they can be answered. Take climate change, for example. With the wealth of climatic data collected around the world over the past few decades, it is not difficult to draw conclusions about current trends in, let us say, global temperature. It has been firmly established by regular and accurate measurements that the level of carbon dioxide (CO_2) in the atmosphere is rising (see Chapter 5) and this may well result in a tendency for temperatures to rise. Indeed, there are indications (from measurements that are somewhat less reliable than those of CO_2) that the overall global temperature is rising and that the elevation is approximately 0.5°C in the course of the past 100 years. Although this figure on its own tells us nothing of regional fluctuations or the variability of annual temperatures within this time period, it does help to demon-

Fig. 1.1 Glaciers, such as the one shown here in Austria, occupied large areas of the land masses in the northern hemisphere only 20 000 years ago. 'Ice-age' conditions have been the 'normal' state of affairs over the past million years of the earth's history.

strate that conditions on earth are not static from one century to the next.

So the starting point for the study of environmental change is the analysis of observations that are documented. Before we can decide whether any changes observed are a matter for concern or not, we must know something of longer-term changes, so that we can set the information into an appropriate historical context. Is a temperature change of half a degree Celsius per century the kind of fluctuation one might reasonably expect in the light of general climatic behaviour? Does temperature swing up and down around some mean figure, or is this current trend a part of a long-term change in climate? Is it, on the other hand, the reversal of a former trend, in which case can we postulate any mechanisms that might account for such a change? Not only are directions of change important, but also rates of change. Is the currently observed rate similar to those that have been found to operate in the past or is it a significantly accelerated or decelerated one?

Scale is also an important factor here because we may find that the current trend is not typical of yet longer-term cycles. Back in the 12th century AD, for example, conditions in the northern hemisphere were probably warmer than they are now, but they have undergone a severe dip in the intervening centuries. The evidence for this comes from historical documents, agricultural information, records of rivers freezing and glaciers advancing, and even from landscape paintings depicting climatic features or their social effects. This being so, is our present rise in temperature just a natural readjustment following a cold spell? Looking back thousands of years rather than hundreds we find evidence for extremely cold conditions over the earth. Only 20 000 years ago the world was deep in the grip of an ice age; our present warmth has actually been with us for a mere 10 000 years. Indeed, the past 2 million years of the earth's history has been dominated by cold conditions, and the warm episodes have been relatively few and far between, accounting approximately for only about 10 000 years of warmth in each 100 000 years, the rest being spent, at least in the high latitudes, under the influence of ice (Fig. 1.1). Since we are now 10 000 years into a warm episode, should we be anticipating a return to the more normal conditions of cold? If the processes active in the past continue in their former pattern this may well be the situation we should be expecting.

So, having set our current observations concerning change into their historical background, we may be in a position to project into the future. Predicting the future is an essential scientific activity and accurate predictions can be very important in many different ways. Short-term weather forecasting, for example, can be of great economic importance to farmers and to shipping, and it may also be of great interest for leisure activities, from mountaineering to sea-bathing. The basis of prediction is the observation of current conditions and recent changes interpreted in the light of a knowledge of past pattern of events and processes. It lies upon the assumption that no newly-discovered factors or new sets of conditions will enter the equation to upset the model.

But sometimes new factors do arrive and must be accommodated in the prediction process. Projecting weather conditions has an analogy in the world of insurance. Here the premiums to be paid for insurance against particular risks are calculated on the basis of the probability of an event (such as a fire, an earthquake or an automobile accident) taking place. But supposing a totally new set of conditions was suddenly to arise. Perhaps a new law is passed allowing all people over the age of 10 years to drive automobiles on the public highway without any restrictions or training. The entire situation regarding road safety has now changed and the probability of an accident each time you drive out onto the highway will have increased, so insurance premiums need to be elevated accordingly to compensate the insurance company for the greater risks.

The same kind of problem can arise in the case of climate prediction. The earth's climate, as will be explained in due course, is strongly affected by the energy budget of the earth, and this in turn is influenced by the physics and chemistry of the atmosphere through which the earth's energy is received from the sun and radiates out into space. If the chemistry of the atmosphere changes, the energy budget will be altered and it can no longer be assumed that the climate patterns that have prevailed in the past will continue to operate into the future. Prediction now becomes a lot more difficult because one can no longer simply project the lines forward; new factors need to be incorporated into the predictive model. This is the problem currently facing climatologists, and their predictions of the future vary according to what assumptions they make about the new factors operating and the way in which they themselves will continue to change.

Meanwhile, biologists and ecologists consider these climatic projections and try to predict the possible outcome as far as plants and animals are concerned. If, for example, the temperature of the earth rose by a further 1°C over the next century, how would this affect vegetation and animal life? Could species continue to survive in their current ranges or would they die out in some areas and have the opportunity of colonization in others? All species have certain temperature limits and, as the pattern of climate over the face of the earth changes, one would expect different species to expand their ranges in some directions and contract from others according to their innate requirements. Some may even be able to adapt through evolutionary development, but this is likely to occur only for those with rapid breeding systems and great genetic variability. It would take a considerable time period for most species

to be able to do this. For the majority of plants and animals much would depend on powers of dispersal. Flying organisms, like many birds and insects, could probably adjust their ranges quickly and effectively, as could plants that produce light, wind-borne seeds in large numbers. But the more cumbersome may find such movements difficult or impossible. The situation is made more complex for such species because our modern world is so fragmented as a result of human activities. Agricultural land, roads, urban settlements and all kinds of development have broken the natural landscape up into a patchwork of habitat 'islands' surrounded by 'seas' of alien conditions. Movements of animals and plants under such conditions is more restricted and the likelihood of local extinction is enhanced.

As in the case of climate predictions, where current changes are introducing new factors which render the simple extrapolation of the past into the future an invalid exercise, similar problems face those involved in wildlife preservation and conservation. New conditions now exist, so one cannot assume that the elk and the oak tree, the bush mouse and the pitcher plant that survived former climate changes by altering their ranges will be able to cope with current changes under new sets of conditions.

Some species will survive because they receive help from our own species. We can be reasonably sure that species such as wheat and maize, sheep and cattle will survive because we shall transport them to areas where climatic conditions are suitable for their growth and production. Agricultural patterns may have to change, however, since areas currently suitable for the growth of one species, such as wheat, may no longer be so, or the crop may become prohibitively expensive because of the need for irrigation or increased pest-control measures. And changes in the pattern of agriculture have inevitable repercussions in social and political spheres.

Climatic changes, however, are only one symptom of a much more complex problem facing our species, namely the build up of waste products from our increasingly abundant and densely packed society. The exhaustion of local resources and the accumulation of waste is not a new feature of human society, but has become increasingly serious and global in extent with increasing human populations. Primitive shifting agriculture, in which a resource exploitation phase was followed by abandonment and population movement into a fresh area, operated under precisely the same ecological rules as our present society except that there was always the option of moving on. That option is no longer available to us, so population control and rational resource use in a sustainable fashion must ultimately be our goal. The isolation of the world in which we live (the 'spaceship earth' concept) is becoming increasingly evident with time and the need for international co-operation is generally recognized. The English 17th century poet John Donne penned the immortal words concerning no man being an island, emphasizing the importance of the individual in society and the interdependence of humanity. It is increasingly apparent that no nation in our present world can be an island, for we each have profound effects upon all others.

Take, for example, a small and relatively self-contained country like Switzerland, apparently isolated in its mountain fastness in the heart of continental Europe. The availability of hydroelectric power means that its consumption of fossil fuels is relatively low for a developed nation and it contributes only 0.2% of global CO_2 output. Per capita emission of CO_2 falls well below the Organization for Economic Co-operation and Development (OECD) average (Fig. 1.2) and not far above the world average, so its record on this score is fairly commendable (Swiss Federal Office of Environment, Forests and Landscape (FOEFL) 1994). Methane emissions from natural gas leaks are low and declining, and the use of chlorofluoro-

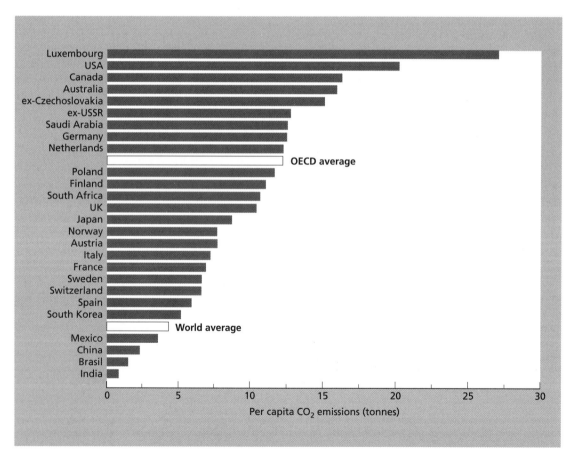

Fig. 1.2 Carbon dioxide emissions (per capita) for a number of developed and developing nations. Switzerland can be seen to lie among the lower ranks of the developed, industrialized nations. (Data from the Swiss Federal Office of Environment, Forests and Landscape (1994)).

carbons (CFCs) will be completely phased out by 1997. So Switzerland can hold its head relatively high among the developed nations as far as the production of greenhouse gases (see Chapter 5) is concerned.

But Switzerland shares the atmosphere with the rest of the world and will import, albeit unwillingly, any pollutant contained within the atmosphere as well as the climatic and environmental consequences of such pollution. The implications of climatic change (as a consequence of greenhouse gas accumulation) for Switzerland could be considerable and perhaps unexpectedly severe. Take the effects of rising temperature on the natural vegetation of the Alps, for example. At present we cannot predict with any degree of certainty just how fast the temperature will rise in the future, but most researchers are convinced that conditions generally will continue to become warmer and we can make projections concerning the effects of warmer conditions on the zonation of vegetation in alpine mountains. Figure 1.3 represents the expected change in vegetation in the Western Alps that would result from a rise of

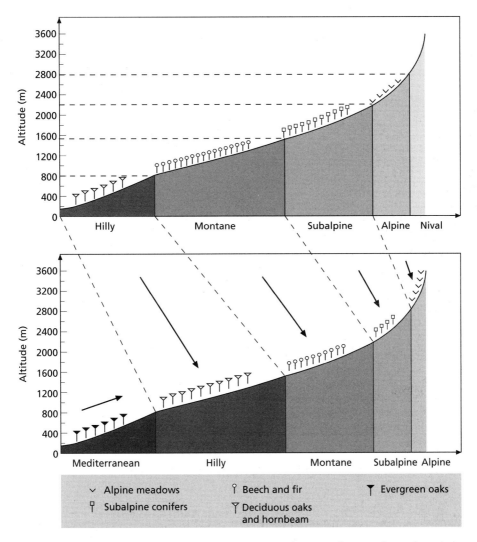

Fig. 1.3 Estimated changes in the vegetation bands with altitude in the Western Alps under conditions of a temperature elevation of 3.5°C (Ozenda & Borel 1995). The upper diagram shows the current situation and the lower is that anticipated under warmer conditions.

3.5°C in overall temperature (Ozenda & Borel 1995). From this it can be seen that each band of vegetation would extend upwards and be outcompeted by other vegetation types from below. The outcome would be approximately a 700 m rise in the altitude of the main vegetation boundaries and a mountain of 3600 m would lose its permanent snow-cap.

A rise in temperature of this order would, of course, affect agriculture and forestry, as well as the natural plant and animal species present in a country that is so rich in wildlife. High-altitude plants and insects that currently occupy the mountain tops above the tree-line may have nowhere to retreat to if warmer conditions permit the larch and spruce trees to invade the high pastures. Reduced areas available to these species is sure to result in local extinctions and lowered biodiversity.

But climatic warming would also have a very direct impact on Swiss economy as a result of its effects on winter sports and tourism, which account for 6% of the nation's gross national product (GNP). Figure 1.4 indicates the projected impact of a 3°C rise in overall temperature on the depth of snow cover at an altitude of 1500 m in the Swiss Alps. Not only is snow depth severely diminished, the length of the snow-cover season declines by about seven weeks or more at this altitude. Low-level, and some medium-altitude

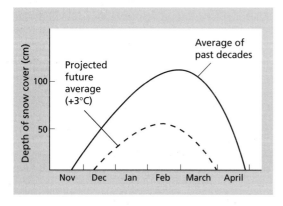

Fig. 1.4 Depth of snow cover through the winter months at an altitude of 1500 m in Switzerland. The solid line represents the current state of affairs and the dashed line is that projected if the average temperature of these months were to rise by 3°C. (Data from the Swiss Federal Office of Environment, Forests and Landscape (1994)).

resorts would be under distinct threat by such a temperature rise, as shown in Fig. 1.5. Ski resorts in the Jura and in the high alpine foothills would certainly be vulnerable to such a change. Already an experience of warmer winters has been affecting some resorts with profits down as much as 50–80% in some seasons. Thus the socio-economic impact of climate change is complex, far-reaching and pervades all nations, the clean and the unclean.

Among the developing nations, Bangladesh may well become a recipient of problems for which it can hardly bear the blame. A rise in global sea levels, resulting from raised temperatures and melting ice-caps, would endanger much of the low-lying coastal zone. Loss of land, life and livelihood are an inevitable consequence for this unfortunate nation. So predictable is the suffering of such low-lying lands that one can develop an entire science of Disaster Management in which responses to flood and cyclone, drought and pollution, can be developed. Although we would all share the unfortunate results of changing climate, it is the poor of the world who would undoubtedly suffer most.

History holds many severe lessons for us. In the past, whole civilizations have been eliminated by climatic change, as in the case of the Maya of Central America, which collapsed from its state as a flourishing, highly-civilized agricultural society in a relatively short period of time, around 750–900 AD. Many theories have been erected to explain the sudden demise of this society, but recent work by archaeologists and environmental scientists in Mexico have shown that it came about almost certainly as a result of an intense period of drought that developed in the region at that time (Hodell *et al.* 1995). Physical and chemical studies of the sediments of Lake Chichancanab on the Yucatan Peninsula have shown that evaporation rates from the lake increased during the 2000 years or so that the Mayan civilization lasted, reaching peak levels at the time of the collapse. We still need to know more about the precise sequence of events that led to the decline of the individual cities, but the circumstantial evidence now points firmly to a climatic change as the prime cause of the death of a civilization. But many other questions can now be raised, such as how the stresses of climate change influenced the social processes within the

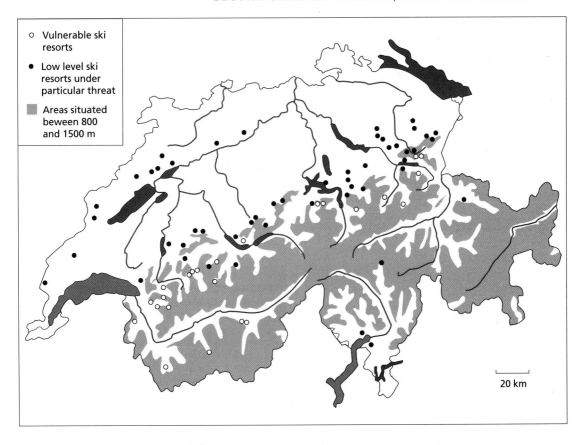

Fig. 1.5 Map of Switzerland showing those ski resorts that would be under threat (solid circles) and could be vulnerable (open circles) if the winter temperatures were to rise by 3°C. As can be seen, the lower-lying resorts could lose their tourist income under such a scenario. (Data from the Swiss Federal Office of Environment, Forests and Landscape (1994)).

agricultural and urban populations of the sophisticated Maya people (Sabloff 1995). Was there a threshold level of population that had been exceeded that rendered the Maya particularly sensitive to the onset of drought? The questions do not end with the demonstration of contemporaneous environmental change; they move on to the social and political problems of that time.

It may be unduly optimistic to suppose that the lessons unearthed by archaeologists will provide us with the information we need to solve our problems as we face the future, but the message of the Maya is that complacency, even in our sophisticated modern technological society, is dangerous. Disaster forecasting and management must be well planned, securely informed and international in scope if such events are to be avoided in future.

Disasters may be avoided, but change is inevitable. It has become customary, for example, to divide the world into 'developed' and 'developing' nations, but this is really a misuse of the terms, since it implies that the developed nations have completed their development (which is far from being the case) and that the developing nations may be given the freedom, and even encouragement, to pursue their development, which is also questionable. The very use of the word 'development' implies change and it is only reasonable to expect that social, economic and technological change within human populations will give rise to further global environmental change. Among the industrialized nations one can observe a rapid evolution through a whole range of industries, such as paper, textiles, and communications. Development in one industry can greatly modify others. In communications, for example, current developments are affecting global requirements for materials such as copper and steel. A communications satellite may weigh only 250 kg, yet it is more efficient than a transatlantic copper telephone cable weighing 150 000 tonnes (Tolba 1992). Changes of this sort in the industrialized nations will impact upon the demand for metals which will in turn influence a range of environmental issues involved in their extraction and transport.

The developing nations, meanwhile, seek to follow the lead of the industrialized world to improve the economic and social conditions of their inhabitants. Yet their share of world manufacturing output since 1980 has remained fairly stagnant at between 12% and 14%. Their industrial development is impeded by the need to service debts to richer nations, by protectionism in the markets of the developed nations and even by the pressure of environmentalists from the developed world to restrict the exploitation of their natural resources, including the rain forests.

Industrialization of these nations, as it proceeds, will certainly generate a new range of environmental problems. So global change studies cannot be restricted to a consideration of climate; our demand for food, for energy, for materials and, ultimately, for space will all add to the pressures for change.

In order to understand the causes of environmental change, therefore, we need to be able to analyze the processes that we see happening around us, physical, chemical, biological and sociological, and set them in the context of their geological and historical development. In order to predict the future we need to be able to project these processes forward, making certain assumptions about their rates, directions and interactions. We cannot claim that this book will give rise to a new generation of prophets and sybils (we have enough of those already), but it should provide a factual basis for a rational and positive approach to the future.

Chapter 2

The Earth as a Planet

A time perspective

The earth is one of the smaller members of the group of planets which make up the solar system. If the earth could be viewed by some remote observers in space, having the technology to see the details of the surface of the various planets, the earth would stand out as having a surface appearance that changes significantly on an hour-to-hour, day-to-day basis (e.g. in cloud movement and the extent of sea ice). On a longer time scale running into weeks and months, they would observe even more striking change, in the amount of surface covered with snow and in the spread of vegetation in response to the changing seasons. If such observers had had the means to extend their observations over much longer periods of past time, they would have seen over the last 10 000 or 20 000 years the retreat of the great ice-sheets off much of the high-latitude land areas of both the northern and southern hemispheres. Accompanying this would have been the changing shapes of the coastlines as the sea level changed in response to the melting of the ice and the warming of the ocean water. If these imaginary observers had been able to study the earth over an even longer time perspective, they would have seen on a time scale of millions of years the movement of whole continents, the rising—and subsequent erosion—of mountain chains, and the repeated spread and retreat of seas over the margins of the continents. This change in the environment of the surface of the earth has been a continuous feature of its 4600-million-year history.

Planetary atmospheres

If we want to understand the processes which have produced this sort of change in the global environment, it is helpful to compare Earth with at least the two closest planets within the solar system. The relevant planets are Mars, whose orbit lies next outside that of Earth, and Venus, whose orbit lies within our own. The major differences in the atmosphere and the character of the surface between Earth and those two planets suggests some of the ways in which all three may have changed with time. Long before the current age of exploration by space vehicles sent out from Earth, astronomers had already learnt a good deal about the composition of some other planets' atmospheres. This was achieved by studying the changes in the sunlight reflected by them produced by the absorption of the gases making up their atmospheres through which the sunlight had had to pass. By this analysis of the spectra of the reflected light of Venus and Mars it was evident that their atmospheres were significantly different from that of Earth.

As seen with optical telescopes from Earth, it was also clear that Venus had dense cloud cover which obscured any view of its surface, and that it evidently had an atmosphere of considerable density. Mars, by contrast, had no cloud cover, and such atmosphere as it had was seemingly very thin. These early observations have been borne out by the much fuller information from unmanned space vehicles. Table 2.1 shows the composition of the atmospheres of those two planets closest to Earth, compared with our own.

Table 2.1 also shows the barometric pressure of the other planetary atmospheres; taking the Earth's as one, the atmospheric pressure at the surface of Venus would be 90, while that of Mars is less than 1/100th that of the Earth. These differences in atmospheric composition have contributed to major differences in the mean surface temperatures of these planets. The mean temperature of the Earth's surface is commonly cited as 13°C (although this figure is of course an average of vastly different temperatures between the poles and the equator, and through the seasons). By

Table 2.1 Comparative data of atmospheres, temperature and pressure on Earth, Mars and Venus. Lovelock suggests that the conditions on Earth would have been more comparable with those on Mars and Venus ('Earth without life' column), if it had not been for the presence of life which has greatly modified the earth's environment ('Earth as it is' column). From Lovelock (1995) by permission of Oxford University Press.

	Planet			
Gas	Venus	Earth without life	Mars	Earth as it is
Carbon dioxide	98%	98%	95%	0.03%
Nitrogen	1.9%	1.9%	2.7%	79%
Oxygen	trace	trace	0.13%	21%
Argon	0.1%	0.1%	2%	1%
Surface temperatures (°C)	477	240–340	–53	13
Total pressure (bars)	90	60	0.0064	1.0

comparison, that of Venus is some 477°C. This would be hot enough to melt many metals—not a promising setting for a space-ship landing! The mean Martian temperature, in contrast, has been calculated as –53°C.

Mobility and variability in the atmosphere and oceans

The most striking feature of the earth's surface showing changes on an hour-to-hour time scale, which would be evident to our space observers, is the appearance of the atmosphere, and especially the movement of the clouds of water droplets and ice crystals. The mobility of the atmosphere, and its resulting capacity to mix and give a high degree of uniformity to atmospheric composition globally, contrasts with the slower-moving action of ocean currents. The mixing of the atmosphere evens out the three constituents that are vital for the survival of all forms of life on earth—oxygen, CO_2 (the substrate for photosynthesis) and water vapour. The nitrogen, oxygen and CO_2 content of the atmosphere remain remarkably constant throughout the world, except for local variations in the CO_2 within vegetation canopy and close to the soil surface. The absolute atmospheric pressure falls off with altitude, so that animals and plants living at high altitude have less available oxygen and CO_2 than do those living at sea level. The amount of water vapour in the atmosphere is by contrast very varied; it is dependent on access to large water bodies over which the air may pass, and also to their temperature and that of the air. The vapour pressure of water rises with rising temperature. Warm air can 'carry' far more water vapour than can cold air. This makes it impossible to cite any meaningful global figures for the water content of the atmosphere.

The oceans are of course highly mobile, but because of the vastly greater density and viscosity of water compared with air, the global composition and temperature of seawater are less effectively mixed by current movement than is the case with the atmosphere. Seas on the continental masses, with variously limited access to the oceans, may become greatly diluted with input of freshwater (as in the Baltic) while other areas where the evaporation exceeds the rate of freshwater input may become far more saline than the open ocean (as in the Arabian Gulf). The capacity of the oceans to store and to move dissolved gases, mineral nutrients and heat energy gives them a role in the global biosphere, on land as well as in the sea. This is dealt with more fully in Chapter 6. The movement of the essential requirements of life between the atmosphere, the oceans and the biosphere is one of the prime reasons for taking a global perspective in considering the problems caused by the impact of humankind on our environment.

Solar energy and the greenhouse effect

Another feature of the atmosphere is its role in regulating the passage of radiant energy from the sun to the earth and to living organisms. Sunlight is of course the ultimate source of virtually all the energy used by living organisms, so making life on earth possible. Only a very small and atypical fraction of the earth's living things obtain their energy from chemically active inorganic substances. An important habitat for such organisms is regions where upwelling of molten volcanic rock in the depths of the ocean causes these reactive substances to be released into the seawater. The harnessing of chemical energy from this volcanic source may well have preceded the harnessing of solar energy by means of photosynthesis in the history of life on earth.

The capacity of the earth's atmosphere to retain heat energy from the sun—the so-called greenhouse effect—enables it to maintain a temperature that makes life possible over a large part of the earth's surface. Without that greenhouse cover, the earth would be a great deal colder than it now is.

The atmosphere plays a further role in partially reflecting and filtering the whole spectrum of radiation that we receive from the sun. The complex of water droplets and ice crystals that form clouds, the dust from volcanoes and blown from desert surfaces and the gases released by forest fires and industrial burning all contribute to modifying the impact of direct solar radiation. Some of the energy is reflected, especially by the cloud cover; some of it is absorbed by the atmosphere, raising its temperature in the process. The energy that reaches the earth's surface may be absorbed by the vegetation, and used in photosynthesis to make energy-rich organic substances, or simply taken up by the ocean or land surface, or directly reflected back. Some of this absorbed solar energy will then be re-radiated, and this is in turn involved in a further pattern of atmospheric absorption and reflection. These processes are summarized in Fig. 2.1.

The effect of passage through the atmosphere on the outgoing (reflected and re-radiated) energy is different from that on the incoming spectrum. Several features of the radiation budget shown in Fig. 2.1 are perhaps a little unexpected. Only about a third of the incoming radiant energy from the sun is actually reflected, from within the atmosphere (from clouds) and from the earth. The greater part is absorbed, both within the atmosphere and by the earth's surface and then re-radiated at a greater wavelength than that of the incoming radiation.

Plate 2.1 (facing p. 136) shows a composite satellite picture of the earth in false colour to show the pattern of absorption within the photosynthetically active waveband by the different parts of the earth's surface. The lowest levels of absorption in this band are from bare rock surfaces (as in sand desert areas such as much of the Sahara and Arabia) and from the ice-caps (as in Greenland and Antarctica). High levels of photosynthetically active light are absorbed by forests, both temperate and in the humid tropics, and in the growing season, by pasture and arable land; snow cover in the northern hemisphere winter drastically reduces this (lower picture in Plate 2.1).

In addition to the reflection and re-radiation of energy, there is conduction and convection exchange between the land surface and the atmosphere. This means that air close to the land surface may be warmed by it (if the air is cooler than the land over which it is passing); the warmer air then rises, carrying the heat away from the surface, and incidentally generating wind movement. This convectional movement is a major source of transport of water vapour in the atmosphere, and hence in the pattern of rainfall, and of the transport of heat energy in the air itself.

It can be seen in Fig. 2.1 that the net energy flow is zero, and indeed this is the basis of all energy budget calculations. That is, we assume that for the short term, at least, there is a steady equilibrium state. This may seem a little puzzling, since we talk of a constant flow of energy from the sun into the earth's living system, the biosphere, as its 'driving force'. One might feel instinctively that the 'greenhouse' process involves more energy being let in than is allowed to escape.

In reality, the 'greenhouse effect' means that the solar radiation has simply raised the mean surface temperature to the point (around a mean value of 13°C) at which the re-radiated energy exactly balances the absorbed energy, retained by our present atmosphere. Should the greenhouse effect become enhanced above its present level, for example by our increasing the atmospheric CO_2 (see Chapter 5), then this equilibrium would be upset. The mean global temperature would then rise until the increased re-radiation of solar energy (which is a function of the surface temperature) brought the incoming and outgoing energy

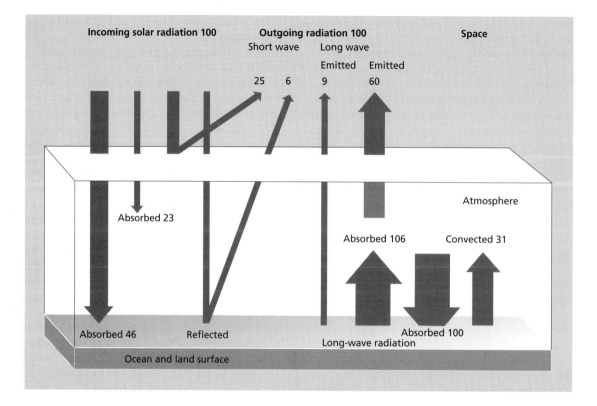

Fig. 2.1 The radiation balance of energy received and emitted by the earth. Incoming radiation is partly absorbed and partly reflected by the atmosphere and the earth's surface. Radiation absorbed at the earth's surface is re-emitted as infra-red radiation of longer wavelength than that received. Note that there is a balance between the incoming and outgoing radiation. Increase of greenhouse gases in the atmosphere would result in that balance being achieved at a higher mean global temperature. From *The Greenhouse Gases*, United Nations Environment Programme (1987a).

back into balance. It is estimated that this greenhouse phenomenon maintains the earth's surface at a temperature some 30°C higher than it would be if we had no atmosphere.

The different constituents of the atmosphere have very different greenhouse potential. It is common knowledge that CO_2 is a more 'powerful' greenhouse gas than the two major atmospheric components of oxygen and nitrogen, for example. Methane is an even more effective greenhouse gas, molecule for molecule, than CO_2. The greenhouse potential of chlorofluorocarbons (CFCs), introduced into the atmosphere by industrial production, is greater still (see Fig. 5.10 and further discussion of greenhouse gases in Chapter 5). It will be evident that the absorption spectrum of any given substance, and hence its greenhouse potential, is a function of the size of the molecule, but that the relationship is not a simple one. Water vapour, an important atmospheric constituent, also has a powerful greenhouse effect, but its role tends to be neglected in the discussion of greenhouse gases. This is because it is hard to quantify in any particular context, since in most parts of the world the water vapour content of the air fluctuates considerably from day to day and even from hour to hour.

Other greenhouses

It is evident from what we have just been considering that the greenhouse effect is essential to the survival of life on earth; without it life, at least as we know it, would be impossible. The concern that has given the greenhouse effect a 'bad press' over the last decade is related not to its existence but to the potential for its increase. This 'enhanced greenhouse', anticipated from the man-made gases released into the atmosphere (most notably CO_2, CFCs and methane), may produce excessive climatic warming, with variously harmful—but also, perhaps, with some advantageous—results. This is considered further in Chapter 5.

The effect on surface temperature of the greenhouse characteristics of a planet's atmosphere is vividly illustrated by comparing the Earth with Mars and Venus (Table 2.1). Mars, with a very thin atmosphere, has a minimal greenhouse effect and generally low temperature, but with enormous abrupt changes as its rotation turns the surface towards or away from the sun. The Earth, in contrast, has a materially higher average temperature, but the day to night changes, at least in most latitudes, is far less drastic than on the nearly atmosphere-free Mars. Venus with its immensely dense atmosphere (and closer proximity to the sun) shows what may fairly be called an excessive greenhouse effect. Its temperature is far above boiling point, and any water present is in its gaseous form; indeed it is in the condition that we would be in if all the oceans and lakes had evaporated and transferred their water to the atmosphere. Its temperature is far above that at which any form of life, as we know it on Earth, could survive. The cloud cover is so dense that there can be little change in the surface temperature between the Venusian day and night. The fact that the surface of Mars has proved to be 'too cold' and that of Venus 'too hot' while that of Earth is 'just right' for life has inevitably prompted reference to the 'Goldilocks effect' of the planetary atmospheres by some American authors.

The surface of the Moon, which like Mars has only a very thin atmosphere, makes a striking contrast with that of the Earth as a result of their differences in greenhouse terms. The memorable NASA photograph of 'earthrise' as seen from the moon's surface (Plate 2.2, facing p.136) puts the Earth with its dark patches of dense vegetation, its oceans and its clouds into striking juxtaposition with the dry, barren and lifeless surface in the foreground.

Components of the earth system

To pursue our discussion of the relationship between the earth and the life inhabiting it, it will be useful to recognize the different constituents of the system. The physical and (largely) solid components of the earth itself have been referred to as the geosphere, which can be seen as complementing the concept of the atmosphere, being the gas which surrounds it. With rather less sharp distinction we can then characterize the hydrosphere, meaning all the water—freshwater, seawater and water vapour—shared between the earth's crust, its surface and the atmosphere. These three units, the geosphere, atmosphere and hydrosphere, are not separate compartments within the earth system, but as we have just seen, are closely interconnected. The hydrosphere is in fact partly distributed within the very substance of the other two, and volcanic eruptions can carry fragments of the geosphere high into the atmosphere! A fourth component of the earth system is of course the totality of living things (including humankind), which has been designated the biosphere. This has in turn invaded all three of the other components; plants and animals have occupied the outermost part of the geosphere, the soil, while a range of life forms have colonized most parts of the hydrosphere and the lower reaches of the atmosphere. Despite the artificiality of the concepts, this terminology of the different parts of the earth system is useful in considering the interactions of these components of our environment.

Carbon in the global system

The movement of carbon between living systems in the sea and on the land, and the great reservoirs of that element in the atmosphere, the hydrosphere and within the geosphere, is one of the fundamental processes associated with life on earth. It is also one of the strongest links between those four components of the earth system. The atmosphere and the waters of rivers, lakes and the ocean are the main reservoirs and pathways of the inorganic carbon, largely in the form of CO_2, available to the biosphere.

The main driving force in carrying carbon from the atmosphere into the biosphere is the photosynthetic fixation of CO_2 by both land plants and the phytoplankton in the oceans. For most animals which inhabit the land, including humankind, the organic matter which makes up the bulk of their dry weight has been extracted from the air by land plants, and consumed either by eating the plants themselves or the meat of animals who have themselves fed on plants. This process is the more remarkable when you consider how relatively little of the atmosphere is actually made up of CO_2—roughly 350 parts per million, or 0.035%. The carbon content of the atmosphere is so thin that it would require the CO_2 in all the air contained in about 10 average-sized houses to produce enough carbon for one human being. The photosynthetic fixation of CO_2 is of course more or less in equilibrium with its return to the atmosphere by respiration, and by the breakdown of dead plant and animal material by micro-organisms in the soil. That slow biological cycling of CO_2 is speeded up by the process of combustion, as in the case of wildfire—the burning of forests and grasslands.

The central role of the carbon cycle in many aspects of environmental change will be taken up in Chapter 5, and its relationship to life in the ocean in Chapter 6.

Water and life: biosphere and hydrosphere

All forms of life, plants, animals and micro-organisms, aquatic and terrestrial, contain far more water than dry matter; a commonly cited proportion is some 95% water by weight for most living organisms, although of course this figure varies greatly between different groups. The distribution and movement of water can be summarized as a 'hydrological cycle' (Fig. 2.2). This shows the amount of water held in the main 'reservoirs' in the form of seawater, freshwater lakes, ice and water vapour in the atmosphere, expressed as a percentage of the total global holding. It also shows estimates of the flow of water along the various pathways ('fluxes) expressed as 10^{15} kg yr^{-1} (these figures need to be multiplied by 1000 to turn them into gigatonnes (Gt) (10^{12} kg), the more commonly used units in handling global data).

As can be seen in Fig. 2.2, by far the largest fraction (about 97%) of the water on the earth's surface is held in the oceans and the epicontinental seas. The remainder is held in the lake and river systems of freshwater (only 0.01% of the total), as groundwater (in cracks and minute spaces within porous rock, about 1%) and as ice in the high latitude ice-caps and montane glaciers (about 2%). The atmosphere is estimated to hold only 0.002% of the global water supply, with about half this as

water vapour, and the other half as water droplets and ice crystals in clouds. However, this atmospheric water represents the principal flux between the oceans and the land-held water of lakes and rivers. The figures of the fluxes and reservoirs of water at the earth's surface are, as would be expected, far greater than those of carbon, if the data in Fig. 2.2 are compared with those of the carbon cycle in Fig. 5.1.

The total evaporation from the global oceans is estimated at 336 000 Gt yr^{-1}. As shown in Fig. 2.2, about nine-tenths of this returns directly to the oceans as precipitation. The remainder joins some 64 000 Gt of water coming off the land surface as evaporation and transpiration, to contribute to the 100 000 Gt yr^{-1} of precipitation which is estimated to occur over land. The difference between that figure and the amount estimated to have been lost by evaporation and transpiration leaves some 36 000 Gt yr^{-1} as river and ice flow being returned from the land surface to the oceans.

The residence time in a given reservoir is theoretically the time that an average water molecule in that reservoir will remain there, before passing through one of the fluxes into another setting. The 'narrowness' of the flux of water through the atmosphere compared with the huge reservoirs of ocean and ice-bound water means that the residence times in these various settings are vastly different. Atmospheric water (vapour or clouds) turns over in an average time of 10 days, while a mean figure for water in the oceans is 1000 years. Since ocean water is not stirred up throughout its depth by current movement, in the same way that the atmosphere is stirred by wind movement, the actual residence time in ocean water is believed to range widely between 10 days and 1500 years.

The freshwater on which humankind depends, both directly for drinking, domestic and industrial use, as well as for crop irrigation, is of course a minute fraction of the whole, and is hence vulnerable to any change in the overall pattern of the hydrological cycle. The importance of the availability of water for plant growth and agriculture is implicit in many of the words that we use to describe the world's major vegetation types—rainforest, desert, wetlands. Much human effort and expenditure of energy goes into the process of irrigation of land areas too dry for crop growth, and into drainage of areas that are too wet. It is

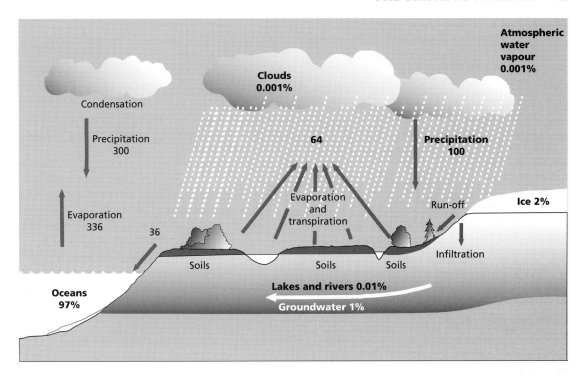

Fig. 2.2 The hydrological cycle; the fluxes are given in units of 10^{15} kg yr^{-1}; the reservoirs as percentages of the global total. From NERC (1989).

estimated that some 5000 additional square kilometres of agricultural land are put under irrigation each year. Ironically, this approximates to the area estimated to have become desertified each year. These and other aspects of the interplay between the hydrosphere and agriculture are considered further in Chapter 11.

Given the ubiquity of CO_2 (albeit at a very low level) in the atmosphere as a carbon source, there are two primary limiting factors which restrict the growth of plants on the earth's surface. These are availability of water (largely as precipitation), and temperature. Light energy is also of course essential, but the other two seem to exert the major constraint in the type of vegetation that can be supported in any location. The distribution of adequate levels of rainfall and temperature produces a close parallel between the climatic pattern and the main 'biomes'—major units of vegetation type—on the earth's surface. The limits of the major vegetation types can be plotted on the two axes of temperature and precipitation,

as in Fig. 2.3. The fact that the major biomes occupy only half of the apparently available 'space' of rainfall and precipitation levels (the triangle in the lower left-hand side of Fig. 2.3) is an inevitable product of the relationship between temperature and rainfall. The vapour pressure of water (and hence what might be thought of as the water carrying capacity of the atmosphere) is a function of temperature. The atmosphere over those parts of the globe with a high mean temperature (principally within the tropics) can carry a maximum load of water vapour. It accordingly has at least the potential of yielding high rainfall—hence the full range of habitats from arid desert to tropical rainforest along the base of Fig. 2.3. In low-temperature areas, generally in high latitudes, the precipitation is limited to the lower part of the range (upper left-hand part of Fig. 2.3). This is largely because a very low air temperature is incompatible with its carrying a high load of water vapour. On a global scale, there are two regions in which vegetation is virtually, or entirely, absent: the low-precipitation, low-temperature areas of the ice-caps, and the low-precipitation, high-temperature areas of arid desert. These are of course the two main areas of low absorption of photosynthetically active radiation recognizable in Plate 2.1.

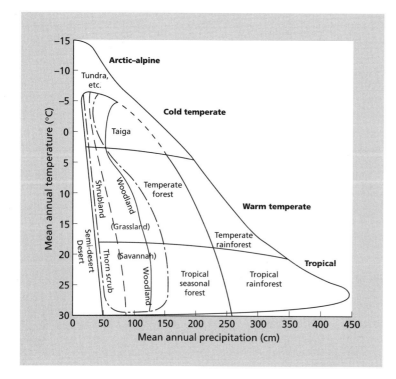

Fig. 2.3 The relationship between temperature, precipitation and the major biomes. All the known biomes occupy the lower left-hand part of the diagram; the upper right, representing high rainfall and low temperature, is not represented in any terrestrial habitat since the capacity of the atmosphere to carry water vapour is closely correlated with temperature. From Cox & Moore (1993).

Earth history

The history of the earth, and of the life on it, is one of unrelenting change. The pattern of the earth's crust, the position of mountains, of epi-continental seas and of the oceans have all changed drastically through the course of geological time. With them the global climate has also undergone major change, from phases with large ice-caps on both poles, as during the successive glaciations of the last 2 million years, to periods free of ice-caps with a globally more equable climate. Against this back-drop of environmental change we see the fossil record of life evolving, punctuated by episodes of widespread extinction. Some of these affected many forms of life more or less synchronously, while others were of more limited impact.

If we are to see the present concerns over the environmental effect of human activities in perspective, we need to see them in the context of the long-term (pre-human) processes of global change. We need to look briefly at the history of life on earth, from its origin to the present complex interactions of humankind with the environment. Of particular interest here is the extent to which physical changes may have influenced the course of evolutionary change, and the ways in which living organisms have in turn brought about changes in the physical environment.

The age of the earth is some 4.5 billion years. Figure 2.4 shows the history of the world, as though condensed into a single year of time past. On New Year's Eve it is customary to contemplate events of the past year. In terms of earth history we can picture the sequence of past events as though we were now gathered on New Year's Eve, at midnight on 31 December, looking back at the 4.5 billion years of earth history, compressed into a single year of time past.

On this scale we have no secure record of fossil remains of any form of life until May of the year of time past. These consist of concentric layers of limestone, forming pillow-like structures (stromatolites) comparable with those formed by microscopic marine algae in certain highly saline sites at the present day. A more satisfactory, but somewhat later, record consists of microscopic cells, preserved in three dimensions in transparent silica rock (Fig. 2.5). Certain chains of cells show close structural similarity to blue–green algae (cyanobacteria) (Fig. 2.5a), of which the living representatives are photosynthetic, and these

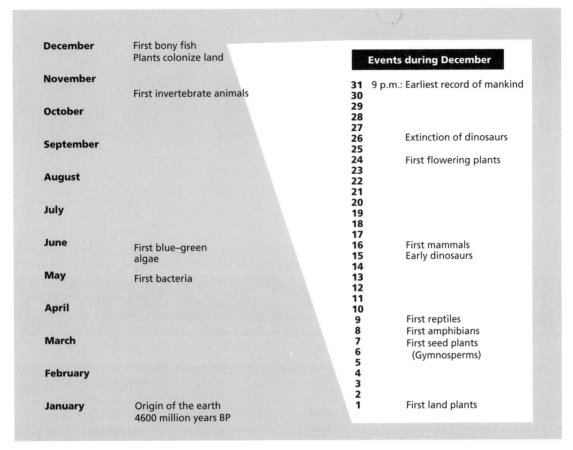

Fig. 2.4 Earth history portrayed as a year of time past. Note that most of the major events involving terrestrial life took place during the month of December, following the colonization of the land at about the beginning of that month.

date back to about July of our year of time past. These well-preserved fossils from the Gunflint Chert of northern Quebec Province in Canada are not the oldest putative cells, but are certainly among the best early records of cellular structure.

The earliest fossil animals large enough to be seen without a microscope appear in the fossil record at around mid-November of our year. Within days, on this compressed time scale, we see the appearance of all the commoner major groups of marine invertebrates. The earliest record of vertebrates comes only a little later, in the last few days of November. That relatively sudden appearance of diversity in the animal kingdom seen in the fossil record of marine life

has been referred to as an evolutionary explosion. The trigger for this relatively sudden spread of new types of animals has been the subject of much debate; it may have been initiated by the passing of some environmental threshold, such as a certain level of atmospheric oxygen. Other explanations include the evolution of the means of making coatings of chitin (the main component of the outer cover of insects and related marine animals) or of shells of calcium carbonate. These forms of armouring would have given animals, for the first time, robust 'hard parts' which were able to survive abundantly in the fossil record. The significance of this early diversity is discussed by Gould (1989) who emphasizes the role of chance in sorting out the competing body-plans of the major animal phyla. This implies that there was no inevitability about the major patterns of animal form, and that those that rose to ecological and evolutionary success, and survived the threat of extinction, did so at least in part through good luck as well as good genes.

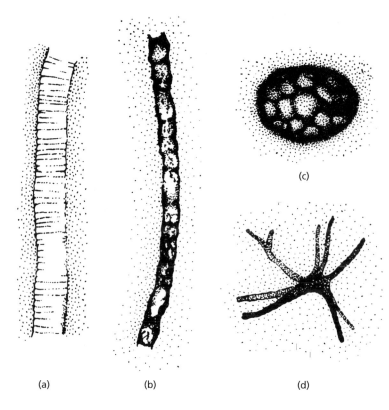

(a)   (b)   (d)

(c)

Fig. 2.5 Some of the microfossils seen in the Gunflint Chert, (approximately 1800 million years old). (a) *Animikiea*; (b) *Gunflintia*; (c) *Huroniospora*; (d) *Eoastrion*. Drawn from photographs by Barghoorn & Tyler (1965). The similarity of the filamentous organism *Animikiea* to living cyanobacteria is one of the best pieces of structural evidence we have that photosynthetic organisms existed at that time. From Stewart & Rothwell (1993).

By the beginning of December of our year, vertebrates were well established in the seas in the form of bony fish, and we get the earliest evidence of both plant and animal life on land, fossilized in deposits formed in lakes and ancient wetlands. Plants adapted to live successfully on dry land, with upright stems, a water-conducting system and aerated tissue structure enabling them to draw in CO_2 from the atmosphere (rather than from water, as their antecedents had done) first appear at that time. The early land plants diversified with the same sort of explosive rapidity that we saw in the early animal groups. Land-adapted arthropods, including arachnids (scorpion-like animals) and collembolans (litter-feeding springtails) appear amongst these early land-plant fossils. Fossil fungi, evidently adapted to life out of water, are also present within the tissue of the fossil plants, and were no doubt involved in the biodegradation of this new source of energy-rich plant material now being produced on the land surface. The structure of some of these fungi even suggests that they may have already developed a symbiotic relationship with the photosynthetic plants, as many of them have now, assisting the

plants in extracting mineral matter from the soil in exchange for a source of food derived from photosynthetic fixation of CO_2.

By the end of the first week of December in our year of time past, plants had evolved from a fern-like life cycle, spreading by wind-dispersed spores, to one in which the seed was the effective unit of dispersal. This eliminated the water-dependent sexual phase of the fern life cycle, in which sperms needed a film of soil water in which to swim to fertilize the egg cell. The dispersed seed contained a small embryo plant with its own food reserve, able to germinate and grow even under the shade of other plants. By this time the vertebrates had effected land colonization, with the first four-legged amphibians able to walk over the land surface, perhaps initially simply to be able to get from one aquatic haven to another.

By the end of the second week in December, the amphibians had been succeeded by the reptiles, with a reproductive system freeing them from dependence on an aquatic environment, just as the seed had done for plant life. The mammals were to follow, and in the plant kingdom, the gymnosperms (primitive seed plants) were to give rise to the angiosperms, the flowering plants. This group was remarkable in using a range of animal groups, but most notably the insects, as carriers of their pollen; this gave them a means of sexual reproduction of unrivalled efficiency. The angiosperms first appear in the fossil record of our year of compressed time on Christmas eve. The appearance and spread of humankind on this time scale would have been accomplished in the last few minutes of the year, and the whole of record-ed history within the last few seconds before mid-night.

This very superficial survey of the time scale of life on earth is perhaps enough just to emphasize the relatively short period within earth history that humankind has been an agent for environ-mental change. We now have to look briefly for evidence in the history of life of direct interaction between living systems and their environment. Most particularly, we need to ask when did life first bring about any changes in the physical envi-ronment? But first, we must see something of the early history of earth before life arose in order to understand the impact that the origin of life must have had on the global environment.

Early history of the earth

Since there are many aspects of the physical sys-tem of the present-day earth which are inade-quately understood, it is scarcely surprising that there is some diversity of opinion regarding the origin of the earth and its early history. It is very generally accepted that the solar system had a sin-gle origin, so that all the planets share a common source. The observed differences in planetary atmospheres, in their sizes and densities must all be products of this history. The early picture of the planets condensing from a single plume of gaseous matter drawn from the sun by a passing large body has given way to a picture of condensa-tion of the planets from a disc of gas and dust without the intervention of any other stellar body (McElroy 1992). Progressive condensation pro-duced numerous nuclei of solid matter, the least volatile components condensing first. The small-er inner planets, of largely solid composition (Mercury, Venus, Earth and Mars) accreted first with the less-dense larger planets (Jupiter, Saturn) forming later in the progression. At a relatively late stage in the process, the earth probably gained

the moon as a condensation of a ring of orbiting minor bodies analogous to Saturn's rings. These may themselves have been generated by a late-stage impact of a smaller planetary body with the earth. As Nisbet (1991) suggests, this impact ori-gin of the moon may have been associated with the earth's axis of rotation being tilted some 20° from its original orientation perpendicular to its plane of rotation around the sun. This interesting possibility links the origin of the moon with two features of the earth's behaviour that have had enormous biological significance. For the moon is the major source of tidal gravitational pull, while the earth's axial tilt is the cause of climatic sea-sonality.

The segregation of the major components of the earth's internal structure into core, mantle and crust (Fig. 2.6) would have taken place as the tem-perature of the accumulating solid matter rose, and became fluid enough to allow gravitational sorting of different elements to take place. Implicit in this version of earth's origin is the con-cept that the stratification of the deep earth was a result of physical segregation at an early stage in its history rather than progressive accumulation of successive layers like the rolling of a snowball.

On this interpretation much of the residual heat in the body of the earth would have been derived from the kinetic energy of the impacting accreted particles, but also from its own radio-activity (Nisbet 1991). Much of the earth's heat at the present day is maintained by radioactive decay of potassium, uranium and thorium which have accumulated in the outer layers of the earth's mantle and crust early in the history of the planet. With the present-day concerns for global warming, it is salutary to remember that radia-tion received from the sun, although of enormous importance to life, is contributing little to the heat energy budget of the interior of the earth. Solar radiation, coupled with our atmospheric greenhouse, has merely maintained a remarkably steady surface temperature, at least in the equa-torial and middle latitudes, which cannot have varied more than a few degrees centigrade of mean annual temperature over the 600 million years that multicellular plants and animals have inhabited the world.

The meteors which from time to time enter our atmosphere and may survive passage through it, to land on the earth's surface as meteorites, give

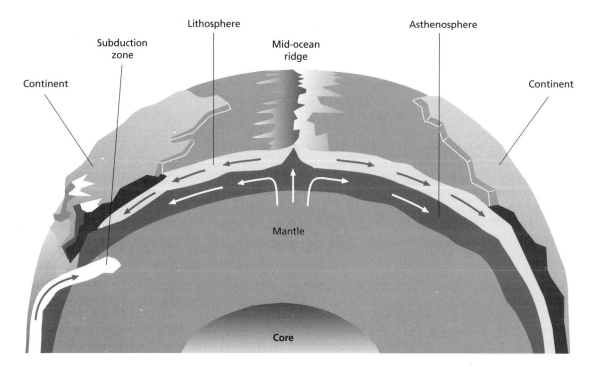

Continent

Subduction
zone

Lithosphere

Mid-ocean
ridge

Asthenosphere

Continent

Mantle

Core

Fig. 2.6 Section of the earth showing the stratified nature of the asthenosphere (the upper part of the earth's mantle), and the more rigid lithosphere and crust above it. New oceanic crust is being formed at the mid-ocean spreading centre. From NASA (1988).

us interesting clues as to the process of formation of the earth. If, as some believe, these bodies represent residual material from the later stages of planetary accretion they may tell us something about events early in earth history, of which we have no direct terrestrial evidence as a result of subsequent physical and chemical changes. There are three broad categories of meteorites: the so-called 'irons', largely of nickel–iron composition; 'stones', largely of silicate minerals; and 'carbonaceous chondrites', which contain some simple organic compounds. The stony meteorites are the commonest, followed by the irons, with the carbonaceous ones the rarest. It is significant that the first two of these three categories correspond to the composition of the earth's mantle and core respectively. The carbonaceous chondrites are noteworthy in containing iron in the form of its sulphide, sulphate or oxide; also, water combined with inorganic crystalline components; but above all, they contain carbon compounds of some

diversity, including amino acids. That is to say, they contain a significant proportion of the group of substances formed from carbon, oxygen, hydrogen and nitrogen that make up the principal components of life. These organic compounds present in meteorites are of particular interest in demonstrating that relatively complex organic substances can be formed from (presumably) non-biotic, extraterrestrial processes.

The origin of life

It is not appropriate for us to review here the wide range of conflicting views on how life arose. It is sufficient to say that the matter is still an area of lively scientific controversy. Even some of the most basic questions are still a matter of debate. However, there is wide agreement that the earth's atmosphere at the time of origin of life was significantly different from that of the present day, and that the early evolution of life had a profound impact on atmospheric composition.

To the early scientists, the idea that life might arise spontaneously (and that indeed it was arising more or less continuously in all sorts of everyday contexts such as rotting food) did not seem at all unreasonable. It was not until the availability of adequate microscopes revealed the role of the microbial basis of decomposition that the fallacy of spontaneous generation came to be established. A critical stage in this recognition lay in the work of Pasteur, who first showed in the middle of the 19th century that micro-organisms do not simply arise spontaneously in a nutrient-rich broth that had been heat sterilized, even when it had access to the air, so long as that access does not allow the passage of air-borne spores. This meant by implication that all forms of life, even the smallest bacteria and other unicellular organisms, come from pre-existing forms and do not simply arise *de novo*.

It is particularly significant that these discoveries were being made at the very time that Darwin (1859) was advancing his thesis on the process of evolution, which sought to explain how life had changed through the history of the earth from simpler to progressively more complex forms. It was implicit in Darwinian thought that if one extrapolated evolutionary change (by whatever process) back through time, all forms of life had not merely shared a common origin, but that life itself must have had a single beginning. Darwin acknowledged the existence of the dilemma that while we have no knowledge of life arising from non-living matter in the present world, this process must have occurred at some time in the past. He briefly contemplated the kind of environment under which life might have arisen, but offered little more than gentle speculation about the process.

One of the major features of the early earth environment which may have been critical to the origin of life is the probability that the atmosphere was of a reducing nature (or certainly lacking free oxygen). This proposition has had great attraction for those contemplating the origin of life. Firstly, it could allow the abiotic formation and accumulation of organic substances which are unstable in the presence of oxygen, while at the same time explaining why conditions now are unfavourable for spontaneous generation to occur. Two scientists, Oparin in Russia and the British geneticist J.B.S. Haldane, came to rather similar conclusions on these lines more or less independently, in the 1920s (Lazcano 1992). They both favoured an early atmosphere of a reducing character in which our present-day composition of nitrogen, oxygen and CO_2 might have existed in varying degrees in their reduced states of

ammonia, water and methane. Although their ideas differed in detail, both visualized ultraviolet light from the sun as an ionizing source which could have given the necessary energy to form simple organic compounds such as carbohydrates, fatty acids and amino acids. These were pictured as accumulating in the world's oceans, stable in a reducing abiotic environment, to form what came to be called a 'pre-biotic soup'. It was suggested that if enzyme systems then arose that were capable of breaking down these substrates anaerobically, they could have used the chemical energy so generated to synthesize new compounds, rather as anaerobic micro-organisms do at the present day. This would have resulted in the release of CO_2 into the atmosphere from the 'fermentation' of the pre-biotic soup (Table 2.2c).

It is now clear that a range of simple organic compounds, such as ethanol, occur in interstellar space, while more complex organic compounds such as amino acids are known from meteorites. Ponamperuma (1992) reports that radioastronomers have identified more than 50 different organic compounds spectroscopically, occurring in interstellar space. This evidence of the formation of energy-rich organic substances outside the earth, without the intervention of living systems, gives a new possible source for the 'building blocks of life'. But as we have noted above, the course of subsequent events remains very uncertain, until we reach a point in the fossil record when recognizable cells first appear.

Some contemporary scientists, interested in the origin of life, believe with Oparin and Haldane that early living systems fed on abiotically-formed organic substrates; that is to say that they had heterotrophic nutrition, like modern fungi and animals. Autotrophic nutrition—either chemosynthetic, relying on inorganic chemical energy, or photosynthetic, harvesting light energy —would then have followed as a later step (Table 2.2a,d). But other workers are sceptical of a pre-biotic soup concept, and believe that early living systems must have been in some way autotrophic. This scenario would involve chemosynthesis preceding and leading to photosynthesis. The discovery of living systems in the totally dark ocean depths, deriving their energy from chemically active volcanic emissions, has given some support to this idea. If valid, this would raise a picture of life perhaps having arisen in the environment

Table 2.2 The principal energy-related processes of living systems and their biochemical basis. Photosynthesis (a) is the energy-storing process of most autotrophic organisms (plants, phytoplankton) that 'make their own food' in this sense from light energy, using CO_2 from the atmosphere. They also use some of this stored energy in various metabolic processes by aerobic respiration (b), consuming oxygen in the process. Heterotrophic organisms (animals and fungi) use aerobic respiration (b) as their way of deriving energy from the other organisms that they consume. Anaerobic respiration (c) is used by many autotrophs and heterotrophs when oxygen is unavailable; this process, used by some plants and bacteria gives a much lower energy yield than aerobically respiring the same substrate. A small proportion of organisms (some bacteria) use chemosynthesis (d) as their primary source of energy, using naturally occurring inorganic substances.

(a) Photosynthesis

$6\ CO_2 + 6\ H_2O + \text{light energy} \rightarrow \text{glucose}\ (C_6H_{12}O_6) + 6\ O_2$

(b) Respiration (and combustion e.g. by wildfire)

$C_6H_{12}O_6 \rightarrow \text{energy} + 6\ CO_2 \uparrow + 6\ H_2O$

Respiration yields 39% of the available chemical energy in the sugar.

(c) Anaerobic respiration ('fermentation')

Alcoholic fermentation (in fungi, plants, some bacteria)

$C_6H_{12}O_6 \rightarrow \text{energy} + 2\ C_2H_5OH + 2\ CO_2$

This yields only 7% of the available chemical energy in sugar.

Lactic acid fermentation (in animals and some bacteria)

$C_6H_{12}O_6 \rightarrow \text{energy} + 2\ CH_3CH\ (OH)\ COOH\ (\text{lactic acid})$

This yields only 2% of the available chemical energy in the sugar.

(d) Chemosynthesis (in bacteria)

Similar to photosynthesis, but energy is obtained from chemical reactions rather than light, for example:

$H_2S \rightarrow S, H$
(ferrous iron)$Fe^{2+} \rightarrow$ (ferric iron)Fe^{3+}
(ammonia)$NH_3 \rightarrow$ (nitrite)NO_2^-

of the mid-ocean ridges where such chemosynthetically-powered communities now exist.

However photosynthesis arose, it would certainly have led slowly to a progressive conversion of some of the atmospheric CO_2 to oxygen. As the oxygen level rose, it would have begun to make possible the oxidative breakdown of organic substrates—that is to say, normal respiratory oxidation, which has a much higher energy yield than the more primitive 'fermentation' by an anaerobic system, which does not require oxygen. This would eventually have produced the range of metabolic processes that we see today, involving oxygen, photosynthetically produced by plants, being consumed by aerobically respiring heterotrophs (animals, fungi and bacteria) as they live on the food and energy source of the plant biomass (Table 2.2b).

The build-up of atmospheric oxygen would have had other important consequences. Firstly, it would have had a very injurious effect on those micro-organisms which had a metabolism totally committed to anaerobic respiration ('obligate anaerobes') (Table 2.2c). These organisms would have been driven from the whole range of environments exposed to the previously oxygen-free air as these now exposed them to atmospheric oxygen. This would have left them with the rather limited range of oxygen-poor habitats (such as organic muds in lakes and estuaries) that they now inhabit. If this was indeed the course of events, as seems likely, it represents the most devastating impact of living organisms on the global environment. This early oxygenation of the earth's atmosphere would have produced an environment of acute toxicity for the anaerobic organisms, which up to that time would have been the prevalent life forms on earth. Indeed, for them this alien gas, oxygen, produced by green plants would have constituted pollution to an intolerable degree. It is irresistable to fantasize that if those early anaerobic bacteria could have effected the kind of pollution control that we attempt to exercise now, they would have introduced legislation to halt the widespread release of oxygen gas by photosynthetic organisms. If they had been successful in this 'pollution' control, it would have drastically altered the course of evolution of life on earth, and indeed spared it from all that ensued, including the arrival of humankind!

Although many of the scientists who are interested in the events leading up to the origin of life accept the general features of the Oparin/Haldane scenario (see for example Ponamperuma, 1992) not all accept the idea of a reducing atmosphere for the primitive earth. Lovelock, for example, in his memorable book *Gaia* (1995) favours a CO_2-rich atmosphere for the early earth (see the column 'Earth without life' in Table 2.1). However, despite the diversity of views on the composition of the primitive atmosphere, most are agreed on two significant features of atmospheric history. Firstly, that the primitive earth had very much less oxygen in its atmosphere than it does now; and secondly, that photosynthetic plant life was instrumental in bringing about the present level of oxygen.

The other important effect of an oxygen build-up, beyond its impact on the metabolism of anaerobes, would have been the generation of ozone in the stratosphere. This would have developed spontaneously, as it does now, under the influence of solar radiation, so forming the 'ozone layer' which plays a significant role in shielding life on the earth's surface from the full strength of the sun's ultraviolet (UV) radiation. The biological consequences of the build-up of stratospheric ozone are considered further in Chapter 8.

Just how important that generation of an ozone UV screen was depends on one's assessment of the threat that 'unscreened' UV light actually represents for living organisms. Some see the penetration of UV light through the photic zone of the upper layers of the ocean as having been an important factor early in earth history, impeding the diversification of the marine plankton, and perhaps even holding back the colonization of the oceans by metazoan life. Others, such as Lovelock (1995), regard the lethal powers of unscreened UV radiation as greatly exaggerated, and attach less importance to the ozone build-up that must have taken place as the atmosphere became oxygenated. These considerations bear greatly on the anticipated ecological impact of ozone thinning (the 'ozone hole') which are explored further in Chapter 8.

Oceans and continents

Having looked at some of the interactions linking the geosphere and biosphere from an early phase of earth history, we need now to look at the geological processes taking place at the earth's surface that have influenced living things. By the time that eukaryotic organisms appear in the fossil record, at around 900 million years ago, the physical environment of the world was already operating much as it does at the present day. That is to say the physical structure of the earth was already composed of relatively deep oceans, and of higher level areas, the continents, of which only the margins, and sometimes parts of their interiors, were flooded from time to time by relatively shallow incursions of the sea. However, we also know that the actual pattern of oceans and continents was very different at that time from the present arrangement. We have ample evidence from geology and geophysics that the continents have moved relative to one another and to the major oceans, throughout the course of earth history, for at least the time during which metazoan life has existed (see Figs 2.8 and 2.9). The science of these movements of the rocks forming the continents and floors of the oceans is called plate tectonics.

Movements of the continents

Geologists recognize two layers within the outer part of the earth: the uppermost part of the mantle, called the asthenosphere, and the crust or lithosphere which lies above it (Fig. 2.7). The mantle behaves as a very viscous fluid, which can be deformed by convective flow; some parts of the mantle are hotter than others, and these rise in convective plumes carrying the more rigid

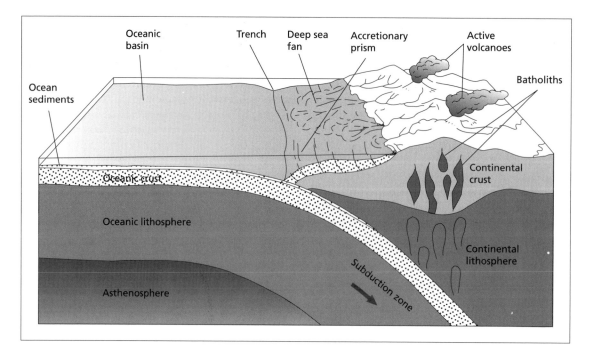

Fig. 2.7 Vertical section through the earth's crust where oceanic lithosphere is being subducted by convectional movement at the continental margin (as is now occurring at the western margin of North America); material deposited in sea-floor sediments, subducted beneath the continental margin, may be returned to the surface millions of years later by volcanic eruption. From NERC (1989).

asthenosphere outwards, horizontally, away from the rising mantle rock. This causes lateral movement of the more or less rigid crustal 'plates' forming the lithosphere; this process is well documented for some mid-oceanic ridges, as in the mid-Atlantic ridge, where rising mantle rock is forming new lithosphere of basaltic lava on either side of the ridge. This is expanding the floor of the Atlantic Ocean, and increasing the distance separating America from Europe and Africa. Geologists distinguish between oceanic plates of the lithosphere, produced by upwelling of basaltic rock beneath the major oceans, and the thicker, lighter bodies of crustal rock which form the continental plates.

In addition to the formation of new crustal rock, as at the mid-ocean ridges, there are other areas where a compensating downward movement carries crustal material into the mantle in a so-called 'subduction zone' (Fig. 2.7). Crustal rock

of the Pacific Ocean plate is being carried downwards in this way under the edge of the continental plate on the west side of North America.

This movement of the continents and ocean floor was first documented by the German geologist and climatologist Alfred Wegener who coined the phrase 'continental drift' to describe the phenomenon. He also realized that the continents had moved with respect to the earth's axis of rotation, and he referred to this as a separate process, 'polar wandering'. We now recognize that the single process of plate movement, of both continental and oceanic crust, moves continents relative to one another and also relative to the poles.

Through the course of geological time, these movements of the great land masses have had a profound influence on the animals and plants living on them. Populations have been separated by the break-up of a continent, while other groups of organisms, long separated geographically, have been brought into juxtaposition by plates 'colliding' with one another. The effect of these processes is evident today in the distribution of many groups of plants and animals.

An early ice age

Throughout the history of life on earth, the continents have moved, sometimes getting nearer to the poles, and so getting colder, sometimes moving into warmer latitudes. Another product of plate movement which has had enormous consequences for living things has been the uplift of mountain chains. This has taken place, for example, where there have been collisions between plates as has occurred in the Alps and the Himalayan–Tibet region. The effect of these major changes in topography not only alters the climate in the immediate vicinity, but by disturbing the pattern of air movement can have much more widespread impact on regional climate.

One of the most far-reaching climatic events of the past took place some 300 million years ago, in the late Carboniferous and early Permian (Fig. 2.9). The present-day land areas of South America, Africa, India, Australia and Antarctica were at that time joined together as a single 'supercontinent' called Gondwana (Fig. 2.8) and the South Pole lay close to its centre. Through the late Carboniferous and into the Permian a major ice-cap developed over the centre of that continent, greater in extent than the ice-sheets which cov-

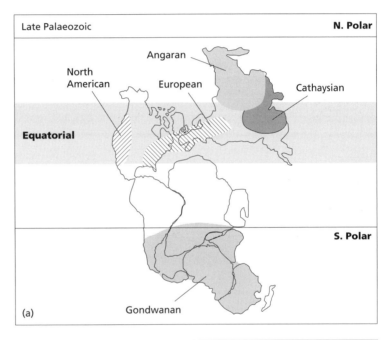

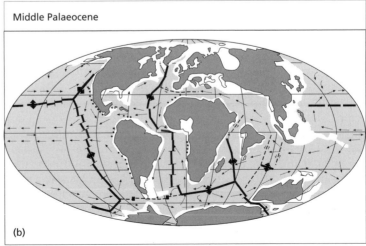

Fig. 2.8 Plate positions in (a) Permian and (b) Palaeocene time. In (a) the South Pole fell within what subsequently became Antarctica, at the core of the Gondwanan land mass. The several provinces characterized by different floras are indicated. By the end of the Permian, Gondwana was supporting gymnosperm forests within a few degrees of the Palaeo-South Pole. In the Palaeocene (b), India was already moving across the equator from its original position in Gondwana, eventually to impact on the rest of Asia. After Rogers (1993).

Era	Period	Epoch	Approximate duration in millions of years	Approximate date or commencement in millions of years BP	
Cenozoic	Quaternary	Pleistocene	2·4	2·4	Million years BP
Cenozoic	Tertiary	Pliocene	2·6	5	
Cenozoic	Tertiary	Miocene	18	23	
Cenozoic	Tertiary	Oligocene	13·5	36·5	— 50
Cenozoic	Tertiary	Eocene	16·5	53	
Cenozoic	Tertiary	Palaeocene	12	65	
Mesozoic	Cretaceous		70		— 100
Mesozoic				135	— 150
Mesozoic	Jurassic		70		
Mesozoic				205	— 200
Mesozoic	Triassic		45		
Palaeozoic				250	— 250
Palaeozoic	Permian		40		
Palaeozoic				290	— 300
Palaeozoic	Carboniferous		65		
Palaeozoic				355	— 350
Palaeozoic	Devonian		50		
Palaeozoic				410	— 400
Palaeozoic	Silurian		25		
Palaeozoic				435	— 450
Palaeozoic	Ordovician		75		
Palaeozoic				510	— 500
Palaeozoic	Cambrian		60		— 550
Precambrian				570	
Precambrian			4000		

Formation of Earth's crust about 4600 million years ago   — 4600

Fig. 2.9 The geological time scale. Note that in the lower part the vertical units are not to scale, and that the large part of earth history is taken up in the Precambrian. From Cox & Moore (1993).

ered much of the northern hemisphere during the last million years (see 'An age of ice', p. 31). The realization that glacially scratched rock surfaces and ancient glacial deposits formed simultaneously on the now widely separated fragments of Gondwana in Carbo-Permian time greatly influenced Wegener and others who advocated 'continental drift' in the early years of this century.

While the ultimate cause of the onset of that glacial phase is still the subject of debate, it is worth taking note of a further dimension to the climate story. We know from the air trapped in the ice cores of Greenland and Antarctica of the last 200 000 years that the CO_2 content of the

atmosphere has gone up and down in phase with the advance and retreat of the ice-sheets. The evidence for this will be reviewed in Chapter 3. It appears that a low level of atmospheric CO_2 (with a corresponding low greenhouse effect) is in some way linked to conditions favouring the extension of polar ice-caps. In this connection it is interesting to compare the past levels of atmospheric CO_2 with the history of global climate. By developing a model of the global carbon budget over the last 600 million years, coupled with other evidence, Berner (1994) has plotted the calculated values of global atmospheric CO_2 levels over the last 600 million years (Fig. 2.10). This shows a major drop in the CO_2

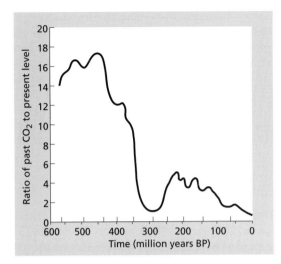

Fig. 2.10 Changes in atmospheric CO_2 through the last 600 million years of geological time; the vertical axis shows the CO_2 level expressed in terms of its pre-industrial level ('RCO_2'), the horizontal axis is in millions of years before present (BP). The main colonization of the land by plants took place around 400 million years ago; the great coal-forming swamps were widespread in the northern hemisphere around 300 million years ago. After Berner (1994).

from the Devonian (c. 400 million years ago), when it was around 12 times its present level, down to about its present level at the time of the Carbo-Permian glaciation of Gondwana (c. 300 million years ago). It appears likely that the burial of plant material in the form of peat (eventually to become coal) through the Carboniferous, over much of the (now) northern hemisphere, played a major role in this fall in the CO_2 level.

An important feature of Berner's CO_2 curve is that a low level of the CO_2 greenhouse effect is seen at the two phases in the last 600 million years in which there have been major glaciations—the Carbo-Permian and the Pleistocene, the Ice Age of the last 2 million years. Whether the low CO_2 is the cause of such an 'ice house world', or was produced by it, will be taken up again in the next chapter.

A greenhouse world

After the Carbo-Permian glaciation, the CO_2 level rose again to about five times the present value, then dropping over the last 200 million years to the relatively low levels of the last 2 million years

(Fig. 2.10). During Jurassic and Cretaceous time (approximately 200–70 million years ago) the world appears to have been in a phase of remarkably equable climate through a wide range of latitudes (Spicer *et al.* 1993). Even well into the late Cretaceous and the ensuing Eocene period, forests extended into the Arctic and Antarctic circles of that time (Creber & Chaloner 1985). By 70 million years ago we see broad-leaved trees of kinds now familiar in the woodlands of temperate Europe and North America, extending into high latitudes in Alaska and Spitzbergen, while trees typical of the present-day humid tropics extended as far north as the British Isles. Palms, which are noteworthy as a family being largely confined to the present-day tropical and sub-tropical latitudes, were growing in southern England, and reached corresponding latitudes in North America. This 'greenhouse world' of high CO_2 level and a globally equable climate contrasts with that of the Carbo-Permian and the Pleistocene ice ages, with extensive and fluctuating ice-caps and steep climatic gradients from the tropics to polar latitudes. As we move on in Chapter 3 to look at the impact of the last Ice Age, it is useful to bear in mind the extent of these earlier climatic changes. If the 'greenhouse world' of the Cretaceous and Eocene Periods was the product of a high CO_2 level, then the environment which that represents is certainly less hostile to life than that of a glacial phase, under a low CO_2 level, as existed only 20 000 years ago.

Carbon burial

Another long-term effect of continental movement involves the cycling of elements, and most particularly carbon. Some organic matter produced by photosynthesis, escapes biological degradation and becomes incorporated and buried in sediments, either on the continents themselves (as in lake sediments or peat bogs) or along the continental shelves of their submerged margins. In this way it is taken 'out of circulation' at least for the time being, in the sense that it is no longer part of the circulation of the carbon cycle (see Chapter 5); this process is referred to as carbon burial. In the oceanic environment the calcium- and magnesium carbonate shells of planktonic plants and animals play a similar role. Such calcareous (limestone) sediments accruing in a marine environment can lock up carbon on a long-term time scale. This is one of the most important ways in which the CO_2 level in the atmosphere has been modified by a combination of biological and geological processes on a scale of millions of years. The implications of this will be considered further when we look at the marine realm in Chapter 6.

The occurrence of wildfire—that is, forest- and grassland-fire—is a further natural process which makes a unique contribution to the long-term cycling and burial of carbon. Wildfire is a process that gives a quick return of biomass back to atmospheric CO_2, without the intervention of biotic processes. There is strong evidence from the occurrence of fossil charcoal that fires have been a feature of terrestrial ecosystems under a wide range of climatic regimes, ever since plant communities have existed as a fuel source. Figure 2.11 illustrates how the process of wildfire is linked to many other processes in the earth system. Although wildfire does of course destroy much living and dead plant material, a certain proportion will become charred in the course of the fire. That is to say, it is heated to the point where volatile constituents are driven off, and if the local lack of access to oxygen then prevents burning from occurring, a residue of charcoal is left. This is remarkably tough physically, it floats (and so is easily water-transported) and is resis-

tant to biodegradation. The residue of charcoal detritus which may be carried into the ocean following forest fire is another route to carbon burial. It has been estimated than on average a forest fire may yield 10% of its biomass as charcoal (Helas 1995). If this figure is borne out, and much of that material is transported out of the ecosystem by erosion and water transport, then wildfire is a significant contributor to long-term carbon burial.

Long-term carbon cycling

The process of carbon burial is not an irreversible one on a scale of millions of years. Carbon in the form of peat, lignite and coal buried in sedimentary rock can, after the elapse of some millions of years, come to be exposed at the surface of the earth's crust through the normal processes of crustal uplift, folding and erosion. Oxidation of this fossil organic matter is in part, simply oxidative weathering, and in part mediated by microorganisms, and both these processes will return the fossil carbon to the atmosphere as CO_2. In a rather different way, carbon in the form of limestone (carbonate rock) exposed at the surface is also subject to weathering. Interestingly, the return of this carbonate carbon back into circulation through the biosphere is itself mediated by CO_2. This, in solution in rain and groundwater,

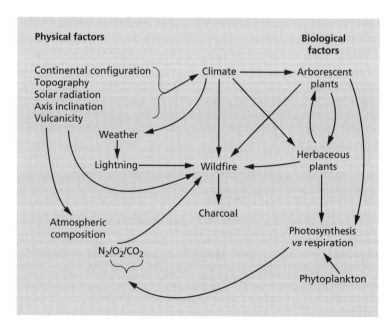

Fig. 2.11 The central role of wildfire in the earth system. The burning of vegetation has been a feature of the biosphere for some 400 million years, ever since plants produced sufficient biomass on the land surface. In any one location the incidence of fire is controlled by the nature of the vegetation, which is itself largely controlled by climate. But a series of other physical and biological factors have influenced and interacted with the occurrence of wildfire in natural communities over past geological time. From Cope & Chaloner (1985).

takes the carbonate into solution as bicarbonate ions. In this form, hard crystalline limestone may be dissolved by rainwater and the carbon in it recycled into the biosphere by passing into the ocean to be used in photosynthesis by phytoplankton or incorporated into carbonate shells.

The weathering of fossil fuel deposits is a slow return of fossil carbon into circulation, occurring as a natural process without human intervention. For the last few hundred years, humankind has been mining and burning fossil fuels, so carrying out the same process but at an enormously accelerated rate. This return of fossil carbon from the vast reservoir in the earth's crust is a very necessary part of the long-term carbon cycle, which maintains the atmospheric CO_2 available to plant life. However, the acceleration of this process by humankind may be a cause for serious concern, and this is explored further in Chapter 5.

The carbon/oxygen balance

The great proportion of all known global carbon is in the form of 'buried' carbon in the sedimentary rocks accumulated through the course of earth history. This outweighs all other carbon in the oceans, the atmosphere and the biosphere by a factor of over 2000. This removal of carbon from the atmospheric reservoir of CO_2, by the process of photosynthesis, means that theoretically at least, for every atom of carbon buried, and so taken out of circulation through the biosphere, there has been a release of an oxygen atom into the atmosphere. (It should be noted that this widely held interpretation presumes that the ultimate source of all atmospheric oxygen was CO_2!) However, even accepting this thesis, there are many problems in relating the amount of buried carbon to the oxygen in the atmosphere, since there are many other elements in the crust able to take up oxygen. These include iron and sulphur, both of which exist in nature in more than one oxidation state. The early history of the earth's atmosphere is widely believed to have involved a stage when oxygen released by photosynthesis began to make a significant contribution to the atmosphere for the first time. Reduced (ferrous) iron at the earth's surface would have become progressively oxidized to the predominantly ferric condition of iron in sedimentary deposits of at least the last 600 million years. Figure 2.12 shows

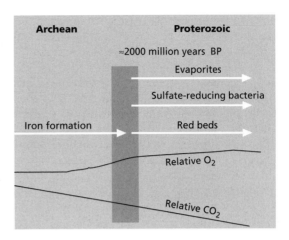

Fig. 2.12 Changes in atmospheric oxygen and CO_2 through the early part of earth history (the Archaean and Proterozoic, the major subdivisions of Precambrian time; see Fig. 2.9). The rising oxygen and falling CO_2, resulting from photosynthesis, caused a significant change in global chemistry at around 2000 million years ago, from a reducing to an oxygenic environment. Although iron-bearing formations exist prior to that time, no oxidized iron in the form of red beds are known earlier than that. After Rogers (1993).

one version of a possible history of oxygen and CO_2 in relation to the prevailing state of iron and other features of the physical environment. The 'iron formation' in the diagram represents deposition of iron, at least intermittently, in its oxidized (ferric) state, and was characteristic of a partially oxygenic atmosphere in Precambrian time. As the atmosphere became increasingly rich in oxygen, such iron formations do not occur later in the geological record. Indeed, it is believed that the present level of atmospheric oxygen was probably reached soon after the great burst of metazoan life some 600 million years ago.

Conclusion

In this chapter we have attempted to review some of the ways in which the physical development of the earth (the geosphere) has changed the environment in which organisms (the biosphere) have undergone their evolution. But it also illustrates how, from the earliest times, living forms have brought about change in their physical environment. This interaction of the geosphere and biosphere has been a consistent feature of earth's history. Humankind, more than any other form of life on earth, has become deeply involved in this interaction; but it is important to understand that 'interference' by life in the physical state of the earth was not a human invention! None the less, we are evidently bringing about change at a rate probably in excess of any that has previously operated. But, as we have seen, the whole of earth history is a catalogue of drastic change. In the following chapters we will try to consider the impact of humankind on the global environment and, vice versa, in the perspective of the past record set out here.

Further reading

Lovelock, J.E. (1995) *Gaia, a New Look at Life on Earth*, 2nd edn. Oxford University Press, Oxford.

Margulis, L. & Olendzenski, L. (eds) (1992) *Environmental Evolution*. M.I.T. Press, Cambridge, Mass., USA.

Pickering, K.T. & Owen, L.A. (1994) *An Introduction to Global Environmental Issues*. Routledge, London & New York.

Rogers, J.W. (1990) *A History of the Earth*. Cambridge University Press, Cambridge.

NASA (1988) *Earth Systems Science*. National Aeronautics and Space Administration, Washington D.C.

Chapter 3

An Age of Ice

The last 2 million years or so of the earth's history are of particular interest to students of global change for a number of reasons.

1 Being the most recent part of geological history, events during this period of time lead up to the present day, and any patterns of environmental change within this time offer the greatest potential for developing predictive models about the future.

2 The last 2 million years has been a time of global climatic instability, the lessons from which are most pertinent to current problems.

3 It is during this time that our own species has evolved and has assumed a significant role in the biological and physical processes of the earth.

The last 2 million years covers the period of time known to geologists as the Quaternary (Fig. 2.9). It succeeds the so-called Tertiary Period and it lasts right up to the present day. Most geologists divide the Quaternary Period into two Epochs, the Pleistocene (covering most of the Period, from its inception right up to the end of the last glacial stage) and the Holocene, or Recent Epoch, covering approximately the last 10 000 years. Many geological boundaries are rather arbitrary and some would regard this subdivision of the Quaternary as particularly so since, as we shall see, there is no unprecedented break in the pattern of environmental or climatic change at the Pleistocene/Holocene boundary that really warrants the separation of the last 10 000 years from the rest of the Quaternary, but the division is a convenient one from the human perspective and so is widely used and accepted.

The commencement of the Quaternary is even more contentious than its subdivision, but the Period contains much evidence of continued global cooling similar to that experienced through the Tertiary. The long-term cooling of global climate over the last 70 million years seems to have occurred in steps rather than smoothly and the arguments about the beginning of the Quaternary centre upon which step in the cooling process is sufficiently clear, strong and recognizable to form a definitive horizon for the establishment of the boundary between the two Periods. The alternative proposals will be discussed later in this chapter.

Geology and the Ice Age

In 1837 the geologist Louis Agassiz presented to the Swiss Society of Natural Sciences in Neuchatel a theory that the glaciers, still to be found occurring in the mountains of his native Switzerland, had once been much more extensive, covering large areas of Europe during what he termed the *Eiszeit*, or Ice Age. The concept of widespread catastrophe resulting from a Biblical Flood was then widely accepted, but such a major role for ice in the shaping of landforms was quite a new innovation and met with considerable scepticism. Many of the deposits of materials now known to have been transported by glaciers were considered at that time to have resulted from the Flood catastrophe and were given the general term 'drift'. But Agassiz pointed out that some of the transported materials found hundreds of miles from their places of origin were in the form of massive boulders, 'erratics', unlikely to have been carried by water flow. The British geologist Charles Lyell had already pondered this problem and had come to the conclusion that such massive structures could have been water-borne only if frozen into drifting icebergs on the surface of the flood waters.

The idea of glacial movements in the past was not totally new. Similar thoughts had been put forward in the late 18th century by James Hutton, a Scottish geologist often regarded as the founder of geological science. His major contribution to geological thinking was the principle of 'uniformitarianism', later to be developed by Charles Lyell. This principle claims that the forces which shaped the earth in the past are essentially the same as those operative today, and they operate on the whole in slow and continuous processes that effect great changes over long periods of time. The idea was diametrically opposed to 'catastrophism' such as that required by global flood. There are occasions, however, when catastrophe has been an important part of earth history as, for example, in the case of the supposed bolide that collided with the earth at the Cretaceous/Tertiary boundary. But even this type of event is not a total break with present-day processes. Asteroids still strike the earth and the possibility of a large collision remains realistic. The occurrence of a global flood, on the other hand, demands the operation of physical and geological laws quite contrary to those observed today for there is insufficient water on the planet to cover the entire land surface.

Adopting uniformitarian ideas, it was natural that geologists in Switzerland, familiar with the power of glaciers and with evidence of their past motions in the form of moraines of rock detritus, scratches along the rock faces they had passed and 'U'-shaped valleys carved by ice movements (Fig. 3.1), should begin to form mental reconstructions of times when ice dominated their landscape. Even in Germany some were daring to suggest that polar ice had once extended into that country, but few scientists took seriously such radical suggestions as those of the German geologist Reinhard Bernhardi in the 1830s. But by 1840, Louis Agassiz was confidently proclaiming that ice had once covered not only northern Europe, but also northern parts of Asia and America.

The ideas of Agassiz took root in America and provided an explanation for many aspects of its superficial geology, and his acclaim there led eventually to Agassiz taking up a chair of geology at Harvard in 1846. But the controversy about the geological role of ice action continued until well into the 1860s.

Even with the increasing acceptance of the Ice Age theory, however, there were still many questions to be answered. When had the Ice Age occurred? Had there been more than one Ice Age? How far did the great ice-sheets extend? Agassiz had tended to assume that there was only one Ice Age, but in the eastern part of southern Britain

Fig. 3.1 A glaciated 'U'-shaped valley near Obergurgl in southern Austria. The passage of a former glacier has steepened the valley sides and rounded its floor.

(East Anglia) Joseph Trimmer discovered that there were two distinct layers of drift (now termed 'till' to denote that they are deposits resulting from glacial action). This was later increased to four as a result of the work of Archibald Geikie who had also found evidence of warm episodes between ice advances in the form of plant fossils between the tills of Scotland. The concept of the 'interglacial', warm stages between glacials, began to emerge.

A formal statement of the number and sequence of glacials, however, had to wait until 1909 when A. Penck and E. Bruckner published their account of the Alpine glaciations. Again, they recognized four ice advances and they named them after four of the Alpine tributaries to the Danube River, namely Günz, Mindel, Riss and Würm, in that temporal order. The names became widely adopted in Europe and the concept of an Ice Age with four glacial episodes, or stages, became an idea almost as firmly established as the Biblical Flood had been a century before. Recent work, which will be described below, has shown that this sequence is an oversimplification of the actual glacial history of the Quaternary Period.

Evidence for cold and warm stages

Ice scratches on rocks provided clear evidence of past ice action, but discerning the number and nature of former glacial advances requires more precise information about the origin and the direction of ice flow. Erratics, the rocks carried by moving glaciers and deposited by them when they melt, have often been a useful source of information in this respect. But even more valuable are the tills, the masses of detritus left behind during glacial retreat. Occasionally these may overlie one another, reflecting sequences of glacial activity, and the alignment of stones within the clays and silts can provide an indication of the direction of flow of different ice-sheets. In the eastern part of the British Isles, for example, a series of glacially derived tills has been described and the orientation of stones within them shows that the sequence of glaciers approached the region from different directions (Fig. 3.2).

In addition to the differences in their directions of flow, glaciers also varied in the distances they covered, some extending further from their origi-nal sources than others, depending on the cold-ness, the availability of precipitation (especially snow, the raw material that feeds glacial growth) and the duration of the cold episode. The greatest spread of a glaciation is recorded geologically by a terminal moraine—an unsorted agglomeration of rocks and rubble that marks the outermost fringe of a glacial advance. The maximum extent of three major glaciations of Britain is shown in Fig. 3.3.

Areas beyond the limits of a particular glacia-tion may still endure extremely cold conditions (termed *periglacial*) and these may also leave their permanent mark upon the geomorphology (the form of a landscape) of an area. Under these very cold conditions, soils were often underlain by permanently frozen subsoils (*permafrost*) and the upper layers, charged with meltwater in the

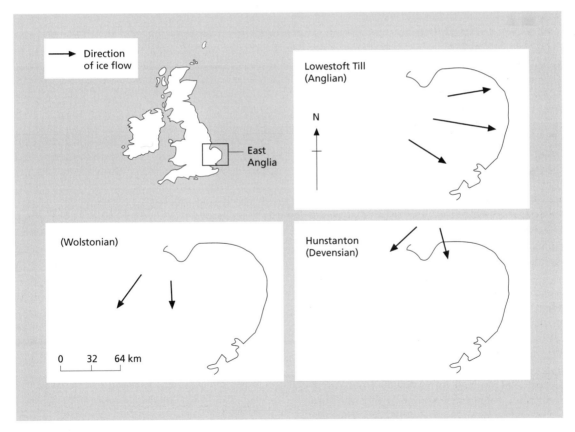

Fig. 3.2 Map of eastern Britain (East Anglia) showing the presumed direction of ice flow in the last three major glaciations, based upon the alignment of stones in the tills. The earliest (Anglian) entered East Anglia from the west while the later two (Wolstonian and Devensian) approached from the north, the former penetrating further south (see Fig. 3.3). After West (1977).

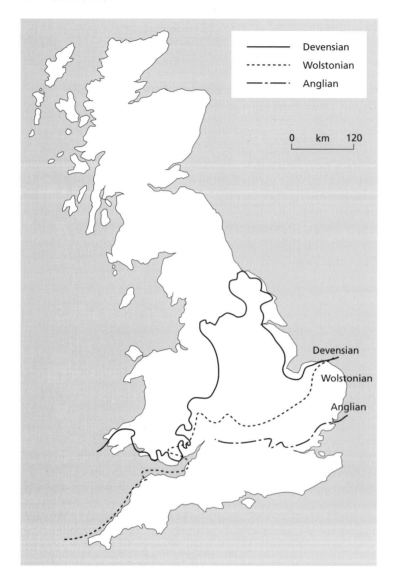

Fig. 3.3 Map showing the maximum extent of the three major Quaternary glaciations of the British Isles based on geomorphological evidence, such as moraines. The oldest (Anglian) penetrated furthest south in the east of the country. The following (Wolstonian) advanced less far in the east, but further in the west. The final (Devensian) glaciation was generally less extensive. After West (1977).

summer, may have become mobile on slopes and sludged into depressions and valleys. The freezing of surface waters in winter would lead to the development of polygonal patterns of ice wedges (Fig. 3.4) penetrating into the underlying soils, and the scars of such structures may still be detected in modern soil sections from former periglacial regions. Dry conditions in these periglacial regions, coupled with high winds and lack of vegetation cover, often led to wind erosion of fine-particled soils (called *loess*) and these were carried considerable distances and then deposited in thick layers, reaching depths of over 30 m in

Kansas and over 100 m in parts of eastern Europe and Asia. Wind strengths and directions can sometimes be determined from such deposits. In China, for example, these loess deposits have proved particularly valuable for the reconstruction of climatic history covering the entire Quaternary Period (Kukla 1987).

Glacial and periglacial sediments of these types may be interspersed with evidence of warm episodes in the form of lake muds, peats, or fossil soils rich in organic matter, and these also play an important part in reconstructing the climatic sequence and the record of former vegetation

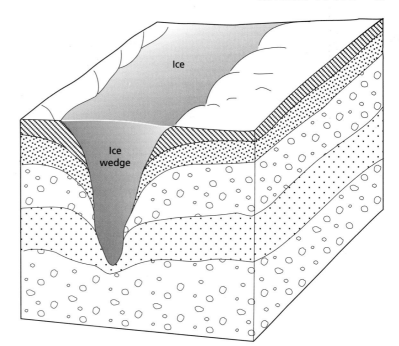

Fig. 3.4 Diagrammatic section of an ice wedge. These structures develop under repeated freeze/thaw cycles in tundra environments. The casts they produce in the soil profile after melting become infilled and may be preserved by burial, thus providing evidence of former periglacial conditions.

cover and animal populations. Where these deposits are stratified (i.e. retained in an evident sequence of layers) it may be possible to follow the sequence of the fossils they contain (leaves, fruits, pollen grains, beetles, molluscs, etc.) that indicate climatic changes. For example, a band of sediments sandwiched between two tills (Fig. 3.5) may contain fossils that indicate a climatic sequence from arctic to sub-arctic to boreal (e.g. coniferous forest) temperate (e.g. deciduous forest), and then back through boreal to sub-arctic to arctic before entering the next glacial phase. Where the episode recorded has been adequately warm and sufficiently long-lasting to permit the development of a temperate biota (plants and animals), it is termed an *interglacial*. If the warmth, or the duration, has been limited and only boreal conditions have been achieved before climatic reversal has set in, then the episode is termed an *interstadial*.

Just as different glaciations can sometimes be recognized on the basis of the rocks in their tills and the directions of their flow, it may be possible to identify interglacials on the basis of the fossils they contain. Early interglacials in Europe, for example, contain far more species of trees than do the later ones (Tallis 1991; see Fig. 3.6) and it is

sometimes possible to identify the interglacial on the basis of its flora. This can only be achieved, however, within limited geographical areas since the vegetation of past interglacials may be assumed to have varied with geographic location just as much as that of the present warm stage.

On the basis of such records of past cold and warm stages it has been possible in some parts of the world to reconstruct quite a detailed sequence, but this is possible only where older materials have not been destroyed or eroded by more recent glaciations. Some detailed and relatively full reconstructions have been achieved in The Netherlands, where at least seven major cold stages have been proposed on the basis of geological sediments. Data from a series of lake sediments on the High Plain of Bogota in Columbia, however, are perhaps the most complete of any terrestrial (land-based) records of the Quaternary sequence. A 357 m deep core from this site is considered to date back 3.5 million years, well back into the Pliocene (Hooghiemstra 1984). At this site approximately 10 cold stages can be detected in the last 2 million years. The initial suggestion of four cold episodes, based on early British and Alpine research, therefore seems to have been an unacceptably conservative estimate.

Fig. 3.5 A coastal exposure of interglacial sediments. The dark organic materials were formed under warm interglacial conditions and the clays above were deposited by glacial action. (West Runton, Norfolk.)

Fig. 3.7 A wave-cut platform in southern England. A former cliff-line is evident on the left of the picture and dates from an interglacial when sea level (relative to land level) was higher than at present.

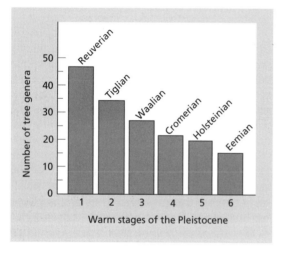

Fig. 3.6 Diagram showing the number of tree species in successive interglacials in north-west Europe. The pressures placed upon trees by each glacial stage resulted in European extinctions of some trees, while others became more restricted with each glacial event and failed to reappear in northern Europe. Data from Tallis (1991).

Sea levels in the Quaternary

One consequence of the expansion of ice-caps and glaciers over the face of the earth is that more of the world's water will be locked up in this form, and that means that less is left behind in the oceans. A situation where the sea level drops relative to the land as a result of the removal of water in ice is termed a *eustatic* change in sea level. But when ice masses form over the land they result in a warping of the earth's crust such that the land surface is depressed relative to the sea level. This is termed an *isostatic* change in the relationship between land and sea levels. The coastline of any particular land area during the Quaternary is essentially the product of these two factors, but there is the added complication that some of the consequences of such changes can take a long while to equilibrate (as in the case of crustal warping, for example) so the study of relative sea levels can be very complex.

Evidence suggests that the relative sea/land level was at least 100 m lower in some parts of the world during the height of the last glaciation than it is at present and this would have had wide-ranging consequences both in terms of exposed land areas over which plants and animals could have spread and migrated and also in its effects on ocean currents and hence heat exchange around the planet.

There is also evidence, however, indicating that sea levels have been higher at certain times in the past. One source of such evidence is provided by wave-cut platforms and raised beaches (Fig. 3.7) that mark former boundaries between land and sea. These must relate to former warm stages when more of the world's water was in the sea and less in the ice-caps than is currently the case.

The relatively rapid climatic changes of the past 2 million years mean that the establishment of a stable equilibrium between sea volumes and crustal warping is unlikely to be established

Fig. 3.8 The island of Hailuoto (Finland) in the northern Baltic Sea is still rising from the waters as the land surface continues to recover from its former downwarping under the ice cover of the final glacial stage. Sand dunes develop on the exposed land surface and pine invades. In the far distance is the receding Baltic Sea.

before new changes occur. It is clear that changes are still taking place today as a result of the end of the last glacial stage, as can be seen in the rising land surfaces of the northern Baltic, where the crust is still recovering from the downwarping caused by its former ice cover (Fig. 3.8). Current sea level changes will be discussed in Chapter 6.

The palaeoclimatic record of the oceans

Evidence from many sources, as we have seen, suggests that there have been several periods of extreme cold in recent geological time, but constructing a precise documentation of the underlying climatic changes has proved very difficult. The records from the land (fossil-bearing sediments, etc.) are difficult to correlate from one geographic area to another and complete records going back throughout the last 2 million years are extremely rare because of the tendency of each cold stage and its associated ice advance to obliterate many of the sediments of earlier stages. Only sites outside the extreme limits of glaciation can be expected to produce a complete record (such as that of Bogota). To obtain a fuller picture we can, however, look to the oceans.

Ocean sediments, particularly those from deep waters, are more likely to have remained undisturbed by glaciation than those on land. The sediments are derived from a variety of sources, including the eroded material from major rivers, aerial input of dust (including that of volcanic origin), detritus carried by icebergs and ice rafts, and the skeletons or shells of sea creatures that sink to the bottom after their death. Among the latter, microscopic organisms are particularly important because they live in large numbers in the surface waters (the plankton) and many of them have persistent shells or scales that are not consumed by bacteria after their death, but sink intact to the sea floor. These fragments, such as the valves of diatoms, the tests of foraminifera and the cysts of coccolithophorids (see Chapter 6), are preserved as fossils in the sediments and provide a stratified record of past planktonic communities which, in turn, reflect the physical conditions of the oceanic surface waters in the past. In particular, they may provide a record of past ocean temperature.

The foraminifera have proved especially useful in tracing past climates. These are microscopic protozoa, related to amoebae, that live within shells (tests) constructed from lime (calcium carbonate, $CaCO_3$). The tests are coiled, rather like snail shells, and are sufficiently distinctive in their structure to allow fossils to be identified even to the level of species (Fig. 6.6, p. 100). Many species of foraminifera live at the bottom of the sea (*benthic*) in shallow water, while some occupy the surface waters of the open ocean (*pelagic*). It is these planktonic, pelagic forms that are most useful in climate reconstruction because the sediments of shallow, coastal regions are often disturbed and strongly affected by river discharge containing much eroded material from land surfaces.

Some foraminifera prefer warm waters in which to live, while others perform better in cold water environments so, provided that these different species can be identified accurately from their fossil tests, it should be possible to reconstruct climate change from the contents of different species in ocean sediments. This is a widely used technique and has proved very useful for this purpose; it has even been employed as a means of demarcating the onset of the Quaternary Period (see West 1977). In a geological section from Vrica in Calabria, Italy, there is a change in ocean sediment from a black organic mud to a clay and at this transition occurs the first appearance of a foraminiferan, *Cyclotheropteron testudo*, which is a cold-loving species. This combination of a change in sediment type and the arrival of a low-temperature indicator organism led the International Commission on Stratigraphy (a panel responsible for agreeing definitions of geological periods) to accept this event as the onset of the Pleistocene cold period (replacing the somewhat warmer Pliocene). The divide was dated at a approximately 1.8 million years ago.

Such definitions are clearly important and necessary, but they inevitably cause heated debate among scientists with opposing views. One objection to this definition of the Tertiary/Quaternary boundary is that it is based on a suspect identification of a foraminiferan. The fossil identified as *C. testudo* is regarded by some as a previously undescribed species closer to the extant *C. wellmani*, which is not a cold-loving species (Jenkins 1987). If this taxonomic criticism is correct, then the boundary is clearly unsatisfactory; indeed the very fact that it is open to doubt of this kind brings its value into question. Deep-sea cores from the Atlantic indicate that severe cooling and possible ice expansion was occurring at least as far back as 2.5 million years ago and some workers feel that this is a more appropriate marker for the Pliocene/Pleistocene transition. But evidence from Norway (Jansen & Sjoholm 1991) suggests that glaciers large enough to reach sea level were present by 5.5 million years ago, so the onset of the Quaternary cannot be regarded as other than a continued cooling in a long history of such events.

Quite apart from the disputed identification of this one important foraminiferan fossil, one should ask the question whether any conclusion about the changing climate of the past should hang on the behaviour of a single species. It is far more satisfactory to base such conclusions on whole communities of organisms (or their fossils). In the case of foraminifera, for example, studies have been undertaken to correlate current assemblages of foraminiferan tests in surface sediments in the Atlantic with present-day conditions and to use these as keys to the interpretation of past conditions. The geologists, Imbrie, Van Donk and Kipp (1973), collected over 60 samples of surface sediments and analysed their foraminiferal content; they used multivariate statistical techniques to look for grouping within their samples and correlate particular assemblages with envi-

ronmental factors, such as winter temperature or water salinity. From these data were derived a series of equations (termed 'transfer functions') that permit a fossil assemblage to be interpreted in terms of these physical conditions. A fossil core from the Caribbean was analysed and interpreted in this way and provided very good correspondence to other sources of evidence for past temperature changes, so the use of whole assemblages of fossils, rather than individual 'indicator' species, seems a more reliable approach to climatic reconstruction.

The use of whole assemblages of fossils, such as foraminifera, can thus produce reliable data about climatic changes over the past 2 million years, but additional sources of information are needed to supplement and confirm such findings.

Oxygen isotope evidence

One of the most powerful tools for the reconstruction of global temperatures over the past 2 million years has been the analysis of oxygen isotopes in marine sediments and in ice cores. The technique is based upon the fact that the element oxygen can exist in a number of forms (isotopes), two of which, ^{16}O and ^{18}O, are stable and therefore persist in sediments and in ice. Of the two, ^{18}O is heavier and is also much rarer: only about 0.2% of the oxygen in circulation is in this heavy form. The heavy isotope is capable of combining with other elements in the same way that the normal isotope does, so it can combine with hydrogen to form water $(H_2^{18}O)$, or with calcium to form calcium carbonate (lime) $(CaC^{18}O_3)$.

In the oceans, water molecules with the heavy oxygen isotope do not evaporate and enter the vapour form as readily as those with the lighter isotope, so the water vapour above the oceans contains a lower proportion of $H_2^{18}O$ than the water left behind. This process is termed *fractionation*. Any of these heavier molecules that do reach the atmosphere also condense more easily, so they soon re-enter the ocean, while the lighter $H_2^{16}O$ continues on its way, ultimately to fall as precipitation at some distance, perhaps even over the land. In high latitudes and over high mountains some of this precipitation may fall as snow and may become incorporated into glacial ice masses, where the water molecules may be retained for considerable periods of time (hundreds of thousands of years in the case of major ice-sheets like the Greenland and the Antarctic ice-sheets).

In the oceans, some of the oxygen from water molecules combines with carbon dioxide dissolved from the atmosphere to form hydrogen carbonate (bicarbonate) ions and these combine with calcium to produce calcium carbonate (lime) that becomes incorporated into the tests of foraminifera and other planktonic organisms. These, as we have just seen, eventually accumulate on the ocean floors as sediment. This means that both in ocean sediments and in terrestrial ice

masses a stratified deposit of material is laid down in which a permanent record is retained of the relative amounts of ^{18}O and ^{16}O present in the medium at any given time in the past. Because of fractionation, however, ^{18}O will always be richer in the ocean sediments.

If global temperatures fall, even less $H_2{}^{18}O$ is able to evaporate into the atmosphere and also more ice is laid down over the surface of the world, so proportionally more ^{16}O becomes locked up in the terrestrial ice and ^{18}O becomes even more concentrated in the oceans. So in cold conditions the ice will contain less ^{18}O and the

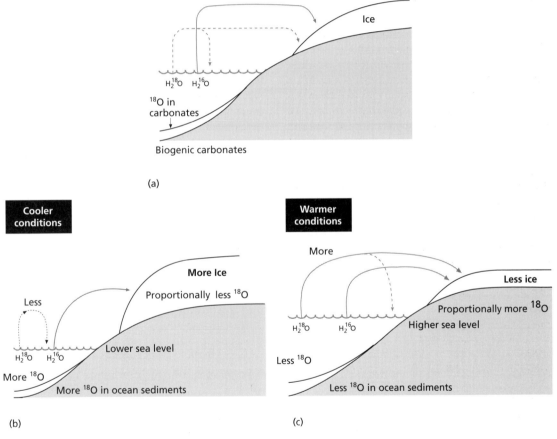

Fig. 3.9 Diagram showing the behaviour of two isotopes of oxygen, ^{16}O and ^{18}O, and their circulation through the atmosphere, oceans and ice-caps. The lighter and commoner isotope, ^{16}O, when combined in the water molecule H_2O, evaporates more easily than the heavier one, so proportionally more of it is transported in the atmosphere, precipitates and accumulates in ice. Under warm conditions rather more ^{18}O reaches the ice and global ice volumes are also smaller; both processes mean that there is less ^{18}O in the oceans and therefore less entering the sediment. In colder conditions more ^{18}O remains in the oceans. The ratios of the two isotopes in ocean sediments or in ice cores can thus provide a record of past temperature (remembering that the ratios behave inversely in the two media).

ocean sediments will contain relatively more (see Fig. 3.9).

If temperatures rise, however, rather more $H_2{}^{18}O$ escapes from the oceans and the melting of glacial ice brings more $H_2{}^{16}O$ back into the oceans. So warm conditions result in relatively more ^{18}O in ice and proportionally less in ocean sediments (Fig. 3.9).

We have in the oxygen isotope ratio of ice cores and ocean cores, therefore, a measure that can be correlated with past climate and that can be used for long-term climatic reconstruction within the Quaternary. It is important to note that the absolute levels of the isotopes are not needed for this process, but the ratio of the two oxygen forms, or their variation in relation to a standard measure. In research of this type the ratio is compared with internationally agreed standards (see, for example, Lowe and Walker 1984 for further details). The system can be calibrated on the basis of current water temperatures and oxygen isotope ratios and this permits a precise interpretation of past curves. An increase in the $^{18}O{:}^{16}O$ ratio in marine sediments of 0.07%, for example, corresponds to a fall in water temperature of approximately 1°C.

Cores of marine sediments from many parts of the world and of ice from major ice-caps have been collected and their oxygen isotope profiles determined. The results have been in good general correspondence and it has proved possible to erect a series of zones, or oxygen isotope stages. An example of an analysed core is given in Fig. 3.10 and the upper 23 stages are marked upon it. This core covers the entire Quaternary Period and a number of important features are immediately apparent.

1 There is no sudden event that can be used to define the beginning of the Quaternary. A cooler episode began at about 1.8 million years ago and may correspond to foraminiferal, changes already referred to.

2 Climatic oscillations during the Quaternary have evidently been frequent (in geological terms) and numerous. The early conception of four (or even ten) major cold events is thus demonstrated to be an oversimplification of the true situation.

3 The mean temperature around which variations occur has fallen during the course of the Quaternary. Global conditions in the long term continue to become colder.

4 The pattern of fluctuations has changed during the course of the Quaternary. In the early stages (from depth 1300 cm to 1000 cm) the wavelength of fluctuations is relatively short and their amplitude is relatively low. Warm episodes alternate with cool ones at high frequency. Above 1000 cm (approximately the last million years) both the wavelength and the amplitude of oscillations has increased. Cold episodes have become colder and last longer.

5 In the most recent cycles there is a tendency for the cold episodes to reach their deepest troughs immediately before the warm stages, leading to a rapid swing into warmer conditions, followed by a stepwise regression into cold once more.

6 The present warm episode (termed the Holocene) has every appearance of being just one more event in a series. There is no reason to suppose that the sequence of fluctuations has come to an end with the present interglacial.

The oxygen isotope record has been divided into a series of zones, or stages, and these are

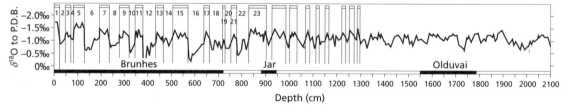

Fig. 3.10 A 21 m core from the Pacific Ocean showing the ratio of the two oxygen isotopes ^{18}O and ^{16}O relative to a standard measure (termed PDB). Increases in ^{18}O (that is, less negative values) relative to the standard indicate colder conditions, while decreases (more negative) indicate warmer climate (see Fig. 3.9 for the mechanism by which this operates). The peaks and troughs of ^{18}O levels are marked as oxygen isotope stages and are numbered from the most recent downwards (23 stages are labelled here). The dark and light bars refer to geomagnetic reversals (see Fig. 3.11). The full length of the core covers about the last 2 million years. Data from Emiliani (1972).

labelled numerically starting from the top of the sequence. So the present (Holocene) warm stage is labelled '1'. Other warm stages in the past can be seen in stages 5, 7, 9, 11, 13, 15, and so on. Matching the wiggles from different ocean sediment cores allows this nomenclatural system to be applied world-wide. It is also possible to match these oxygen isotope stages with those exhibited in cores from the ice-sheets of Greenland and the Antarctic, but the ice-core records fluctuate in an inverse manner (warm episodes have relatively more ^{18}O rather than less in the ocean sediments).

One aspect of this type of study that needs further explanation is the method of dating. The correlation of cores and their interpretation depends upon a reliable system of dating and the methods available and their limitations will now be considered.

Dating methods

The methods available for assigning a date to geological materials can broadly be divided into absolute and comparative approaches. Both are useful, but comparative methods need to be calibrated by reference to an absolute method before they can be applied with confidence. The absolute methods will be considered first. A more detailed discussion of many dating methods is found in Lowe and Walker (1984).

Potassium–argon dating

The radioactive potassium isotope ^{40}K is often present in rocks and like most such isotopes it decays in a regular fashion with time. ^{40}K can decay in two ways, resulting either in the formation of the calcium isotope ^{40}Ca, or argon ^{40}Ar. The isotope ^{40}Ca is relatively common and can arise in a number of other ways, so it is of no value for dating, but the transformation of ^{40}K to ^{40}Ar does provide a useful radiometric clock if the argon is retained. The age of the material is thus determined by measuring the accumulation of argon. Unfortunately, many sedimentary rocks lose argon, so the method cannot work with them, but argon is retained well in volcanic rocks and these can be effectively dated on the basis of their ^{40}K:^{40}Ar ratio.

The method has been widely used in the dating of volcanic rocks from the late Pliocene and early Pleistocene in East Africa, where sediments containing the remains of fossil hominids often alternate with lavas that can be dated in this way. The method has provided an absolute basis for some comparative dating methods such as reversals in the earth's polarity (see below).

Uranium series dating

There are several unstable isotopes of uranium that decay with varying half-lives. In some situations the decay products of these isotopes may build up and provide a means of age estimation. In corals, for example, uranium may be accumulated and subsequently decay to thorium and protactinium. The principle of measurement in this case is thus the increase in the product of decay rather than the loss of the unstable isotope itself (contrast radiocarbon below). Because of this there is no theoretical limit to the age of materials that can be dated.

Radiocarbon dating

This method depends on the regular decay of the radioactive isotope ^{14}C with time. The isotope is actually produced in the upper atmosphere where free neutrons generated by cosmic rays from space bombard nitrogen atoms and convert them into the ^{14}C carbon isotope. This then decomposes by radioactive decay (emitting β radiation) and the decomposition follows a very regular and predictable course with half of the total of carbon atoms decaying back to nitrogen every 5568 years (plus or minus 30 years). This value is known as the 'half-life' of ^{14}C.

The ^{14}C isotope cannot be distinguished from ^{12}C (the common form) by plants, so radioactive molecules of carbon dioxide ($^{14}CO_2$) are taken up by plants in photosynthesis and some is incorporated into their structure. Most of this will re-enter the atmosphere on the death of the plant as a result of microbial decomposition, but some may become preserved as peat, or desiccated in dry climates, and the radioactive decay will continue. Measurement of the ^{14}C left in an organic material will indicate its age; the lower the level the older the sample. Such measurement may be carried out by direct observation of how many atoms decay in a given time, or the precise concentration of ^{14}C can be determined using an accelerator (cyclotron). The latter method is more accurate, especially for old material where very few radioactive carbon atoms may remain.

There are several problems associated with radiocarbon dating, one being the very limited timespan over which it can be used. Any samples with an age in excess of about 40 000 years cannot be dated with accuracy because the levels of remaining ^{14}C are too low for secure determination. The method is also based on the assumption that the production of the ^{14}C isotope in the upper atmosphere is constant. Suspicions that this was not the case arose as a result of radiocarbon datings of objects that could be dated by alternative and undisputed methods, such as the use of wood samples whose age could be calculated by counting annual growth rings, or samples of material from ancient Egypt where a very precise chronology is known from historical records. Dates were found not to agree, and the difference between dates generally increased with time. Eventually, preserved wood samples from the long-lived tree *Pinus longaeva*, the bristlecone pine, from California allowed a calibration of the radiocarbon time scale going back many thousands of years (Fig. 3.11) and it was found that divergence from agreement was quite irregular, presumably depending on variations in the rate of formation of ^{14}C in the past. Such curves are now being constructed for other parts of the world to ascertain whether all of the smaller variations ('wiggles') are widespread and statistically significant. On the basis of such curves it becomes possible to calibrate radiocarbon dates and to express them as 'solar', or calendar years.

There are other problems associated with radiocarbon dating. An organism may take up carbon from an ancient source and incorporate it into its body. Calcium carbonate in lakes, for example, can produce carbon dioxide (CO_2) that is absorbed by aquatic plants and gives them a very considerable apparent age. This is termed hard water error. There is even a record (Schell 1983) of an old squaw, or long-tailed duck, being shot in North America with an apparent age of 1300 years because it had been feeding upon eroding peats of great antiquity!

The nuclear testing of the 1950s released considerable quantities of ^{14}C into the atmosphere and this has also influenced the validity of the method for very recent materials. Radiocarbon dates are usually expressed as years 'before present' (BP), where 'present' is taken to be 1950.

These methods of absolute dating have proved most useful in producing a time framework for Quaternary studies. One other is becoming increasingly important.

Conventional radiocarbon dates in
radiocarbon years BP

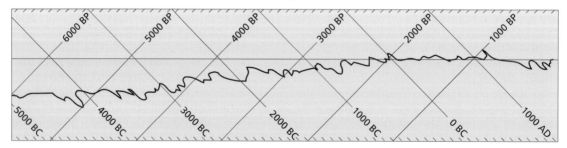

Bristlecone pine dates in calendar years

Fig. 3.11 Calibration of the radiocarbon time scale has been achieved by using the technique of dating slices of wood of known absolute age derived from the annual rings of ancient trees, in this case the bristlecone pine (*Pinus longaeva*). The horizontal line shows the expected correlation if there is no variation between the radiocarbon years before present (BP, where present is taken to be 1950) and so-called 'solar years' or calendar years. Two points emerge: (i) the curve is somewhat erratic, or wiggly; and (ii) the curve departs quite strongly from the expected value in samples older than about 2000 years. Radiocarbon 'years' and solar years are therefore not identical and radiocarbon dates need to be calibrated. From Lowe & Walker (1984).

Amino acid racemization

Amino acids are the biochemical building blocks of proteins. All amino acids are constructed on the same basic plan, centred on a carbon atom. Like all carbon atoms, this has four locations for the attachment of other atoms or groups of atoms, and the four locations in the case of an amino acid are occupied by an acidic carboxyl group ($-COOH$), an amino group ($-NH_2$), a hydrogen atom (H) and an organic chain ('R') that varies from one amino acid to another. The overall configuration is thus:

$$\begin{array}{c} H \\ | \\ R\text{-}C\text{-}COOH \\ | \\ NH_2 \end{array}$$

This molecule, of course, is actually three-dimensional and it can occur in two possible forms depending on the locations occupied by the four attached groups. In living organisms, for some as yet unexplained reason, only one form is found, the so-called 'L-isomer'. The proteins of dead organisms normally decay rapidly, but in certain circumstances, such as in bones or mollusc shells, they may survive long after death and a gradual process of change occurs in which the configuration of groups in amino acids change and the L-isomer partially alters to the D-isomer. This change is time-related and it is possible to estimate the age of a bone or a shell on the basis of the ratio of D- and L-isomeric forms. The reversion rate varies with the amino acid, but for isoleucine, for example, the half-life is about 15 000 years (approximately three times that of radiocarbon). This offers an opportunity to date suitable materials at least as far back as 200 000 years, which makes the method particularly useful for the last interglacial stage.

Magnetic polarity reversals

The earth's geomagnetic poles periodically reverse, or move through 180°c. This leaves its mark in geological sediments since part of their content becomes magnetized by the earth's field and direction of the field is thus recorded for posterity. The record of field reversals is now thoroughly known and the horizons at which such reversals have occurred in the past have been dated using the potassium/argon method (Fig. 3.12).

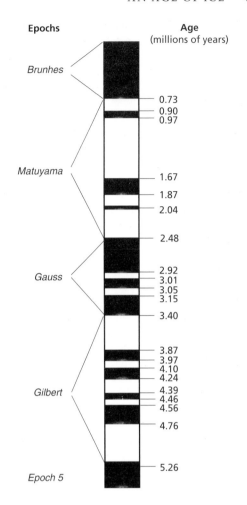

Fig. 3.12 Periodically, the polarity of the earth's magnetic field is reversed and this switch is recorded in magnetized particles in rocks and sediments. This diagram shows the pattern of reversals during the last 5 million years, together with their dates and names. Such reversals, which have been securely dated by absolute radiometric techniques, are most valuable as datum horizons for fixing a time scale to ocean and ice cores, as in the case of the Pacific core shown in Fig. 3.10. Adapted from Lowe & Walker (1984).

The documentation of polarity reversals does not, therefore, provide an absolute dating technique, but a comparative method by which the age of a particular horizon can be ascertained. These reversals are shown on the oxygen isotope diagram in Fig. 3.10 and it can be seen that a cold episode coincides with the commencement of the Olduvai magnetic reversal event, dated at 1.87 million years ago. This marker horizon has also made this particular point in geological history attractive as an acceptable commencement for the Quaternary.

The last glacial/interglacial cycle

Geologists and palaeoecologists (ecologists primarily concerned with the history of ecosystems) have now accumulated a very large body of data concerning the vegetation and the animal life of the different glacial and interglacial stages in different parts of the world. It would not be appropriate to attempt a summary of all this information here, but some examples of the results of such research will be presented as they form a necessary background for the understanding of current climatic changes, both recorded and projected. The data are also useful in displaying the response of animals and plants to climatic changes in the past, which helps us to understand how they may respond in the future.

Biological information is available in the form of fossils retained within sediments (lake muds, clays, etc.) and these come in various forms and sizes. Megafossils (or macrofossils) are derived from parts of plants and animals that are large enough to be examined without the use of a microscope. They include bones, mollusc shells, beetle wing cases, fruits and seeds, twigs, leaves, and so on. Being relatively large, their density in a sediment is limited, so even in rich deposits considerable volumes of material often have to be sifted to obtain large quantities of fossils. They are, however, often so distinctive in their form that they can be identified with considerable precision. This is made easier, of course, in the case of such recent material as that of the Quaternary, because many of the species represented or their close relatives are still extant (living at the present time). Unless they were deposited in very energetic conditions (such as in the alluvial sediments of rivers), megafossils usually have their origin relatively close to the site of deposition.

Microfossils consist of those fossil materials that are so small they need to be examined and identified with the aid of a microscope. They include algal spores and cysts, diatoms, foraminifera, protozoan cysts, microscopic crustaceans, pollen grains, moss and fern spores, etc. Those which are air-borne in their dispersal, such as plant spores and many pollen grains, may travel

considerable distances before deposition, perhaps even thousands of miles, and may have been carried by streams into lakes even after their initial aerial dispersal. So an assemblage of fossil pollen grains in an old lake sediment may represent a considerable catchment area of vegetation. Other microfossils, however, like diatoms, are more likely to be local inhabitants of the lake in which they have become fossilized. The regional representation of vegetation provided by fossil pollen, together with their abundance in fossil lake sediments (densities of tens of thousands per cubic centimetre are frequent), their toughness and the distinctive sculpturing of their outer coats (Fig. 3.13), make them ideal for reconstructing broad changes in plant communities (see Moore *et al.* 1991).

Pollen is so abundant and usually so well preserved in lake and peat sediments that it is possible to extract sufficient numbers of fossils from narrow bands of sediment to allow counts of many hundreds to be made and detailed accounts of vegetation assemblages to be constructed. Close sampling through a core of stratified sediment can then permit the observation of changes in the pollen content (and hence vegetation) through time. The data are usually displayed in a pollen diagram, as shown in Fig. 3.14. The vertical axis represents depth, which is related to time in a sequence with the oldest materials at the base.

Pollen diagrams of this type convey in a succinct manner the history of vegetation in an area over the time period covered by the analysed core of sediments. They must, of course, be interpreted with caution, for many factors influence the relative abundance of pollen apart from the density of

Fig. 3.13 Pollen grains. Photograph by M.E. Collinson.

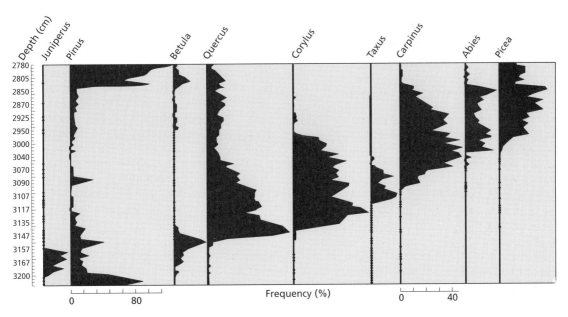

Fig. 3.14 Pollen diagram from Ribains, a mire site near Lyon in France. Only part of the full diagram is shown, covering the last interglacial episode. The vertical axis represents depth (related to age) and the columns show the relative frequency of pollen grains of various tree types at different levels. From Reille and de Beaulieu (1995).

that plant in the regional vegetation. Some plants produce more pollen than others, some types disperse more effectively, some live closer to wet sites, so are more likely to be over-represented, and so on.

Pollen diagrams of this type are now available from a wide range of sites all over the world and cover many periods of Quaternary history. This particular diagram comes from a deep depression in southern France near Lyon, called Ribains, and was analysed by Maurice Reille and Jacques-Louis de Beaulieu of the University of Marseille (1995). The hollow in which it lies was excavated by the last but one glaciation ('Riss') and was occupied by a lake during the last interglacial. The final glaciation ('Würm') did not extend as far as the site, so sedimentation continued uninterrupted right through to the present day, but only the sediments of the last interglacial are shown here (from 27 to 32 m in depth).

The basal part of the deposit mainly contains pollen of pine (*Pinus*), together with grasses and other herbaceous plants (full details of the non-arboreal pollen are not given here because they are

too complex to represent briefly, but mugworts, sedges, rock-roses, saxifrages and joint-pines are frequent and suggest a diverse, open glassland). The overall picture is an open forest tundra, and this was undoubtedly the vegetation of the final stages of the penultimate glaciation in this part of France. Forest canopies then became more closed. The detailed analysis of the tree pollen indicates that pine was replaced by a succession of other tree types, firstly a deciduous group passing from birch (*Betula*) to oak (*Quercus*) to hazel (*Corylus*) to yew (*Taxus*), and hornbeam (*Carpinus*), then back to a coniferous assemblage with fir (*Abies*), spruce (*Picea*) and pine once more as the next cold period began.

The entire sequence provides a picture of vegetation responding to a climatic cycle in which sub-arctic vegetation is replaced by boreal forest, then temperate forest, before reverting via boreal forest to forest-tundra conditions once more. The sequence thus follows that typical of an interglacial in the temperate zone.

Some pollen records for the last interglacial of two widely separated sites (California and Norway) are shown in Fig. 3.15. The full pollen stratigraphies of the sites are far too complex to show in detail because dozens of different pollen types are recorded, some very specific to certain parts of the world. A comparison between California and Norway is possible, however, if we consider only two pollen types, namely oak

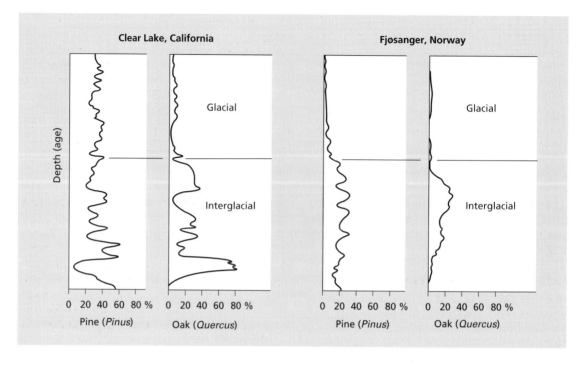

Fig. 3.15 Simplified pollen diagrams showing the history of pine and oak through the last interglacial and the final glacial in California and Norway. In high latitudes (Norway), pine is lost during the glacial stage, while in lower latitudes like California (and in southern Europe, Fig. 3.14) it survives. Data derived from Adams (1985) and Mangerud *et al.* (1979).

(*Quercus*) and pine (*Pinus*). Obviously, different species of these two genera are represented in America and Europe, but the general similarity of response for these two plant groups allows some comparison. In both of these locations oak pollen becomes much more abundant during the interglacial and then decreases in the glacial stage. Evidently this represents an invasion and an expansion of populations of oaks during the warm stage and a local extinction during the cold stage (as was the case also in the south of France, Fig. 3.14). Pine behaves rather differently, however. In California and in France it is reduced in abundance during the warm stage, presumably because of increased competition from deciduous species like oak, and increases during the cold stage. In Norway, on the other hand, which lies considerably further north, pine is able to expand only during the warm stage and becomes locally extinct during the glacial. The fact that oak extended as

far north as the Norwegian site (near Bergen), which lies at a latitude of 60°N (roughly the same latitude as Anchorage, Alaska), during the last interglacial is itself a consequence of the influence of warm ocean currents that pass into this region of the North Atlantic from the tropics.

The pattern of vegetation change during interglacials must be expected to vary with latitude, as shown here. One other obvious question is whether the last interglacial was warmer or cooler than our current one. Evidence from Europe suggests that it was warmer. The fossil bones of hippopotamus are widespread in southern England and many plants were present and others apparently were capable of fruiting further north during that interglacial than is now the case. The biological evidence suggests that the climate in Britain during the last interglacial was more continental, with warmer summers than during our current interglacial (West 1991).

Another question concerns time scale. Putting precise dates on the last interglacial has proved difficult because it lies beyond the range of radiocarbon dating. The use of amino acid racemization techniques has proved locally useful (Miller *et al.* 1979), as has the employment of uranium series dating (Lambeck & Nakada 1992). The latter work has been successfully employed in the dating of interglacial coral growths in Bermuda and Western Australia. Since corals are unable to grow above the level of extreme low water, fossil corals are useful tools for the recording of past sea levels and in some cases it has proved possible to date them using uranium decay techniques. Results from these corals place the last interglacial between 135 and 120 thousand years ago.

The last interglacial and the final glacial stage are well marked in the oxygen isotope record of the oceans and the ice-caps, as can be seen in Fig. 3.16. The final six oxygen isotope stages are marked in this diagram and, because of its complexity, stage 5 has been subdivided into five substages, a–e, of which the lowermost, stage 5e, can be identified with the height of the last interglacial. The characteristic mentioned earlier of a stepwise regression to very cold conditions followed by rapid reversion to warmth is very clear here, and the later substages in stage 5 show an alternation of cool and warmer episodes; but the warmer ones do not reach sufficiently high levels to be regarded as interglacials in their own right. These substages are represented in terrestrial sediments by interstadials where boreal vegetation was often established within what are now temperate latitudes.

The very strong decline in temperature seen at the transition from stage 5a to stage 4 is of particular importance. At this horizon in the sediment cores of the Indian Ocean there occurs a layer of volcanic ash that can be traced to a massive volcanic explosion on the island of Sumatra, caused by the eruption of the volcano Toba 73 500 years ago. Work on this event, which is widely recorded by ash layers in sediments, suggests that a billion tonnes of fine ash may have been thrown to

heights of over 30 km (Rampino & Self 1992). This could have led to a 'volcanic winter' as a result of the shielding of the earth from solar radiation and could have reduced global temperatures by 3–5°C for several years. At high latitudes the cooling could even have been as great as 10–15°C, which would have led to a permanent snow cover over Quebec and Labrador (Ramaswamy 1992) and possibly also in Scandinavia. This may well have triggered the start of the next round of global glaciation as the most recent ice age began in earnest.

Following this explosive start to the last glaciation, however, the attainment of maximum glacial advance was not a smooth affair. The climate became rather warmer once more between about 60 and 35 thousand years ago and then temperatures declined until the glaciation became most severe at about 20 to 18 thousand years ago. At that time the northern glaciers extended well south of the Great lakes region of North America and south to Washington on the west coast (Fig. 3.17). In Europe the glaciers covered northern Ireland and Britain and extended south into Germany and east into Russia, being centred on Scandinavia. Glaciation in the southern hemisphere affected New Zealand and many of the higher mountainous parts of Africa.

The question of the instability of the climate at this time is an important one if we are to draw general lessons about climate change from this particularly extreme (and very recent) episode in the earth's history. To what extent did climate vary between centuries and millenia and how quickly did climate change? Once again we must turn to the oxygen isotope record for information on these points, but we have to question its reliability when dealing with relatively short time-

Fig. 3.16 Oxygen isotope record of the last interglacial (stages 5a–e) and the most recent glacial episode (stages 4, 3 and 2). In contrast to ocean cores (see Figs 3.9 and 3.10), higher values for ^{18}O indicate warmer conditions. The onset of the glaciation (stage 5a/4 boundary) coincides with a major volcanic eruption, Toba in Sumatra. Data based on Martinson *et al.* (1987).

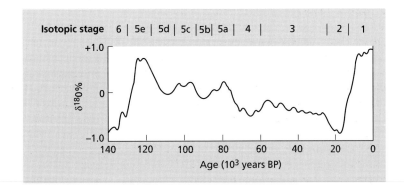

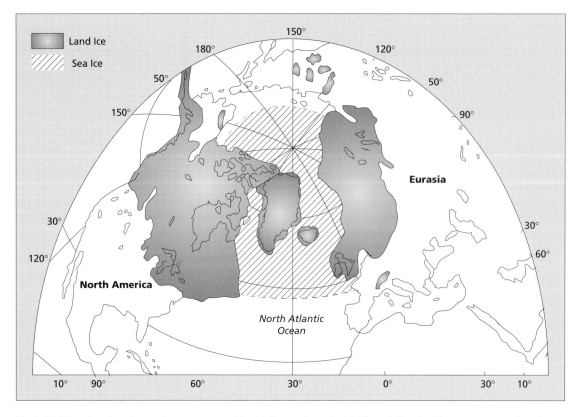

Fig. 3.17 Map showing the maximum extent of ice in the northern hemisphere following the most recent glacial advance about 20 000–18 000 years ago.

scale events. Very detailed analyses of the record from the ice cores, for example, show very fine-scale fluctuations of quite an extreme nature (Fig. 3.18). To what extent are these a result of 'noise', random fluctuations in the system of measurement? Or, since ice moves laterally in glaciers, could such short-term fluctuations be a consequence of ice shifting and destroying the strict stratification of ice layers in the core? One way of testing this is to conduct repeat analyses of cores from other sites and see if there is any correspondence between the fine-scale wiggles. If so, it suggests that we are dealing with real climatic fluctuations.

In Fig. 3.18 the results are presented from a large team of workers (Johnsen *et al.* 1992) researching on the Greenland ice-cap. They have now completed the analysis of four separate cores from different parts of Greenland (Fig. 3.19) and are in a position to compare the results. The cores

shown go back about 35 000 years, covering the time of the last glacial maximum. The upper part of the diagram (the last 15 000 years) shows a very complex series of fluctuations associated with the final stages of the glaciation and the beginning of the Holocene. We shall look at these in detail later. Below this there is a pattern of quite strong peaks and troughs in the ^{18}O isotope level and the peaks have been labelled 2–11 from the top. The number 1 refers to the complex of peaks in the upper part of the diagram. The important question is whether these wiggles are simply artefacts of the sampling or disturbed horizons in the sedimentation of the ice, or whether they are real reflections of brief warm intervals in the generally cold environment.

Until 1992, data was available only from the sites at Dye 3, Camp Century and Renland, all of which were away from the summit of the ice-cap and therefore potentially subject to possible disturbance by the lateral movement of glaciers. The later analysis of the Summit site has provided almost perfect conditions for ice deposition and its oxygen isotope profile gives us a sound basis on which to test the validity of the fluctuations in the other three cores. As can be seen, the correlation is extremely good, even in the detailed form of many of the peaks, which can now firmly be regarded as genuine warm events. The cold troughs were about 12–13°C lower than the present day and the peaks of the interstadials were about 5–6°C lower than present. The warmer interstadials lasted between 500 and 2000 years and usually began abruptly. The lesson to be learned from these sharp fluctuations is that climate, even before any human influence could have been present, could swing quite sharply through 6 to 7°C shifts and sustain them for centuries. This baseline situation makes the predictive modelling of climate change extremely difficult.

The cold episodes recorded in the ice-cores of Greenland during the last glacial also left their record in the ocean sediments of the North Atlantic. In 1988 the geologist Hartmut Heinrich first described six layers of debris in the Atlantic sediments that he could only account for by proposing that armadas of icebergs were periodically released into the North Atlantic. In his honour these episodes have become known as 'Heinrich Events'. These events seem to have taken place roughly every 10 000 years through the glacial stage, and they are correlated with cold episodes in the ice-core record, so it is tempting to suppose that these iceberg releases triggered rapid global climate change. Or could it have been the other way round, with a cold event causing the release of icebergs? As yet the complexities of cause and effect have not been unravelled (Broecker 1994).

One further question about this climatic instability needs to be raised. Did it extend back into the last interglacial stage? Early evidence from ice cores suggested that it did, but this evidence was questioned because the material analysed was close to the base of the ice-sheet (about 2800 m deep) and problems of ice-wastage, slumping and contamination could not be ruled out. Evidence from ocean

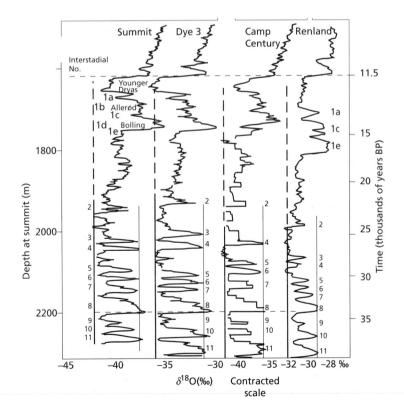

Fig. 3.18 Oxygen isotope profiles derived from ice cores taken from four sites in the Greenland ice-cap (see Fig. 3.19). These are high resolution analyses of cores covering only the last 35 000 years. A series of brief warm interludes (interstadials, numbered 1–11) is apparent in all four cores and the close similarity in their form and pattern suggests that these are genuine and relatively widespread climatic variations, not simply the products of irregularities in sedimentation. Climate during the glacial evidently fluctuated erratically. From Johnsen *et al.* (1992).

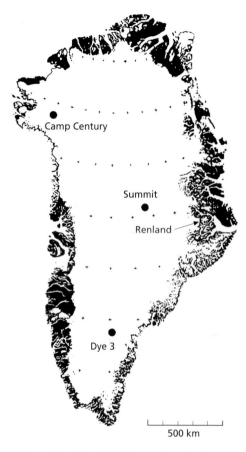

Fig. 3.19 Map of the Greenland ice-cap showing four stations from which ice cores have been recovered. From Johnsen *et al.* (1992).

cores in the North Atlantic has provided more reliable data and suggests that the last interglacial stage was much more stable climatically than the subsequent glacial (McManus *et al.* 1994).

Reasons for rhythms

What determines the evident rhythm of the glacials and interglacials and the shorter time span oscillations of the interstadials? Many theories have been put forward and it is probable that many of them are important influences on the pattern and the extent of climate change. Different factors interact with one another and a combination of effects may result in the crossing of certain thresholds in the way the oceans or the atmosphere behave and so result in quite consid-

erable and rapid climatic swings. Only a clear understanding of the processes at work will enable us to predict with any degree of confidence the consequences of current trends in climate and our responses to them.

Milankovitch cycles

In the 1920s a Yugoslav geophysicist, Milutan Milankovitch, developed and elaborated a scheme of astronomical variables that seemed to correspond in broad terms to the climatic patterns of the Pleistocene. He recognized three different types of variation in the earth's behaviour that occur at three different time scales. When superimposed, the three behaviour patterns come surprisingly close to the observed climate changes.

The first and largest scale of pattern is caused by a regular change in the shape of the earth's orbit around the sun. Sometimes the orbit is almost circular and at other times it is much more elliptical. The time taken for a complete cycle to take place (circular–elliptical–circular) is 96 000 years (Fig. 3.20). The effect of this orbital change on the earth's climate will mainly be on its seasonality, for an elliptical cycle will produce a stronger contrast between summer and winter climate than a circular one.

The second source of variation concerns the angle of tilt of the earth upon its axis. The polar axis of the earth is not vertical with respect to the sun, but is tilted at an angle varying between 24.5° and 21.5° and the precise angle of tilt influences the pattern of energy distribution falling on the surface of the earth. A full cycle of tilt change, from extremely tilted to weakly tilted and back to strongly tilted, takes 42 000 years.

The third variable involves a wobble of the earth on its axis of rather shorter duration, about every 21 000 years. This wobble again influences the relative contrast between summer and winter, by determining whether a particular hemisphere is experiencing either winter or summer when it approaches that part of its elliptical orbit that brings it closest to the sun (see Fig. 3.20).

Much of the variation in the earth's climate can be explained in terms of these three variables when they are superimposed upon one another. The overall pattern of glacial and interglacial, for example, follows a cyclic pattern with a wavelength of about 100 000 years.

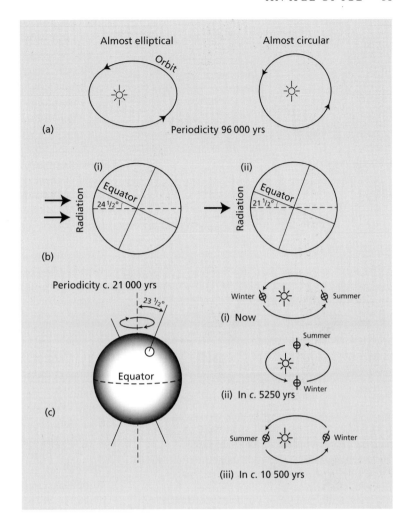

Fig. 3.20 Variations in the earth's orbitals and axis angles that affect the incidence of solar radiation and hence may modify the global climate (Milankovitch cycles). (a) Eccentricity of the orbit; (b) axial tilt; (c) precession of the equinoxes (right) resulting from the wobble of the earth's axis (left). From Lowe & Walker (1984).

Volcanism

Mention has already been made of the explosive eruption of the volcano Toba on the island of Sumatra during oxygen isotope stage 5 (p. 47). This may have acted as a trigger that caused a major decline in global temperature by injecting dust into the upper atmosphere and thus cutting out a significant input of solar energy to the earth. There is now a considerable body of evidence to show that a major eruption can cause a global change in climate, possibly involving reductions in temperature lasting several years subsequently. This means that periods of high volcanic activity could experience generally cooler climates and this hypothesis is borne out by data from the Greenland Ice Sheet in which acidity of the ice

(contributed to by acidic sulphates from volcanoes) is in good agreement with records of glacial extent in mountain areas during historic times (Porter 1986).

The importance of volcanism, as with several other possible mechanisms of climate change, is that it may create a new set of conditions in which overall cooling trends are set in motion or enhanced.

Solar energy output

The output of energy from the sun is not constant, but varies in a series of regular patterns (Crowley & North 1991). Solar magnetic cycles of 11- and 22-year wavelength are well known and there is also a suggestion of an 88-year cycle in sunspot

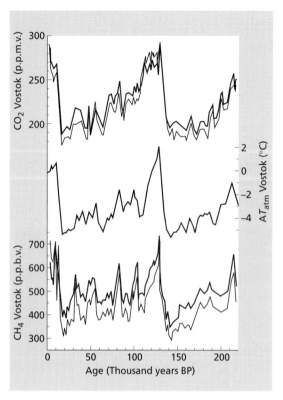

Fig. 3.21 Levels of carbon dioxide (CO_2) and methane (CH_4) trapped in air bubbles in the Vostok ice core from the Antarctic, compared with the suggested temperature variations based on oxygen isotope ratios. From Jouzel *et al.* (1993).

activity, which may also correspond to patterns of variation in the generation of ^{14}C that complicates the use of this technique in dating. Whether such changes in solar output are sufficiently large to account for any significant climate change on earth is still a matter for debate. Actual changes in the energy output of the sun are very small; calculations suggest that it could account for changes of only about 0.1°C, far less than those experienced during an interstadial like the Younger Dryas (see Chapter 4). Nevertheless, some of the minor changes in glacial movements and other climatic responses during the last 10000 years do seem to be connected with solar variability.

Carbon dioxide and methane

Cores through the Greenland and Antarctic ice-caps, referred to previously with respect to oxygen isotope analysis, have also provided information about the abundance of the gases carbon dioxide (CO_2) and methane in the atmospheres of the past. Bubbles of air trapped within the ice at different depths are regarded as fossils of past atmospheres and they can be extracted and analysed for their gaseous components. Carbon dioxide is a small, but important, component of the earth's atmosphere which will be considered in greater detail in Chapter 5. In the ice cores it can be seen (Fig. 3.21) that the quantity of this gas varies quite considerably with depth (equivalent to age) and when these variations are compared with oxygen isotope reconstructions of temperature, there is a close correspondence between them (Lorius *et al.* 1990; Jouzel *et al.* 1993). High peaks in CO_2 occur in recent times (last 10000 years) and during the warm period of the last interglacial (about 130000 years ago) and low levels prevailed about 20000 years ago at the height of the last glaciation.

Methane is an even smaller component of the atmosphere, but its abundance can be seen to mirror very closely both the CO_2 content and the temperature inferred from oxygen isotope studies (Fig. 3.21). Even short-term events are clearly marked in the methane curve (Chappelaz *et al.* 1990). The methane record in ice cores has now been taken back over 200000 years.

The question that arises from this set of data is whether there is any causative link between the two curves. Just because two variables follow a similar pattern does not mean that one causes the other. In this case, however, it is known that increased atmospheric CO_2 could cause climatic warming but, equally, global warming could result in a faster generation of atmospheric CO_2 by, for example, enhanced microbial activity in soils or CO_2 release from oceans. So we are left with a situation in which the correspondence is so close that one cannot determine which variable might be forcing a response form the other. It is impossible to tell from this diagram which event precedes the other, a rise in temperature or a rise in CO_2. There remains the possibility, however, that atmospheric CO_2 levels have influenced past global climates. In the case of methane, current opinion is that the atmospheric methane level is determined by global climate rather than the reverse (Chappellaz *et al.* 1990), but the existence of a feedback mechanism (i.e. higher levels of methane cause increasing warmth) complicates the issue.

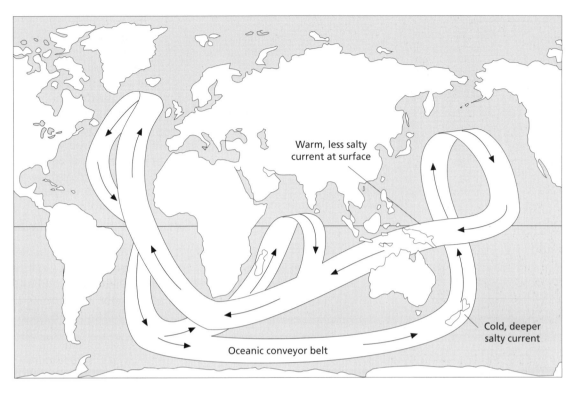

Warm, less salty
current at surface

Cold, deeper
salty current

Oceanic conveyor belt

Fig. 3.22 The oceanic 'conveyor belt' carrying warm, low-salinity surface waters from east to west into the North Atlantic and deep, higher salinity, cold waters from west to east into the Pacific.

Oceanic/atmospheric circulation

The earth's climate is strongly affected by the way in which the waters of the ocean carry heat from one part of the globe to another. Warm ocean currents moving into the North Atlantic from tropical regions, for example, pass their energy on to the air masses of northern latitudes which greatly modifies the climate of the adjacent land areas. If currents of this sort were reduced in their strength it could result in profound changes in the climate in these parts of the world and could well cause repercussions around the entire globe. Before the possible importance of such a mechanism can be assessed, however, one must first examine the causes of the circulation of the oceans. Evaporation of water from the warm northern Atlantic and its atmospheric and subsequent river transport across Europe and Asia into the Pacific leads to the generation of saltier waters in the Atlantic and relatively more dilute waters in the Pacific. The Atlantic salty water is denser

and sinks, forcing its way as a deep current southwards through the Atlantic basin and eventually around the southern tip of Africa where part moves northwards into the Indian Ocean and the remainder continues its journey east into the Pacific. Here it is both warmed and diluted by incoming fresh water and becomes less dense, as a result of which it is forced to the surface and begins its westerward movement back into the Atlantic (Fig. 3.22).

The close interaction between salinity and ocean cycling leads one to consider the implications of large volumes of freshwater being locked up in ice during the glacial stages. This would have resulted in smaller ocean volumes consisting of more saline water. The geographical pattern of freshwater discharge during warmer episodes could well have had important effects on the strength and direction of the salt water circulation systems that may well account for some of the more localized and more rapid fluctuations in climate during the later Pleistocene and early Holocene, such as the Heinrich events and the Younger Dryas episode; a period of intense cold that lasted from 10 800 to 10 000 years ago (for details, see Chapter 4).

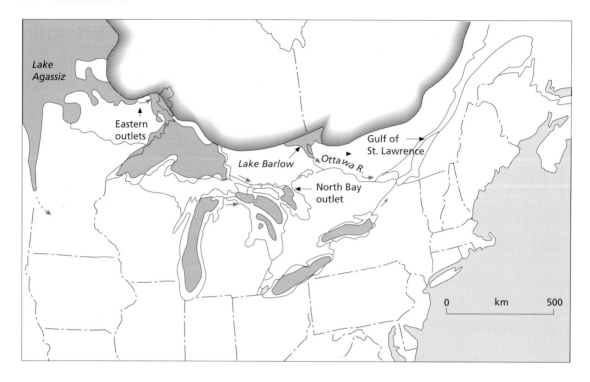

Fig. 3.23 The retreating Laurentide Ice Sheet at 9500 years ago. Lakes formed by the melting ice drained into what is now the Gulf of St Lawrence.

As the great North American Laurentide Ice-Sheet began to melt, some 18 000 years ago, most of the meltwater (as indicated by the oxygen isotope analysis of ocean sediments of the area) seems to have flowed southwards into the Gulf of Mexico (Broeker *et al.* 1989). Warming of these low-density waters in the tropical regions would have set in motion the Atlantic oceanic circulation patterns described above. But as the Laurentide Ice Sheet retreated northwards, what is now the St Lawrence discharge route became exposed and the flow of freshwater was diverted into the North Atlantic (Fig. 3.23). This would have produced a body of freshwater in the region that could have effectively shut down the oceanic circulation system and prevented the further development of a warm North Atlantic seaboard (see Fig. 4.2). In this way the Younger Dryas cold event could have been rapidly generated in the region, but its influence would also have been felt globally via the oceanic and atmospheric circulation system. If the oceanic heat-exchange system

were switched off, for example, the atmospheric circulation pattern of the equatorial regions would be effectively isolated from those of higher latitudes and this would give rise to drier conditions in North Africa at the time, which is precisely what is observed (Street-Perrott & Perrott 1990). The very rapid climate changes (within a decade or less) projected by Alley and co-workers (1993) certainly suggests that a threshold, or trigger mechanism, is important in determining the shifts in climate in the late-glacial.

Some objections have been raised to this explanation of sudden global climatic change, however. Perhaps the most serious of which is that the cold Younger Dryas event does not seem to correspond in its timing with the maximum discharge of meltwater (Fairbanks 1989). But it may be the case that the initial development of a meltwater plume into the North Atlantic was the trigger for the event, besides which radiocarbon dating of this period of time is somewhat insecure because of the 'radiocarbon plateau' described in Chapter 4 (Fig. 4.3).

What is particularly informative about this period and about this mechanism is that it depends upon cataclysmic events, such as the bursting of ice-dammed lakes and the development of major new drainage channels. It is important that we should be aware that such short-term catastrophes may cause a sudden shift in global climate as it runs counter to the general teaching of geological uniformitarianism developed last century in response to the then-pervading ideas of Biblical Flood mechanisms in geology. There are mechanisms available by which the earth's climate can change both rapidly and radically. Thresholds can be crossed which result in major shifts occurring very rapidly and this fact is a salutary one to bear in mind when we consider current climate change.

Ice-sheet instability

Since the salinity of the ocean, particularly the North Atlantic, is important in determining climate changes, one must consider the question of what might influence salinity. As we have seen, ice-sheet melt down can be a major factor, so the question of the stability of the ice-sheets assumes great importance in determining the earth's climate. The launching of the Heinrich iceberg armadas clearly demonstrates a link between ice-sheet instability and climatic change events.

The explanation of such episodes could lie in the thermal blanket effect of the ice cover, for geothermal energy from the earth's core would build up beneath the continental ice-caps over time and eventually result in the whole ice-sheet becoming unstable. Massive collapse would then result in iceberg production around the continental margins and a reduction in oceanic salinity, with all its consequences for oceanic circulation patterns, interaction with the atmosphere and global climate.

The fact that these Heinrich events have had an impact on global vegetation has now been demonstrated by studies of the pollen stratified in the sediments of Lake Tulane in Florida over the past 50 000 years (Grimm *et al.* 1993). A wet/dry cycle of climate is reflected in the alternation of pine with oak/ragweed pollen assemblages, and the timing of the wetter pine periods corresponds with Heinrich events as reflected by large-particle sediments in the North Atlantic.

The discovery of this set of complex interactions has important implications for the current problem of global warming (Maslin 1993). If the expected rise in atmospheric temperature resulting from the greenhouse effect should push the ice-sheets over a stability threshold we could see as period of iceberg production and a reduction in the salinity of the North Atlantic. Such a change could have dramatic consequences for the global climate.

The human factor

Having dismissed the possibility of any global impact of human beings up to the time of the last glacial maximum, we should consider briefly the progress of human evolution up to this point so that the dismissal can be justified. We need to know what stage human physical and cultural evolution had reached by the last glaciation before the potential influence of our species can adequately be assessed.

Human evolution has been studied both by analysis of current similarities and differences between modern humans and related animals, such as the apes, and also from the fossil record, which is obviously very fragmentary and incomplete. On the basis of genetic studies of apes and humans it is apparent that the similarities between the species are very close. Only about 1.2% of our genetic makeup differs from that of a chimpanzee and about 1.4% differs from that of a gorilla. So the chimpanzee is regarded as our closest living relative. The difference between the chimpanzee and the gorilla is about 1.2% (Lewin 1984). On the basis of such studies it can be estimated that our line of evolution diverged from that of the African apes about 20 million years ago (Andrews 1992): in other words, at about 20 million years ago we shared a common ancestor with these animals.

The fossil record from the late Pliocene and early Pleistocene presents a confusing array of bone fragments and primitive stone tools. Archaeologists and palaeontologists have often had to change their concept of what constitutes our own direct line of evolution (the genus *Homo*) as further pieces of evidence have come to light. Some of the earliest evidence for the hominids (the line of ape evolution that included our own ancestors) come from the African sites of Hadar in Ethiopia and Laetoli in Tanzania (between 3 and 4 million years old), and when the fossils were first described it was considered that some belonged to the genus *Homo*. But subsequent work has led to

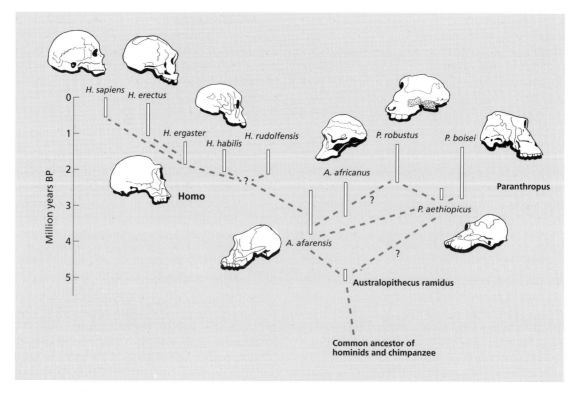

Fig. 3.24 A scheme suggesting the possible inter-relationships between fossil hominids in the last 5 million years (late Pliocene and the Pleistocene). The diversification of the hominid group in the last 2 million years is apparent. Based on the scheme of Wood (1994).

the conclusion that all the material is referable to one species of ape, called *Australopithecus afarensis*. Among these finds is the well publicized 'Lucy' (Simons 1989). It is possible that *A. afarensis* dates back to 6 or even as far as 9 million years ago (Andrews 1992).

This species appears to have been replaced in the fossil record by two or more species, a relatively slight species (*A. africanus*) and a heavier built one (*Paranthropus. robustus*). The robust skeletons from eastern Africa are often regarded as a separate species *Paranthropus boisei*. These last two species are sufficiently different from the presumed parental *A. afarensis* to warrant being placed in a separate genus, *Paranthropus*, and we shall adopt this terminology. This evolutionary diversification took place over the course of about 2 million years and this group of australopithecines survived until a little more than a million years ago (Fig. 3.24).

In comparison with our own genus, *Homo*, these australopithecines (including *Paranthropus*) had smaller brains and relatively short arms. Their hip structure suggests that they were bipedal but they had a short stride and were probably not very speedy. It is unlikely that their larynx structure or their limited brain complexity would have permitted language as we know it, and there is no evidence to suggest that they used fire (Wood 1992). This latter fact, of course, would also limit their capacity to modify their environment to any great degree. They were probably omnivores, scavenging rather than hunting for any meat in their diet. Whether *Paranthropus* was able to make and use stone tools is still a matter of dispute (Lewin 1988).

The genus *Homo* was undoubtedly a contemporary of some of these australopithecines, bones often being found in the same rock strata. Difficulties in defining the genus, particularly on the basis of small scraps of bone, have made it especially hard to determine exactly when the genus made its first appearance. Andrew Hill and colleagues (1992) have worked on fossil material they attribute to *Homo* from Kenya and they have produced evidence for a date of about 2.4 million years ago, but their results have been widely criticized and opposed (for example by Tobias 1993). If correct, it means that our genus arose in parallel to the *Paranthropus* genus, possibly from a common ancestor, perhaps another australopithecine (Fig. 3.24). It is also worthy of note that this date of about 2.4 million years ago has already been quoted as a time of climatic cooling and is a strong contender for being accepted as the datum for the commencement of the Quaternary (see p. 38). It is interesting to speculate whether changing global climate, perhaps through its influence on vegetation, placed new selective pressures on our ancestors that actually spurred on our evolution.

The earliest member of the genus *Homo* has been named *H. habilis*, but there may have been more than one species present in its early history. Bernard Wood (1994) has proposed an evolutionary scheme shown in Fig. 3.24, where some additional species of *Homo* and *Paranthropus* are inserted. Most palaeontologists still favour the idea that human origins were in Africa, although the evidence from primitive stone tools in Asia and Europe (Ackerman 1989) continues to raise the question of the geographical birthplace of our genus. There are difficulties in determining whether these supposed primitive tools were indeed worked by early man; there are difficulties in dating them, and also there arises the question of whether *H. habilis* may have moved out of Africa and spread into neighbouring continents as long as 2 million years ago. These problems have yet to be solved.

From this species, or group of closely related species, emerged *Homo erectus*, which is a species associated with a very recognizable culture of stone tool manufacture, termed the Acheulian. This consists of a complex array of tools including handaxes, cleavers, picks, biface scrapers, etc. and replaces the former primitive 'Oldowan' tool technology. *Homo erectus* first arose in Africa about 1.7 million years ago (Asfaw *et al.* 1992) and, judging by the stability and long persistence of the stone technology, it proved a very successful species. Its success may also be appreciated by reference to the expansion of its range, spreading to southern Africa, north Africa, northern Europe (north Germany), eastern Asia (Peking Man) and Java (Java Man) over the next 1.5 million years. The anatomical stability of the species is remarkable, an almost complete skele-

ton from Kenya dating from 1.6 million years ago (Simons 1989) being clearly of the same species as Peking Man, dating less than 0.5 million years ago. The spread of *H. erectus* took place as the intensity of ice age sequences became greater during the last million years (see Fig. 3.10).

The brain sizes of both *H. habilis* and *H. erectus* were small (relative to body size) in comparison with modern humans, but were considerably larger than those of the australopithecines. Skeletal evidence of *H. erectus* is scarce, but one measurement of the height of a 12-year-old male is about 1.64–1.68 m (5′4″–5′6″), which is surprisingly tall. On the basis of associated fossils (elephant, horse, rhinoceros, giraffe, hyaena, etc.) they appear to have evolved in a savanna environment and their possession of fire would have given them a degree of control over the environment not previously achieved. It is difficult to trace the precise impact of fire on the contemporaneous environment, but in south-eastern England, at Swanscombe, a *H. erectus* (or possibly early Neanderthal—see below) population was present in the last-but-one interglacial (the Hoxnian). Pollen diagrams from the other side of the river Thames at that time show an interesting loss of trees and replacement by grass vegetation in the middle of the interglacial, and it is tempting to associate this episode with the presence nearby of human populations and their use of fire (Turner 1970).

There can be little doubt that *H. sapiens* arose from the *H. erectus* stock. Two lines of development are evident, with very little indication of genetic interchange between them. The distinctive heavy skull characteristics and pronounced brows of the Neanderthal line, or subspecies (*Homo sapiens neanderthalensis*), made their appearance in the fossil record before 230 000 years ago and remained confined to western Eurasia, from Britain to the Middle East. These people were culturally advanced and their stone tools (Mousterian culture) are distinctive. There is evidence to suggest that they buried their dead (Diamond 1989), so they may have practised some form of religion, and they may well have been capable of speech (Arensburg *et al.* 1989).

The origins of the more successful subspecies, *Homo sapiens sapiens*, is still a matter of conjecture, but genetic evidence (Stringer & Andrews 1988) suggests that our race of the species arose in

Africa and emigrated into Europe and Asia probably between 150 000 and 300 000 years ago. The
two subspecies were certainly contemporaneous,
not part of a linear evolutionary sequence. In
some sites, such as Israel, there is evidence for
coexistence in an area, but this could have been
episodic, one race periodically replacing the other
(Stringer *et al.* 1989). The spread of *H. sapiens* was
remarkably wide ranging and rapid, occurring
mainly during the final glacial stage. South Africa
had been reached by about 80 000 years ago (Grun
et al. 1990); northern Australia had human populations by about 60 000 years ago (Roberts *et al.*
1990) and much of Polynesia by about 40 000
years ago. The colonization of North and South
America probably took place via the Bering Land
Bridge around 30 000 years ago. Theories about a
'three wave' invasion and about the possibility of
a separate and earlier arrival in South America
have not been supported by genetic studies of
modern native American populations. Genetic
uniformity through North and South America
implies a single wave of invasion from Asia.

The extinction of the Neanderthal subspecies
of humankind has never been fully explained.
Neanderthals certainly survived until about
35 000 years ago, just before the final advance of
the last glaciation. The loss seems to have begun
in the Middle East at about 40 000 years ago and

then spread westwards into Europe. The massive
brows and skull features of the Neanderthals contrasted strongly with the erect forehead of the
sapient race, and these features are lost from the
fossil history quite rapidly about 36 000 years ago
(Mercier *et al.* 1991), suggesting that the subspecies was not lost by interbreeding with *H. sapiens sapiens* (Lewin 1984). One must assume that
they were simply replaced and that our own
species was successful in a competitive interaction. There is no evidence for 'Palaeolithic geno-
cide' and the dates suggest that (in the words of
Stringer & Grun 1991) they 'went out with a
whimper, not a bang'. The only hominid to
emerge from the final glaciation was thus our own
species and this fact, as we shall see, was to prove
particularly important for the entire globe during
our current interglacial.

Further reading

Cox, C.B. & Moore, P.D. (1993) *Biogeography: An
Ecological and Evolutionary Approach*, 5th edn.
Blackwell Scientific Publications, Oxford.

Lowe, J.J. & Walker, M.J.C. (1984) *Reconstructing
Quaternary Environments*. Longman, London.

Mannion, A.M. (1991) *Global Environmental Change*.
Longman, London.

Tallis, J.H. (1991) *Plant Community History*. Chapman
& Hall, London.

Chapter 4

After the Ice

As we have seen in Chapter 3, the last 2 million years of the earth's history have seen some extreme climatic fluctuations in what, geologically speaking, is a relatively short period of time. Only 20 000 years ago ice-sheets covered not only Antarctica and Greenland, but also much of what is now Canada, northern Europe and parts of northern Asia, together with more isolated glaciers in Africa and New Zealand. The whole world was in the grip of its most recent age of ice. Clearly, there have been massive changes in climate during relatively recent times and it is only by understanding the rates and the patterns of these changes that one can set the current concerns about climate change in a realistic perspective.

The last 2 million years have shown that we should not expect climate to be static; it is constantly changing. The period has also shown us that the current warm climate of the earth is not typical, but must be regarded as one of those relatively infrequent episodes in which the Ice Age is temporarily set on one side and the higher latitudes can bask for a while in welcome warmth and energetic productivity on the part of vegetation. But the lessons of the past point quite clearly to a pattern in which a return to the Ice Age seems inevitable and the timing of past patterns imply that we should already be well on the way in that direction. How does the current debate about a 'Greenhouse World' fit into such a picture? How will the two opposing trends resolve themselves? And how may we expect the living world of plants and animals (including ourselves and our domesticated species) respond to continued change?

Once more we must turn to the past to provide a setting in which such questions can be posed. We need to know more details about current underlying climatic trends, their patterns, their rates and, if possible, their causes. For how long has the human impact on global environments

been important? We also need to understand more about the ways in which plants and animals have responded to climate change in the recent past. Have species evolved in response to such change, or have they merely changed their distributional ranges? These are the questions that will be faced in this chapter.

Climatic flutters

Towards the end of the last chapter it was pointed out that the temperature curve reconstructed from oxygen isotopes in the Greenland ice cores was far from smooth during the last glacial advance. Figure 3.18 showed that the cold stage was punctuated by a series of minor warm episodes (interstadials). The period between about 25 000 and 40 000 years ago, immediately before the last major ice advance, contained about nine such events. Some of the intervening cold episodes have been correlated with iceberg rafting in the Atlantic—the Heinrich Events. The same diagram shows that such fluctuations in temperature continued after the final ice advance (at about 20 000 years ago), but take a somewhat different form. The oxygen isotope records suggest that there was a steady build-up in temperature, accelerating around 13 000 years ago, then entering a series of flutters before plunging back down again. Following this very cold incident the present warm interglacial began in earnest.

These flutters, and especially the very cold interruption to climatic warming (that has come to be known as the Younger Dryas for reasons that will shortly be explained), have been the focus of a great deal of recent research by climatologists, geophysicists, oceanographers and palaeobiologists. The main interest in these flutters is that they provide a clue both to the underlying causes of rapid climate change on earth and also to the biological responses to such rapid changes.

The potential application of the knowledge gained from such studies to the prediction of the outcome of current climate change are very evident.

The late-glacial climatic flutters were first described on the basis of fossil lake sediments discovered in Denmark where dark organic muds, denoting high lake productivity under warm conditions, were found to alternate with biologically sterile clays with angular fragments of rock that had clearly been sludged into lakes as a result of extremely cold, periglacial conditions leading to soils becoming mobile as they were subjected to seasonal freezing and thawing. Unlike the organic bands, the clays contained few fossils, but some plant remains were present, mainly the leaves and fruits of the arctic alpine mountain avens (*Dryas octopetala*), and this fossil plant led to the naming of the supposed climatic fluctuations that had

given rise to these clay layers as Younger, Older and Oldest Dryas episodes. The intervening organic bands were named after the locations in which they were first described, namely Allerød (near Copenhagen) and Bølling (Fig. 4.1). Of these various names, the Younger Dryas has proved the most durable and is now used globally for the major cold event that interrupted the warming process since the last glacial maximum.

A number of questions surround the Younger Dryas and its associated climatic events. What is the precise chronology of the changes? How rapidly did climate change? How warm were the milder episodes? How extensive was their impact in geographic terms? What caused the changes? How did animals and plants respond to the changes? Many of these questions are still not fully answered, but within them lie some important clues to the climatic and biological balance of

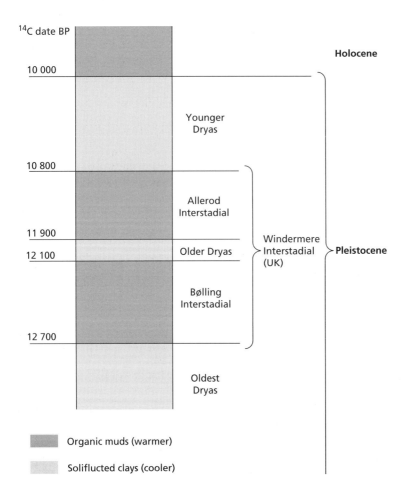

Fig. 4.1 Geological stratigraphy of the end of the Pleistocene showing the effects of climatic fluctuations upon sediments in north-west Europe. The Younger Dryas event (between about 10 800 and 10 000 years ago on the radiocarbon time scale) was a time of intense cold in the North Atlantic region and many lake sediment sequences that had showed increased productivity and hence organic sediments showed a return to inwashed clays and gravels resulting from soil erosion (solifluction) under the changed climate.

the earth, so it is appropriate that we should look at them in some detail.

The geographical extent of the Younger Dryas cooling is still being determined. After its discovery in Denmark, the Allerød/Younger Dryas sequence of sediments was soon described from Ireland (Jessen & Farrington 1938), where fossil remains of the giant Irish Elk were found associated with the Allerød deposits, and from many other parts of northern Europe. In North America, on the other hand, the apparent traces of sediment changes in lakes and the small variations in the patterns of vegetation as depicted in the pollen sequences in these sediments led to controversy and some confusion. It was even postulated that the event was an essentially European phenomenon and had no impact on the climate of North America (Mercer 1969). Even within Europe, the impact of the event was very uneven, being most marked in the north west and becoming increasingly weak as one moves south east into the Alps (Watts 1980). The event is, however, quite strongly apparent in the late-glacial lake sediments of the western Pyrenees. Circumstantial evidence began to point very strongly at the North Atlantic

as the centre of the episode. Attempts have been made to plot the extent of the cold polar waters (the Polar Front) and the extent of sea ice during the late-glacial (Ruddiman & McIntyre 1981) and the evidence suggests that the sea ice limit had retreated to a line between Newfoundland and Iceland during the Allerød, but advanced as far south as Portugal once more during the Younger Dryas (Fig. 4.2).

As the scientific interest in rapid climatic fluctuations has become more apparent in recent years, more effort has been devoted to the global search for the effects of the Younger Dryas. As a result, the event is now clearly recognized in the sediments of the eastern seaboard of North America (Wright 1989) and the detailed analyses of sites in New Brunswick and Nova Scotia have led to the detection of even finer variations in the climate of these times, particularly a pre-Younger Dryas cold event (Levesque et al. 1993) along this eastern seaboard. In the Mediterranean region of Europe, from Italy (Lowe 1992) to Bulgaria (Huttunen et al. 1992), there is pollen evidence of a chilling of climate during the Younger Dryas and in Israel the same time period is marked by a

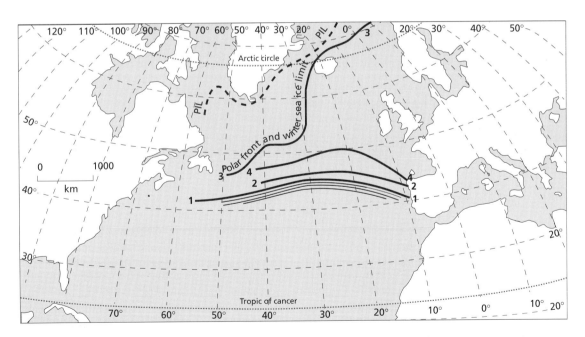

Fig. 4.2 The position of the 'polar front' (cold waters bearing winter pack ice) at various times in the late Pleistocene and Holocene. (1) 20 000–16 000 years ago; (2) 16 000–13 000 years ago; (3) 13 000–11 000 years ago; (4) 11 000–10 000 years ago (the Younger Dryas); (PIL) present limit of winter pack ice. From Lowe & Walker (1984).

dry phase in which the water level of the Dead Sea is thought to have fallen by perhaps as much as 300 m (Yechieli *et al.* 1993).

In the South Atlantic, the evidence is rather weak, but this could be due to a lack of investigated sites and a lack of detailed analysis of the appropriate sediments. Ledru (1993) has conducted some detailed studies of lake deposits in central Brazil and has found pollen evidence for the disappearance of monkey puzzle (*Araucaria*) forest between 11 000 and 10 000 years ago, which could be due to the climatic fluctuations of the Younger Dryas. But the analysis of beetle fossils from Chile (Hoganson & Ashworth 1992) failed to detect any changes that could be related to the event.

Within the Pacific Basin, the Younger Dryas has been detected in the ocean cores of the South Sulu Sea (Kudrass *et al.* 1991), and in the lakes of Alaska (Engstrom *et al.* 1990), though some would regard the evidence in Japan and the North Pacific as weak (Heusser & Morley 1990). In southern Asia there is evidence of vegetation change associated with the Younger Dryas event in Tibet (Gasse *et al.* 1991) and in the loess deposits of China (Zhisheng *et al.* 1993). The oxygen isotope record from the Dundee Ice Cap in Tibet, however, does not supply clear evidence of a climatic fluctuation at that time (Thompson *et al.* 1989).

The overall evidence suggests that changes resulting in the Younger Dryas episode were indeed centred around the North Atlantic, but the repercussions of the event were felt in many parts of the world.

The next question relates to timing. Not only is it important to determine precisely when the Younger Dryas occurred and how long it lasted, but also it is vital that we know how rapidly the changes observed during the episode took place. The oxygen isotope curves indicate that the Allerød warming involved a temperature rise of at least 4 or 5°C and that the decline into the Younger Dryas was at least as great, if not greater, than that. How fast can this degree of climate change occur? In order to answer this question we must return to the problem of dating, for without an accurate series of dates for our sediments we are unable to determine rates of change. We need to establish firm time boundaries for the climatic events of the late-glacial (Broecker 1993).

In Chapter 3 we looked at various techniques for dating geological materials and one of the most useful for dating organic material from the last 40 000 years has been radiocarbon dating. A disadvantage with this method, however, was the need for calibration if the dates were to be translated into calendar, or solar years (see p. 42). Radiocarbon dates are now available for many of the sites in which the late-glacial climate fluctuations have been studied and these are summarized in the dates quoted in Fig. 4.1. But the dates given are in radiocarbon years before present (BP, i.e. before 1950). Two approaches have been made towards the calibration of these late-glacial dates, one using tree rings (dendrochronology) and the other using the annually laminated sediments (varves) of certain lakes which permit a precise back-counting of years into the late-glacial.

The best set of tree-ring dates covering the final part of the late-glacial sequence is based upon fossil pine and oak wood from south-central Europe, largely from south Germany (Becker *et al.* 1991). This extends back to 10 030 radiocarbon years BP. As can be seen from Fig. 4.3, the correction needed is about 1000 years, the radiocarbon dates being that much younger than the calendar dates ('dendro-years'). This places the end of the Younger Dryas at about 11 000 calendar years ago. A much more serious problem, however, is that there are parts of the calibration curve that appear flat, i.e. a spread of calendar years would all be represented by the same radiocarbon date. This is probably due to a change at that time in the rate at which ^{14}C is generated in the earth's atmosphere (see Chapter 3). As Fig. 4.3a indicates, ^{14}C was probably falling off in its atmospheric concentration at that time. Unfortunately, the 10 000 BP radiocarbon date, conventionally and conveniently taken to mark the end of the Pleistocene and the commencement of the Holocene, lies at this very point, so precise determination of the rate of climatic change at this important horizon cannot be achieved using radiocarbon dating.

The alternative approach, using annual varves, has been attempted in Switzerland (Amman & Lotter 1989; Lotter 1991). Once again the radiocarbon plateau at about 10 000 radiocarbon years BP was evident plus a further plateau at about 12 700 BP, corresponding to the beginning of the warm Bølling/Allerød interstadial. So, once again, it is impossible to determine precisely how fast the interstadial warming occurred on the basis of

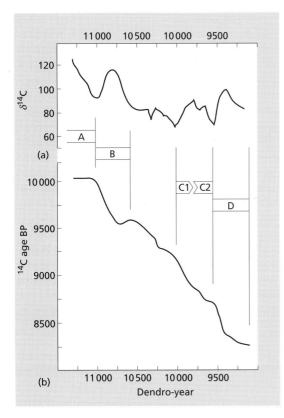

Fig. 4.3 Changes in the level of radiocarbon present in the atmosphere in the past (^{14}C) can lead to problems in the use of radiocarbon dating. (a) Past ^{14}C levels. (b) Calibration curve for radiocarbon years against solar (dendro-) years at the time of the Pleistocene/Holocene transition. Increased atmospheric ^{14}C leads to a plateau in the curve, meaning that the technique cannot produce reliable dates at those times. Four such plateaux are seen, labelled A–D. From Becker *et al.* (1991).

radiocarbon dating. Even allowing for these problems in dating, however, the climatic changes during the late-glacial must be regarded as abrupt and the end of the Younger Dryas also took place very rapidly. Dansgaard *et al.* (1989) estimate from their oxygen isotope cores in Greenland that a warming of about 7°C occurred within a period of only about 50 years at the end of the Younger Dryas stadial. Some workers (e.g. Alley *et al.* 1993; Mayewski *et al.* 1993) have suggested even more rapid rates of change, not only in temperature but also in precipitation. Snow accumulation in the Greenland ice-sheet, for example, may have doubled in only 1–3 years. The rates of temperature change projected for these late-glacial times

are considerably greater than those recorded over the past century (about 0.5°C), which is why their mechanisms and also their effects upon the ecosystems of the time are so important to understand.

If the records of the past can inform us about the nature and rate of change of climate in the past, can it also provide information about the rapidity with which living organisms can respond to such changes?

Biological responses to rapid climate change

Despite all the problems of dating changes at the end of the last Ice Age, it is clear that there were some profound changes in temperature, centred on the North Atlantic but having global repercussions, that took place within fairly short periods of time (perhaps even within decades). The fossils within sediments of these times provide information about the responses of plants and animals to such changes that could be valuable in predicting the consequences of the climate changes that are predicted for the coming century.

When the Allerød interstadial sediments were first described from Denmark, they were recognized both because of the additional organic matter they contained (an indicator of enhanced biological productivity) and the more warmth demanding character of their fossils. Leaves, fruits and pollen grains all display a clear expansion of woodland species in northern Europe, with birch, juniper and pine particularly well marked. But the spread of such plants (the word 'migration' should, perhaps, be avoided for plants since it implies a volitional and repeated movement of a population along a set route that is usually associated with seasonal shifts in animal populations) in response to rapidly changing climate is limited by the dispersal capacity of their fruits and the length of time between germination and breeding maturity. Trees are obviously not likely to be the most rapid organisms in their response because of their relatively slow maturation rates. Herbaceous plants with shorter generation times may well be expected to spread more rapidly, and the late-glacial/early Holocene sediments certainly show a more rapid response on the part of some aquatic and wetland herbaceous species that are well represented in the fossil record. One of the most widely used types of fossil material, however, is the pollen grain, for this is produced in large numbers and often well dispersed. It is also recognizable and usually well preserved in sediments. But it has the disadvantage that those plants which rely on wind for pollination produce much larger quantities of pollen than those which employ insects, and this produces bias in the record and leads to the impression that wind-pollinated species are more frequent. It may even lead to problems of detecting when a particular plant is actually present at a site and when its pollen is being blown from far away.

There are two approaches to the solution of this problem in the study of plant spreading rates. One can conduct detailed analyses of succeeding layers of sediments at a given location and observe the increase in the quantity of pollen sedimenting at the site. (If the sedimentation rate and the pollen concentration per unit volume of sediment are known, then the absolute influx of pollen in terms of number of grains per unit area of sediment surface per year can be calculated). Current surface studies of pollen sedimentation of the species both within and beyond its current range can then aid the interpretation of such data and indicate when the species can assuredly be regarded as present at the site. The increasing abundance of the pollen of the species also indicates the pattern of population expansion and provides a model in which a tailing off can be recognized as the carrying capacity of the particular habitat is reached and also where the competitive resistance to further invasion set up by the existing plant communities may be detectable as some species find it difficult to expand their populations. An example of this approach to the study of plant spread is afforded by the work of Keith Bennett (1983) on the early Holocene invasion of trees into eastern Britain and some of the pollen curves he has obtained are shown in Fig. 4.4.

This approach can provide information concerning the time of arrival and the pattern of population expansion at a site, but to observe the rate and pattern of spread one must conduct similar analyses for a range of sites over a broad geographic area. Margaret Davis (1983) and George Jacobson, Tom Webb and Eric Grimm (1987) conducted this type of study based on pollen stratigraphy in a large number of lake sites in eastern North America and the maps for selected tree taxa are shown in Fig. 4.5. The lines on these maps show the advancing front of the trees at given dates (radiocarbon years). This provides graphic illustration of the rate at which different tree types responded to the postglacial warming of global climiate and one thing is very evident—the response was rather slow. In fact, when one considers the problems trees face in dispersal and regeneration, the spread rates are actually quite surprisingly speedy, ranging gener-

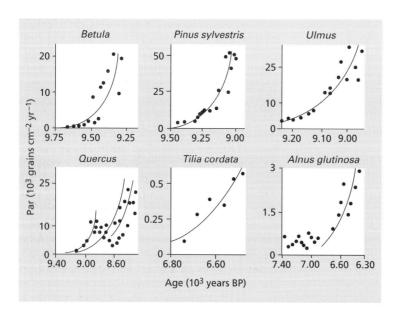

Fig. 4.4 The measurement of pollen accumulation rates in the early Holocene in eastern Britain can be used to reconstruct the population expansion of tree species. From Bennett (1983).

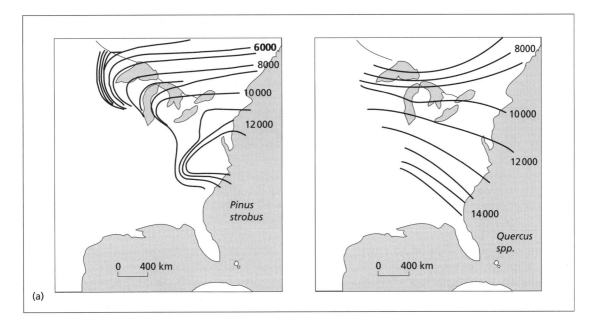

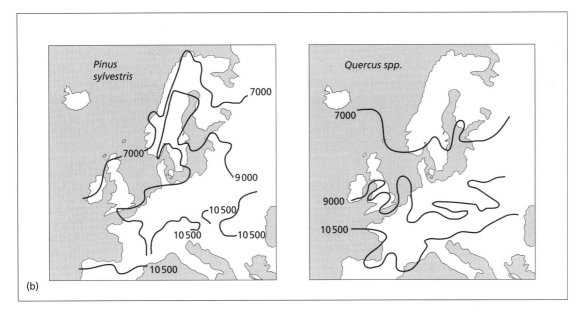

Fig. 4.5 Spread of trees in the early Holocene in (a) North America and (b) Europe determined from pollen analyses of lake sediments over the area. From Davis (1983) and Huntley & Birks (1983). Lines show the probable limits at the times indicated (BP).

ally between 300 and 500 m yr^{-1}. But this is slow when compared to the rapidity of climate change.

This can be illustrated by considering the demands imposed upon a plant species by a climatic shift. A 1°C rise in temperature is approximately equivalent to a spatial range shift of 100–150 km (Roberts 1989). So the 7°C change at the beginning of the Holocene (which may have occurred within only a few decades) might demand a shift in range for a tree species of about 1000 km. A 'fast' tree species with a spread rate of 500 m yr^{-1} would thus take 2000 years to arrive at its new limit, by which time it is quite probable that the climate will have shifted once more. Herbaceous species share with the trees all the problems of dispersal (birches and thistles, for example, have adopted similar methods of overcoming them), but they often have shorter generation times (they may be able to produce seed within a year or two of establishment) and this enhances their potential spread rate. Fast spread is particularly associated with plants that have a very short generation time (such as annuals), with high seed production and good dispersal mechanisms, and particularly with those associated with open habitats where dispersal is more easily achieved. Not surprisingly, such species (including what we might now term 'weeds') were among the ones that spread rapidly into areas in the higher latitudes that experienced the sudden warming in post-glacial times. Their arrival often accompanied or even preceded the arrival of many trees, yet they were equally alien to these regions, for tundra vegetation is practically devoid of annual plant species.

Herbaceous species of the forest understorey may well have found it more difficult to respond to the climatic demands for shifting ranges. Many of these species have large seeds and poor dispersal potential and many rely on vegetative means of increasing their cover or population levels. As a result, some of them were slower in their spread. In Europe this occasionally led to particularly acute problems because the global warming was resulting in melting ice masses and rising sea levels that isolated some coastal regions. The British Isles, for example, became isolated from mainland Europe and some of the species that were slow to spread found themselves excluded from this geographical region (Cox & Moore 1993). Others reached mainland Britain but failed to reach Ireland, including the heavy fruited, slow spreading herb paris (*Paris quadrifolia*). On the eastern seaboard of North America some herbaceous plants evidently spread over land that is now submerged by the sea, as is the case with the skunk cabbage (*Symplocarpus foetidus*). This wet woodland species is found in the western end of the Nova Scotia peninsula but has failed to expand into the east. It is now cut off from the mainland by the high sea level, so its spread route has been lost (Pielou 1991).

The development of barriers to spread can be seen to have interrupted the process on occasions and to leave populations isolated in other cases. Both situations have salutary messages to impart to the modern situation where human-created barriers, such as areas of agricultural land, are the equivalent of past oceans, both blocking the spread routes and isolating fragments of populations from their main ranges. A response to climate change on the part of modern vegetation could thus be constrained by new obstructions.

Another feature which is evident from these fossil studies of plant responses to climate change is the way in which each species acts independently of others. Ultimately each species has its own climatic optimum and tolerance limits, which may be modified by other species growing with it and competing for limited resources. As climate changes, the species spreads into new areas where it is now able to survive and compete effectively with other plants already in residence and perhaps now finding themselves under climatic stress. Meanwhile, it may find itself being out-competed in some parts of its range where formerly it was able to hold its own. Each species thus moves over the face of the continent, advancing in some areas and retreating in others. Occasionally species coincide in their distributions and sometimes they separate. What we now regard as plant associations and communities within our familiar vegetation may be unique to our own particular time, never having had quite their equivalents in the past and never likely to be repeated in quite the same way in the future. Vegetation is dynamic and what we see is a single snapshot in time (Huntley 1991).

To this view of vegetation flux must be added the shorter term cycling of vegetation as patches become opened by minor catastrophe, such as fire, wind or disease. These factors, together with the patchwork imposed by varying topography and soils, leads to a mosaic development of vegetation in which different patches are at different stages of development and possibly even on different developmental courses. It is upon such a patchwork that climatic changes exert their selective pressures (Ritchie 1986).

A further point that needs emphasis is that climate change is not uniform over large areas. A consideration of the Younger Dryas episode makes it apparent that some areas of the earth's surface were affected more strongly than others,

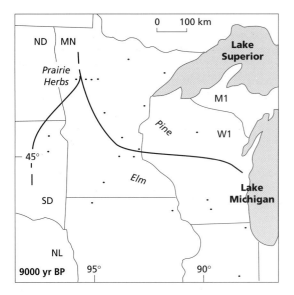

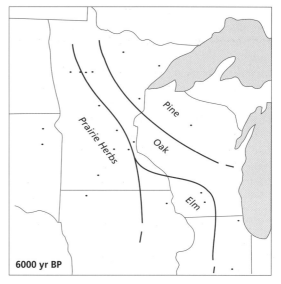

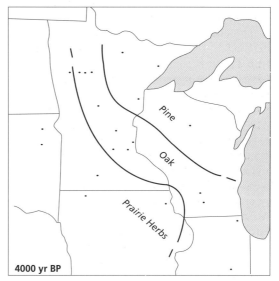

Fig. 4.6 The mid-western United States at three different times in the Holocene, showing the movement of the major vegetation types. The maps are produced from the collation of data from pollen preserved in lake sediments through the region. From Wright (1992).

and the same applied to subsequent climatic shifts. As the Holocene advanced, the vegetation of the North American Mid-West, for example, shows some distinct geographical variations that may relate to climatic thresholds (Baker *et al.* 1992; Wright 1992). By 9000 BP, elm forest in the Dakotas was giving way to spreading prairie grasslands (see Fig. 4.6a). Between 8000 and 6000 BP the prairies spread eastwards across Minnesota (Fig. 4.6b), and they reached north-eastern Iowa and southern Wisconsin only by 4000 BP (Fig. 4.6c). Wright has suggested that in these latter regions, tropical, moist, maritime air masses from the Caribbean were blocking the advance of the dry Pacific air masses from the west whose easterly spread was favouring the advance of the prairies. These tropical air masses were being brought in by continental warming in these southern regions leading to monsoonal conditions with high summer precipitation, thus favouring the maintenance of deciduous forest in the area. According to the Milankovitch model (p. 50), the last maximum of summer insolation was at about 10 000 BP, so at that time the monsoonal effect would have been greatest and since then the effect would have retreated eastwards in the observed manner.

The entire situation, however, would have been complicated by the proximity of the wasting Laurentide Ice Sheet only a few hundred kilometres to the north, which must also have had an impact on local climatic gradients.

The complexity of the relationship between climatic and vegetation change is apparent from these palaeohistorical studies, but the value of such information as a basis for predictions into the future should not be underestimated. The data provide the closest approximation to experimental manipulation that is likely to be achieved.

Climatic trends in the Holocene

Perhaps the most significant fact to emerge from the study of ancient climates, as far as our present problems are concerned, is the rapidity and the frequency with which major climatic shifts have taken place in the past. As studies on the ice cores become more and more detailed and extend further into the past (see the report by GRIP members 1993), it becomes evident that the earth's climate has been extremely unstable especially through the final glacial. The evidence indicates that climatic changes of mean annual temperature of 10°C or so within a decade may have been relatively frequent over the last 100 000 years but, even more remarkably, the climate over the past 10 000 years (our current interglacial, the Holocene) has remained quite surprisingly stable (Fig. 3.18). In this respect it resembles the last interglacial.

The oxygen isotope record indicates a very rapid rise in temperature during the early Holocene and relatively stable conditions (varying only by, at most, 3°C) since that time (Fig. 3.18). By 8500 years ago the temperature was about 0.5°C warmer than the present and by 5500 years ago about 1.5°C warmer. Since this time the temperature has declined. This evidence from the isotopes is generally confirmed by a range of other sources, all of which contribute to our knowledge of detailed, but relatively minor, climatic changes that have taken place in the last 10 000 years. Some of these sources are as follows.

Fossil records of plants and animals

As has been discussed, the response of animals and plants to Holocene climatic warming has been a process of readjusting distribution patterns, with some organisms being able to respond much faster than others. Some of the changing patterns can provide information on climate change. For example, the distribution of hazel (*Corylus avellana*) in Europe extended further north 5000 years ago than it does today (Fig. 4.7) and the painted terrapin (*Emys orbicularis*) also had a more widespread distribution (Fig. 4.8). This might indicate a response to warmer conditions in former times, but the impact of human beings on plants and animals in the last 5000 years always leaves some room for doubt. In North America, the spadefoot toad was found 100 km north of its present limit in Alberta (Pielou 1991) and this is unlikely to have been modified by human intervention.

What is apparent from the temperate fossil record is that sudden, erratic changes in abundance of particular species is rare. It is not totally unknown, however. The hemlock declined sharply in North America about 4850 years ago and elm suffered a similar decline in Europe about 5000 years ago (Tallis 1991). Both of these, how-

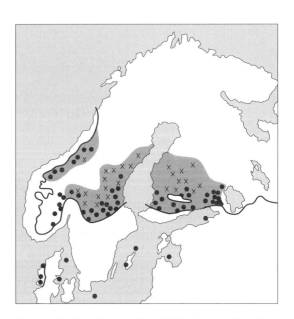

Fig. 4.7 The distribution of hazel (*Corylus avellana*) in Fennoscandia during the mid-Holocene (based on fossil finds of the tree in the past, shown by the symbol ×) and its limits at the present day (• shows the site of isolated present day occurences; — shows its northern limit at present). Its retreat may be a response to cooler temperatures in the later Holocene.

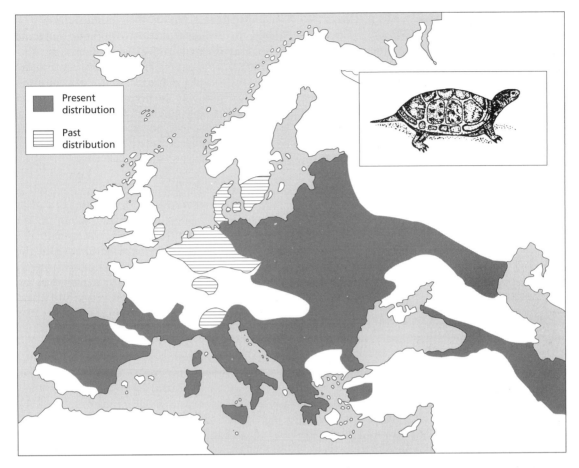

Fig. 4.8 Past and present distribution of the European pond terrapin. Its change in distribution may in part be explained by climatic change since the mid-Holocene.

ever, were probably caused by catastrophic outbreaks of pathogens, although the impact of human activity may also have contributed to the elm decline. But neither reflects any sudden change in climate, because they are not accompanied by marked changes in other climatic indicators.

In the tropics, pollen records indicate rapid rises in temperature at the beginning of the Holocene, together with increased precipitation. In equatorial East Africa, for example, precipitation rose by about 500 mm yr^{-1} at the beginning of the Holocene (Bonnefille *et al.* 1990), and has remained high ever since. Further north in the Sahel, however, the steep climatic gradient has resulted in a greater sensitivity to climate change (see Chapter 11), and the initial expansion north

of the tropical forests has been succeeded by a retreat southwards as aridity increased over the past 7000 years (Lezine 1989). The general picture from the fossil record, therefore, apart from that of climatically sensitive areas, supports the indications of the oxygen isotope record and points to a general stability in Holocene climate, with relatively gradual shifts rather than sudden changes within decades.

Lake levels and sedimentary studies

Some lake levels, especially those in more arid regions which may lack any outlet, respond strongly to changes in precipitation : evaporation ratios and hence can act as climatic indicators. In Africa, for example, the early Holocene humidity is again reflected in the higher levels of lakes (Street-Perrott & Harrison 1985; Roberts 1990). Generally, there is evidence for low lake levels during the last glacial maximum (21 000–12 500 years ago), followed by raised levels due to a more humid climate between 12 500 and 5000 years ago.

The Nile Delta provides evidence of former wet periods which can be detected with precision by examining their mineral components and hence the part of the Nile catchment from which they have been derived (Foucault & Stanley 1989). But here again, one must be cautious in the interpretation of this kind of evidence as human activity by forest destruction and grazing can influence both hydrology and sedimentation in lake and alluvial sediments.

Peat stratigraphy

Peat-forming ecosystems (mires) derive their water supply from rainwater and, for those fed by inflow streams or soaks, from groundwater. Since the formation of peat depends on the restriction of microbial decomposition as a result of waterlogging, changes in water supply may affect the rate of peat formation and the degree to which fossil organic materials are preserved. Those mires fed by groundwater are relatively complex in their hydrology and changes in water supply can vary with patterns of water movement and with changes in vegetation (often caused by human activity) within the catchment. The exclusively rain-fed mires, however, may respond directly to the precipitation : evaporation ratio in the quality of preservation of the peat.

Studies in peat stratigraphy have permitted some palaeoclimatic reconstruction, but they require extensive radiocarbon dating if those horizons in which peat accumulation rate suddenly change (indicated, in the case of increased accumulation, by a lighter, fresher appearance to the peat and poorer decomposition). This approach has been used extensively in Europe with some encouraging results (e.g. Blackford & Chambers 1991; Korhola 1992) and there are hopeful prospects for this line of research.

Timberline studies

The limits of tree growth up mountains or at high latitudes is determined by temperature, so studies of past timberlines (from fossil wood, or pollen in sediments) provides some indication of past climate changes. One must bear in mind, however, that tree lines may respond fairly slowly to shifts in climate, and there is always a possibility that other factors, such as intensive grazing, fire and other human agencies can modify timberlines. A range of timberline studies from the European Alps (Fig. 4.9) shows that they ascended quite rapidly in most sites during the early Holocene, reaching their maximum extent between 3000 and 6000 years ago (Burga 1988), which broadly agrees with other evidence regarding the climatic optimum of the present interglacial. The fluctuations around the general curve are relatively small when compared with the overall shifts through the Holocene.

Treeline studies can also be important in determining how fast an ecosystem can respond to shifts in climate. In northern Canada, for exam-

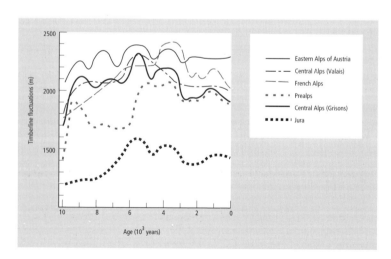

Fig. 4.9 Timberline fluctuations in the European Alps during the last 10 000 years. Note that the highest tree lines occurred largely between 6000 and 3000 years ago. Both climate and human activity may have played a part in this pattern. After Burga (1988).

ple, a northward shift of the treeline occurred in the mid-Holocene (about 5000 to 4000 years ago) that illustrates just how rapidly such responses to climate change can take place. Within about 150 years the treeline had moved at least 100 km further north (MacDonald *et al.* 1993). Such studies provide baseline data for predicting the outcome of future climate changes on the geographical distributions of ecosystems.

Tree-ring studies

The annual rings found in the wood of trees living in seasonal climates has already been mentioned in relation to radiocarbon dating calibration (p.42). These same rings can supply information concerning climate, for their thickness (i.e. how much wood is laid down by the tree each year) is determined in part by climatic conditions (such as temperature and water supply). There are many complications in the interpretation of such information, for the wood accretion can also be influenced by pathogen attacks, grazing animals, human harvesting and local conditions of soil and aspect (Hughes *et al.* 1982). But if a large enough sample of trees is available, covering a relatively wide area, then it is possible to obtain reliable data. Figure 4.10 shows tree-ring studies of spruce growing in northern Canada (Payette *et al.* 1989). A clear depression of growth during the 1600s is evident here and corresponds with the European 'Little Ice Age' (see 'meteorological records', below).

   Tree-ring studies may be limited by the length of life of a tree, unless it proves possible to match up the ring pattern with fossil trees and hence extend the record backwards in time. Long-lived trees have proved particularly useful in such studies, such as the bristlecone pine (*Pinus longaeva*) in North America and the Patagonian cypress (*Fitzroya cupressoides*) in South America. Some

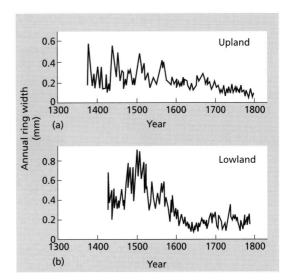

Fig. 4.10 Tree-ring studies of black spruce in northern Canada can demonstrate the response of trees to changing climate. (a) Upland and (b) lowland populations, both at the tree line and therefore sensitive to climatic shifts. Both show clearly the impact of the cold 'Little Ice Age' of the 1600s. From Payette *et al.* (1989).

recent studies of the latter in southern Chile (Lara & Villalba 1993) have resulted in the construction of a proposed temperature curve for the area covering the past 3622 years (Fig. 4.11). The data emerging from this work demonstrate both long-term and short-term temperature fluctuations over the period studied and it can be seen that the amplitude of fluctuations is only about 1°C. The general climatic stability of the late Holocene is once again a feature of these studies.

Art history

Our own genus, *Homo*, has been around for the entire Quaternary Period, but our species, *Homo*

Fig. 4.11 Reconstruction of summer temperatures in southern Chile based upon tree-ring studies of the Patagonian cypress. A period of above average temperatures is apparent between 80 BC and 160 AD. From Lara & Villalba (1993).

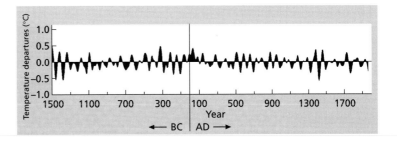

sapiens, emerged only during the last glacial. Agriculture began with the opening of the Holocene, but records that can help with the reconstruction of climate are scarce until recorded history began and even then are of very limited value. Some early information, however, has emerged from the study of art history, particularly the rock paintings of the Sahara Desert (Roset 1984). The paintings are present on rock outcrops over many parts of the Sahara and they depict a wide range of animal life that provides some indication of the contemporaneous conditions.

The oldest paintings probably date from the latter stages of the last glaciation and they show animals of an antelope type, which fits in well with what is known of the steppic vegetation of the area at that time. The early Holocene paintings are of big game animals, with giraffe and rhinoceros, now associated with the savanna vegetation that is currently found in more southerly latitudes and more humid conditions. The paintings of between 7500 and 4500 years ago take on a more agricultural theme, with pictures of cattle. Grazing cattle in the central Sahara would be quite unthinkable at the present day. More recent paintings depict horses and camels, reflecting the onset of more arid conditions. Although the climatic information available from such paintings is limited, it can provide valuable confirmation of other forms of evidence.

Meteorological records

Historical writings, even those of a descriptive rather than a quantitative, numerical nature, have provided a database on which climatic reconstruction has proved possible. In Fig. 4.12, the reconstruction of climate in England over the past 1000 years has been based on historical evidence of a variety of forms (collated and interpreted by Lamb 1972). When this tentative record of climate is compared with those derived from oxygen isotope data, the correlation is agreeably satisfactory. Much of the last 1000 years has been colder than the present day according to these historical records, with a distinct temperature drop around 1300 and further troughs in the early 1600s and 1800s. This extended cool event has often been termed the 'Little Ice Age' and it had considerable agricultural and sociological effects in the temperate zone.

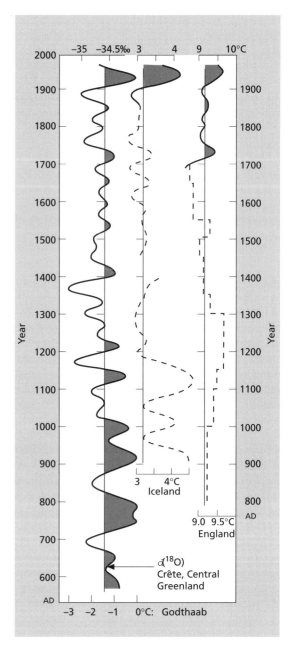

Fig. 4.12 Oxygen isotope curve from the Greenland ice core (left) compared with projected temperature curves based on historical reconstructions from Iceland and England. From Dansgaard *et al.* (1975).

Instrumental records of past climates date back to the 18th century in Europe, but are very limited outside that continent (Jones 1990). The changes of the present century are, of course, known in

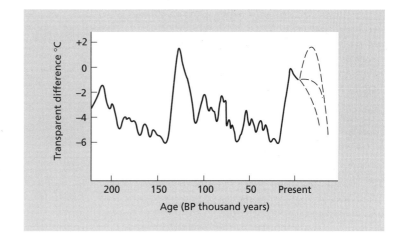

Fig. 4.13 The temperature curve of the past 200 000 years based on oxygen isotope studies projected into the future, with or without the greenhouse effect.

considerably more detail and these will be discussed later, but this and the previous chapter should have provided enough information concerning our current knowledge of the climate of the past for us to be able to place the debate about climate change prediction into its true historical setting. Figure 4.13 summarizes the problem that now faces us. Will we continue to head towards the next glaciation as one might have expected were it not for the 'greenhouse effect'? Or will this human impact on global climate result in a prolongation of the present interglacial, or may it even result in the development of a super-interglacial without Quaternary precedent?

Further reading

Cox, C.B. & Moore, P.D. (1993) *Biogeography: An Ecological and Evolutionary Approach*, 5th edn. Blackwell Scientific Publications, Oxford.

Jacobson, G.L., Webb, T. III & Grimm, E.C. (1987) Patterns and rates of vegetation change during the deglaciation of eastern North America. In: Ruddiman, W.F. and Wright, H.E. Jr. (eds) *North America and Adjacent Oceans During the Last Deglaciation.* Geological Society of America, The Geology of North America, vol. K-3, pp. 277–288.

Jones, P.D. (1990) The climate of the past 1000 years. *Endeavour* **14**, 129–136.

Pielou, E.C. (1991) *After the Ice Age: the Return of Life to Glaciated North America.* University of Chicago Press, Chicago.

Chapter 5

Carbon: Sources, Sinks and Cycles

We saw in Chapters 3 and 4 that the earth's climate, even in the absence of human beings, is quite unstable. Over the past 2 million years, which is a very small proportion of the earth's history, the climate has fluctuated between glacial and interglacial conditions and sometimes the change seems to have been effected in a matter of decades. It is against this background that we must ask the question whether the activities of humankind, including the combustion of fossil fuels and the deforestation of large sections of the globe, may be modifying the composition of the atmosphere to such an extent that the earth's climate may be significantly affected. The problem generates two component questions. First, are the industrial and other processes of modern human society really having a substantial effect on the atmosphere; and second, is there any evidence to suggest that atmospheric changes really are having (or will have) an impact on climate?

The carbon cycle

In Chapter 2, the concept of the global carbon cycle was outlined and it was explained that certain products of human industrial activities could affect the energy balance of the earth by acting as a thermal blanket through which the short-wavelength energy from the sun could penetrate, but which would absorb the long-wavelength, heat energy reflected from the earth's surface. Some of these gaseous by-products of human life seem to be building up in the atmosphere and enhancing this 'Greenhouse Effect', as remarkably predicted by John Tyndall of Manchester, England, back in 1861. We need first to look at the evidence for this build-up to determine whether our species can really have such a profound effect on biogeochemical cycles. As recently as 1959 an eminent scientist attached to the Shell oil company wrote in the journal *New Scientist*, 'Nature's carbon cycles are so vast that there seem few grounds for believing Man will upset the balance' (Matthews 1959). By 1970 it was widely recognized that such statements were unduly complacent; the microclimatic effects of urbanization had been clearly demonstrated and the possible role of carbon dioxide (CO_2) in future climate change was being discussed (Landsberg 1970).

The carbon cycle is illustrated in Fig. 5.1 and estimated figures are provided that indicate how much carbon is considered to be present in the various 'pools' or 'reservoirs' where it resides and how quickly it moves from one reservoir to another. Another set of information that can be useful in understanding the carbon cycle is the residence time for carbon in any particular pool. This is an estimate of how long an average carbon atom can expect to reside in a pool before being transferred to another. For some reservoirs, such as the fossil fuels, which are ultimately derived from the photosynthesis of plants that may have lived hundreds of millions of years ago, the residence time is likely to be extremely long. The residence time for carbon in a short-lived organism, however, such as a ragweed plant or a mouse, is extremely brief in comparison.

Atmospheric carbon

From the point of view of current global change, the most important figures to know are those which affect the size of the atmospheric reservoir.

The most obvious information required by environmental scientists is the level of CO_2 in the atmosphere and its variation in the course of time. In 1958, at the very time when complacency regarding the human influence on the carbon cycle was being expressed in the scientific literature, a recording station was set up in Hawaii at Mauna Loa. The site was chosen because it was situated far from areas where there might be a

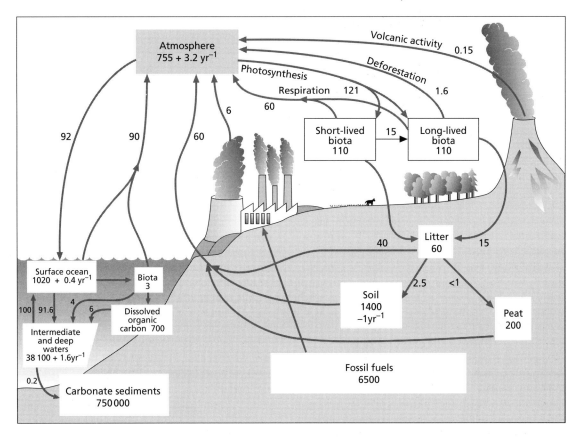

Fig. 5.1 The global carbon cycle. Units in boxes (reservoirs) are in gigatonnes (Gt) = 10^9 tonnes = 10^{15} g of carbon. Figures beside arrows denote flux rates in Gt yr^{-1}. Based on a variety of sources, particularly Butcher *et al.* (1992), Green *et al.* (1993) and Schimel (1995).

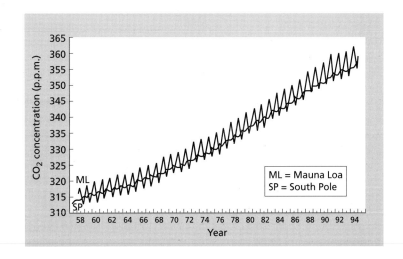

Fig. 5.2 Mean monthly concentrations of atmospheric CO_2 at Mauna Loa, Hawaii (ML), and at the South Pole (SP) since 1958. Units are parts per million by volume (Keeling *et al.* 1995).

strong local influence on atmospheric carbon levels, such as industrial complexes or concentrations of population. A recording station was also set up at the same time at the South Pole. Within a few years it became evident that there was an upward general trend in atmospheric concentration of CO_2 and the current developments are shown in Fig. 5.2 (Keeling *et al.* 1995).

The figures used in this graph are parts per million (ppm) by volume in dry air. (The measurements are obtained by measuring the absorption of infrared wavelengths of radiant energy by CO_2 and since water vapour also absorbs strongly in the same wavelengths the samples must be dried first.) The overall changes have now been documented for over 35 years and the upward drift is quite clear, but in the early years the trend, amounting to a change of only about 1 ppm yr^{-1}, was difficult to detect against the background variation. The mean figure for the atmospheric concentration of CO_2 in 1958 was 315 ppm but, as can be seen from the diagram, there is a cyclic change in concentration with season at the Mauna Loa site, amounting to about 4 ppm between winter (low photosynthesis, so high CO_2) and summer (high photosynthesis, so low CO_2). In parts of the globe where seasonal differences in the growth of vegetation are more pronounced, this amplitude can be even higher, and at the South Pole, far removed from dense vegetation, the amplitude is low, but still detectable. Note in Fig. 5.2 that the southern hemisphere peaks and troughs alternate with those of the northern hemisphere because of the difference in growing season.

Just as there is a seasonal variation in CO_2 concentration, so there is a daily variation, amounting to about 40 ppm amplitude, depending on where the measurements are taken. This is because photosynthesis absorbs CO_2 only during daylight, whereas the production of CO_2 by plants, animals and microbes continues day and night. Obviously, this short-term, but very pronounced variation cannot be shown on a diagram like Fig. 5.2, but it is not difficult to imagine the huge background 'noise' it would impose on the relatively simple pattern shown there.

Carbon dioxide concentration in the air also varies with height above the ground simply because the soil contains almost exclusively non-photosynthetic organisms, mainly invertebrate animals, bacteria and fungi feeding on the detritus falling from the photosynthetic organisms, the living plants. The soil 'respires', generating CO_2 but does not photosynthesize, so recorders close to the ground will pick up higher concentrations of CO_2 than those elevated above it. There may be as much as 12 ppm difference between soil level and a height of 14 m in a tree canopy during the day. Just as the soil respires, so does a city, and we can expect similar variations in CO_2 concentration over the face of the earth depending upon the density of human populations and their industrial and agricultural activities.

Against this background of variability it becomes evident that long-term trends in concentration of the order of 1 ppm yr^{-1} are quite difficult to detect, especially over short periods. As time has progressed, however, the upward rise in atmospheric CO_2 has clearly been sustained and is actually increasing with time. Once the rise was firmly established the question of the increasing capacity of the atmosphere to retain infrared radiation (heat) became an urgent one, and the possibility of a 'Greenhouse Effect' was recognized.

Long-term trends

Having established that there is an upward trend in atmospheric CO_2 the next question relates to long-term variability. How long has the upward trend been proceeding? Is it part of a longer cycle? What is the 'baseline' level of CO_2 from which it is building? To answer these questions we need to look back again to the ice cores, which supplied information concerning CO_2 levels in the glacials and interglacials (Chapter 3). Can we use the same approach when studying the history of atmospheric CO_2 over the last few hundred years? The answer is yes, and the results are shown in Fig. 5.3.

Work by Raynaud and Barnola (1985) on bubbles of air trapped in ancient glacial ice from Antarctica showed that atmospheric CO_2 levels between 1650 and 1850 AD were about 260–280 ppm, so the upward trend has evidently been taking place for around two centuries. Subsequent, more detailed analyses have mapped the rise with greater precision, as shown in Fig. 5.3, and it is clear that the rise is still accelerating, having begun in the late 18th century (Gribbin

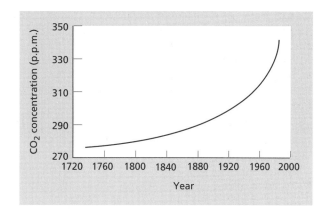

Fig. 5.3 The changing concentration of CO_2 in the atmosphere since the early 18th century, based upon analysis of air bubbles trapped in ice. The most recent values are derived from atmospheric measurements at Mauna Loa.

1990). The initial rise is too early to be accounted for by the burning of fossil fuels. As we see from Fig. 5.4, fossil fuel consumption was very low until late in the 19th century, but forest clearance was proceeding apace, and this may well have been the source of additional carbon in the atmosphere during the early stages of the rise (Houghton 1991).

Sources of atmospheric carbon

Having established the reality and the history of the upward trend, we next need to look more carefully at the sources of this accumulating CO_2 in the atmosphere. To do this we need to calculate with as much accuracy as possible the carbon con-

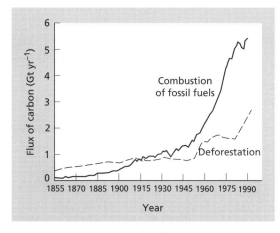

Fig. 5.4 Estimated annual global emissions of carbon in Gt yr^{-1} derived from fossil fuel burning and deforestation. From Houghton (1991).

tent of all its global reservoirs and the rates of movement (flux) between these reservoirs, especially those directly involved with the atmosphere. These are usually considered in terms of the element carbon, for this is the critical element in the process, since oxygen is in far greater abundance, especially in the atmosphere (about 20% by volume compared with the 0.04% of CO_2). When sugars are oxidized by plants and when coal or oil is combusted in power stations, CO_2 is produced. The supply of oxygen from the atmosphere is plentiful, so carbon is the important element in the transfer.

Figure 5.1 shows the general outcome of such calculations, many of which (such as the carbon resources in fossil fuels) are still conjectural, but the flux rates are becoming known with increasing reliability. If we concentrate for the moment on the movements of carbon to and from the atmosphere, it will help to clarify the possible source (or sources) of the current rise. This is shown in Fig. 5.5.

Knowing the rate of rise in concentration of CO_2 in the atmosphere it is possible to calculate how much extra carbon accumulates in this pool each year, and this comes to about 3.2 billion tonnes (10^9 tonnes = 1 gigatonne, Gt = 10^{15} g) (Schimel 1995). One source that is reasonably well documented and therefore can be calculated with some accuracy is the combustion of fossil fuels. For the years 1989–1990 this has been estimated at between 5.5 and 6.0 Gt (IPCC 1992), most of which is derived from industrial activity in the northern hemisphere (Fig. 5.6). Obviously, this total varies from year to year; in 1987 the figure was 5.7 Gt. It also varies with social and

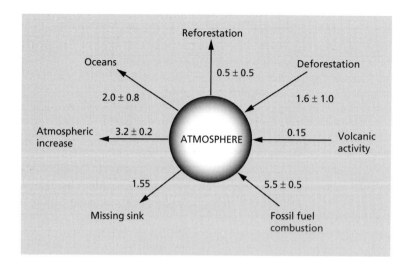

Fig. 5.5 Atmospheric budget for carbon showing sources and sinks with estimated fluxes. Figures are in Gt yr^{-1}. The 'missing sink' could in large part be accounted for by vegetation biomass growth as a result of CO_2 fertilization (estimated at 1.0 + 0.5). (Data based largely on Schimel 1995).

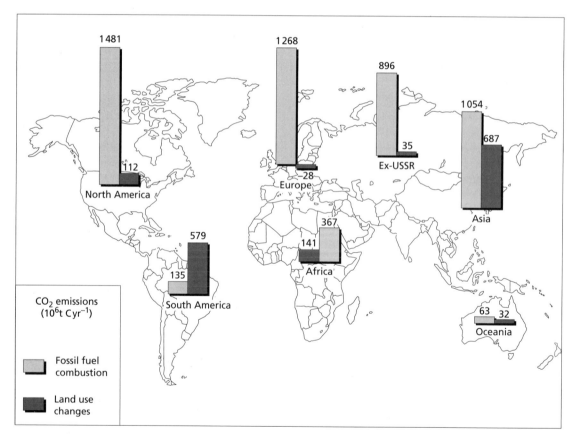

Fig. 5.6 Emissions of CO_2 to the atmosphere as a result of human activities during 1980. Two types of emission are recognized: those resulting from the burning of fossil fuels and those resulting from land-use change (e.g. forest clearance). The negative value of the latter for Europe implies that this continent acted as a carbon sink during the year. From UNEP (1991).

political factors, a recent example being the war-time combustion of oil wells in Kuwait in 1991. In that particular example, the calculated emission was about 0.065 Gt, only little more than 1% of the total annual emission from fossil fuel combustion in the world, so it did not represent too serious an addition to the atmospheric load of CO_2.

Fossil fuel combustion is not the only way in which human activities are adding to the atmospheric load of CO_2, as has been seen from historical studies. Human modifications of the world's vegetation also contribute to atmospheric CO_2, especially by forest destruction and converting the cleared land into a vegetation bearing lower amounts of biomass (the living and standing dead material in an ecosystem). The net loss of carbon from the destroyed vegetation finds its way into the atmosphere either rapidly, as when burning is employed, or more slowly when microbial decomposition gradually degrades the waste organic matter. It is difficult, however, to obtain reliable estimates of how much carbon is transferred to the atmosphere in this way, partly because of the problems in estimating the area (and therefore the biomass) destroyed in this process. Estimates for forest destruction in the late 1980s in Brazil alone, for example, vary between 1.4 and 8 million hectares per annum.

The other major problem in estimating carbon transfer to the atmosphere on this route is that there may be less obvious movements of carbon to the atmosphere, such as the oxidation of organic matter from soils that accompanies forest destruction. This is obviously difficult to quantify and varies considerably from one ecosystem to another.

There are also some other land management processes that are difficult to assess in terms of carbon flux, for example the burning of tropical grasslands. This is a regular feature of the human management of this biome, but although it may result in the transfer of large quantities of carbon from biomass to atmosphere, the grasses regrow and presumably absorb an equivalent mass of carbon in the process. Nevertheless, the carbon flux from the burning of about 750 million hectares of savanna grassland each year is considerable, and if the process is then followed by overgrazing, the carbon recovery will not take place and a net contribution of carbon to the atmosphere results (see Levine 1991). The accuracy with which fires can

be detected and their impact quantified has increased very considerably with the development of remote sensing techniques. Over Africa, for example, the use of night-time satellite imagery has permitted detailed study of the patterns of burning, the season at which it takes place and the area affected (Cahoon *et al.* 1992). It has also proved possible, by analysing radiation from the earth at specific, fairly long wavelengths (about 3.5–4 µm) to record fires even during the daylight, and using this technique has provided sufficient resolution to be able to map stubble burning from individual fields in Britain (Scorer 1993).

One consequence of these problems in estimating carbon flux from land-use changes is that projected values have a wide range of error. The estimate of the Intergovernmental Panel on Climate Change (IPCC 1990) for the year 1980 was 0.6–2.5 Gt yr^{-1}. The estimate for 1990 was 1.1–3.2 Gt yr^{-1}, confirming that this source of atmospheric carbon is rising. The net average emissions over the decade proposed by IPCC and used in Fig. 5.5 is 1.6 Gt yr^{-1} (Siegenthaler and Sarmiento (1993) prefer a figure of 1.9 Gt yr^{-1}). This process of biomass destruction is also important in that it contributes compounds other than CO_2 to the atmosphere, including methane and nitrous oxide, both of them greenhouse gases in their own right. This aspect will be considered later.

One natural source of CO_2 that deserves to be mentioned is volcanic activity. The estimate of global CO_2 emission from subaerial and submarine volcanoes is about 0.13–0.175 Gt yr^{-1} (Gerlach 1991), about 10% of which seems to come from a single volcano, namely Mount Etna. This total from volcanic activity is relatively small when compared with fossil fuel burning (only about one-tenth as great), but is still a significant input.

Sinks of atmospheric carbon

Fortunately for the balance of carbon on earth, there are also paths by which carbon is removed from the atmosphere. One of these is by solution in the oceans. Sea water is slightly alkaline and this assists the solution of CO_2 where it may form hydrogen carbonate ions. These can be used by the phytoplankton as a source of carbon for photosynthesis and some of the carbon fixed in this way will sediment into deeper waters as the dead bodies of the plankton sink downwards. Hydrogen carbonate ions may also combine with calcium to generate calcium carbonate (lime), especially in forming the lime-impregnated outer coats on some of the planktonic organisms. These also sink after the death of the organism, so provide a means of transferring carbon away from the upper layers of the oceans (see Chapter 7). The limiting factor in the movement of carbon along this path is the actual transfer of CO_2 into the water at the interface between sea and atmosphere. Under rough conditions and in colder temperatures the rate of solution is enhanced. In the warmth and low wind speeds of the equatorial regions the ocean may even become a source rather than a sink of carbon to the atmosphere. But the overall net movement of carbon on a global scale is from atmosphere to ocean with a flux of about 2 Gt yr^{-1}. Once again, there are great difficulties in determining a reliable figure because of experimental problems and the very wide amplitude of background variation.

In theory one should be able to calculate the flux of carbon from atmosphere to sea by multiple observations on the difference in concentration (strictly, partial pressure) of CO_2 in the atmosphere and that in the sea, for this is the driving force of solution rates. But again, the variability of these values is very considerable. The global mean partial pressure of CO_2 in the surface waters of the ocean is about 8 ppm (compared with about 350 ppm in the atmosphere), so it is clear that carbon will be inclined to move into the water. But water temperature and local biological respiration processes can elevate the CO_2 partial pressure in the water to 80 ppm or more, so reducing the flux. Simple measurements, therefore, provide only a very imprecise estimate.

There is also the question of oceanic circulation to consider, for this could be important in carrying carbon from one hemisphere to the other. Most fossil fuel burning (about 95%) is carried out in the northern hemisphere (Fig. 5.6), therefore atmospheric concentrations differ and so does the atmosphere/ocean flux. Then there is the problem that some of the oceanic carbon may have entered from dissolved organic carbon in river outflow rather than directly from the atmosphere, so calculations become very complex and

the outcome is usually dependent upon the assumptions made in these calculations. A flux value of about 2 Gt yr^{-1}, however, seems to be of an acceptable order.

Just as deforestation represents a source of atmospheric carbon, reforestation (plantation and regrowth) constitutes a sink. Currently, this is small in comparison with destructive activities, but it is important in some parts of the world and, overall, probably accounts for an uptake of 0.5 Gt each year.

The carbon budget eventually derived from all these estimations, shown in Fig. 5.5, is out of balance unless there is another sink for carbon apart from the oceans. On the basis of the figures used in this diagram, 7.25 Gt of carbon arrive in the atmosphere each year, 3.2 Gt remain in residence there and about 2 Gt are taken up by the oceans. This leaves a further 1.55 Gt that must have been removed by some other sink. The most likely explanation for this 'lost' carbon is that the vegetation of the earth that has not been removed by humankind is actually growing in biomass and taking up carbon. This is very difficult to prove, but it is not an unreasonable supposition. It is possible that the raised CO_2 in the atmosphere is itself contributing to a higher productivity on the part of the remaining vegetation because CO_2 is very often the limiting factor in photosynthesis. It is also possible that the nitrogenous compounds that are released into the atmosphere and into waterways by human activities might raise the productivity of certain vegetation types previously suffering from a lack of nitrogen. So the human additions to the atmosphere may actually be providing a fertilization effect, increasing the carbon sink provided by plant productivity.

One approach to the study of changing vegetation biomass and its relationship to the carbon cycle is the close study of changes in the seasonal amplitude of CO_2 in the atmosphere. If the photosynthetic biomass of the earth were getting bigger, it would progressively take more CO_2 out of the atmosphere each growing season and return it in the dormant season, leading to an increased amplitude in the seasonal cycles of CO_2 as shown in an exaggerated form in Fig. 5.7a. Conversely, if the biomass is declining, then the seasonal amplitude will decrease (Fig. 5.7b). There is an added complication to this approach, since the seasonal changes in sea temperature may also contribute

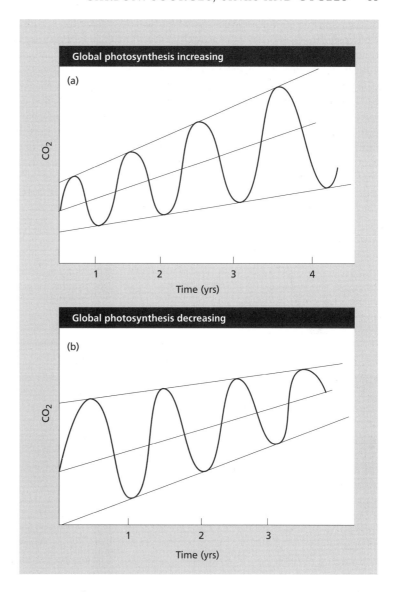

Fig. 5.7 Hypothetical curves for annual variations in atmospheric CO_2 under conditions: (a) in which global photosynthesis is increasing, leading to a greater drawdown of carbon from the atmosphere each growing season; or (b) where global photosynthesis is decreasing, leading to a reduced seasonal fluctuation in atmospheric CO_2.

to the cycles, but these would act in opposition to the vegetation cycles (cold water would absorb more CO_2), so would only tend to dampen the oscillations. Hall and co-workers (1975) first used this approach, but their analysis of the Mauna Loa data showed a very steady 'hemispheric metabolism' (production vs. respiration in the northern hemisphere). More recently, D'Arrigo and co-workers (1987) have found an increasing amplitude in cycles, suggesting that there is an increase in biomass growth and they regard the boreal forests as particularly important in their contribution to this process. There is also evidence that

the sink effect is greater in the northern than in the southern hemisphere, which again points to terrestrial vegetation as the missing sink since the bulk of the world's vegetated land masses lie in the northern hemisphere (Denning *et al.* 1995).

Some additional evidence for increased biomass growth in recent times comes from studies of tree rings, and these were also used by D'Arrigo *et al.* in their work. They studied white spruce (*Picea glauca*) tree rings at four sites on the northern limits of forest growth across Canada and they found that tree-ring width in any given season corresponded very closely with the seasonal drop in atmospheric CO_2 during that growing season. This suggests that the drop is actually caused by the growth of forest trees and that the increasing amplitude of seasonal atmospheric CO_2 fluctuations is due to increasing biomass and growth of the forests. This supports the idea that the 'missing sink' for carbon is indeed biomass growth (see also Keeling *et al.* 1996). It also points to the importance of the boreal forests, currently under increasing threat of destruction especially in Asia, in relation to the earth's carbon balance.

In Europe, where almost all forests are heavily managed, it is possible to calculate changing forest resources in some detail, and these figures also indicate an increase in forest biomass over the past 20 years (Kauppi *et al.* 1992). They suggest that European forest growth alone is responsible for a carbon uptake of perhaps 0.12 Gt yr^{-1}, largely as a result of the fertilization effect of nitrogenous pollution in the atmosphere.

It is worth pausing at this point to consider the role of vegetation in the carbon cycle because this is an area in which some very loose thinking is often evident. While vegetation is growing in bio-

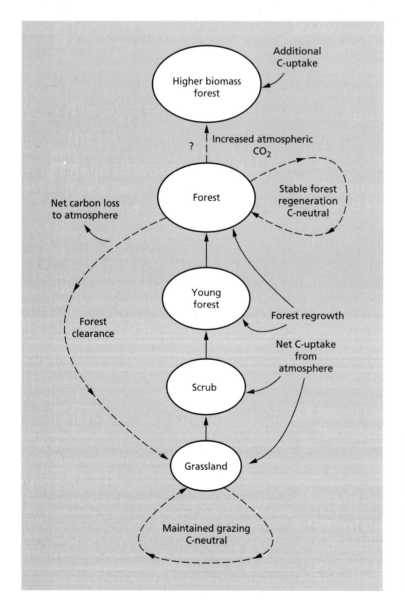

Fig. 5.8 Schematic diagram to illustrate the way in which forests act as a carbon sink only while they are increasing their biomass. This may occur during the course of ecological succession, or as a response to some form of fertilization as, for example, when atmospheric CO_2 levels are raised. Stable, equilibrium forest (or any other ecosystem) is carbon neutral.

mass it provides a sink for atmospheric carbon. Once it reaches maturity its biomass stabilizes and the vegetation assumes an equilibrium, respiring as much carbon back into the atmosphere as it absorbs in photosynthesis. Some of this respiration, of course, is derived from animals (herbivores, carnivores and detritivores) and also, even more important, from microbial decomposers (fungi and bacteria). But at equilibrium the total ecosystem respiration equals the total photosynthetic productivity, so the ecosystem becomes 'carbon neutral'. For this reason, the great forests of the world (particularly the equatorial rain forests and the boreal coniferous forests), although extremely valuable in very many ways, cannot be regarded as carbon sinks unless they are growing in bulk. As we have seen, it is possible that they are growing in bulk, but it is also likely that the regrowth areas of the world, where former cultivated land is now being permitted to redevelop forest, accounts for a significant proportion of the net global uptake of carbon, rather than those forests that are already mature. These relationships are shown in Fig. 5.8.

What, then, is the potential for carbon storage in vegetation in the future? Is it an adequate sink to buffer the possible effects of the greenhouse? Clearly, it is not an adequate sink to cope with current CO_2 emissions, otherwise there would be no rise in its atmospheric levels. The future of vegetation as a sink depends upon its continued growth (as explained above) and this is unlikely because of the pressures human populations will continue to exert upon forests—both for timber resources and for additional agricultural land. It would be unwise, therefore, to rely upon the biomass sink as the solution to the greenhouse problem (Schlesinger 1990; Smith & Shugart 1993).

Peatlands are unusual ecosystems in that they accumulate organic carbon in dead but undecomposed form, and hence act as a carbon sink, but the amount of carbon captured each year is relatively small, mainly because most peatland areas in temperate and boreal regions have relatively low productivity. The amount of carbon stored in this way, however, estimated by Gorham (1990) to be between 180 and 277 Gt, is very substantial and could become an important source of atmospheric carbon if oxidation processes took place. If the waterlogging of the peatland is reduced and if the temperature is raised, microbial decomposition would take place and the carbon would be returned to the atmosphere. Global warming and/or human activity could thus transform a sink for carbon into a source, and there is some evidence to suggest that this has already begun to occur in the arctic tundra of the North Slope of Alaska (Oechel *et al.* 1993). Such peatlands are also an important source of another greenhouse gas, methane, which will be discussed later. This means that land reclamation schemes in which wetlands are drained for agricultural development (as has taken place, for example, over much of the Everglades region of Florida) is environmentally damaging both by the removal of species-rich habitats and by contributing to the imbalance in the carbon cycle.

The rising curve of atmospheric CO_2 (Fig. 5.2) has also been analysed for fluctuations that cannot be accounted for by known inputs into the atmosphere. These have shown some regular variations with a cycle period of about 10 years and again at a 3–5 year interval. These may be a consequence of climatic cycles resulting from variations in solar activity and the periodic development of the El-Niño Southern Oscillation respectively (see Chapter 6). There are also less regular anomalous features in the atmospheric CO_2 curve, such as the plateau observed in 1992–1993 (detectable in Fig. 5.2). This could again be a climatic effect, this time due to the eruption of Mount Pinatubo that resulted in the suppression of a carbon source by lowering temperatures. But many features of the atmospheric phase of the carbon cycle still need to be researched before acceptable explanations of such variations can be made.

Climatic consequences

Whatever the precise sinks and sources of atmospheric carbon prove to be, the fact that the atmospheric reservoir is growing is undisputed, and it is reasonable to argue that a continued growth in atmospheric CO_2 will contribute towards a greenhouse warming. Indeed, there is a significant positive relationship between atmospheric CO_2 concentration and average global temperature over the last 30 years (Fig. 5.9) (Schneider 1989; Kuo *et al.* 1990). But one must bear in mind that many other factors are also at work, includ-

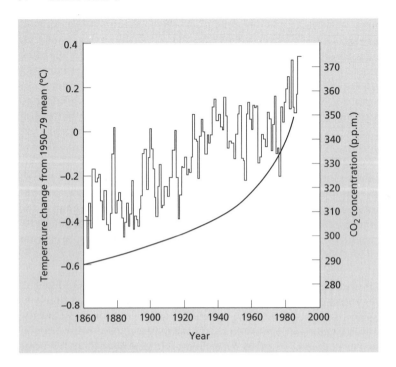

Fig. 5.9 The general global temperature trend over the past two centuries follows a similar course to that of atmospheric CO_2 levels. The temperature is recorded as deviation from the 1950–1979 mean value. From Schneider (1989).

ing the possibility of increased cloud generation as a result of a warmer sea surface (and also as a result of other human pollutants, such as sulphur dioxide), hence less sunshine will reach the earth's surface. There are also the background climatic changes related to the orbital and other variations discussed in Chapter 3 that currently seem to be leading us back towards a period of cold. Solar changes, such as sunspot density, remain important in determining climatic fluctuations and their influence can clearly be demonstrated in some current climate changes (Kelly & Wigley 1992). Volcanic activity can seriously disrupt global climates for a period of years, but fortunately the human equivalent, nuclear war (which was once investigated in climatic terms because of the possibility of its generating a so-called 'nuclear winter' in which temperatures could fall by 10–20°C) (Schneider & Thompson 1988; Turco *et al.* 1990), has now become less of a threat than it was a few years ago. The possibility must also be borne in mind of quite sudden alterations in oceanic and atmospheric circulation patterns that are currently unpredictable because of our ignorance of their controlling factors. We know from historical studies that conditions

can change very quickly when certain thresholds are crossed (see Chapter 4).

If, however, global warming were to take place, how would it modify the carbon cycle? Warmer surface waters would not only produce more water vapour, but would also dissolve less CO_2, because the gas is more soluble in cold water. The result could be a shift in the equation leading to more CO_2 being retained in the atmosphere. How a climatic warming would affect global vegetation will be discussed in Chapter 8, but its general effect upon global biomass is extremely difficult to predict. One cannot use history as a means for projection into the future in this case, because the estimated global biomass before the development of industry is about 610 Gt of carbon, compared with the modern estimate of 560 Gt (Siegenthaler & Sarmiento 1993). So biomass has fallen as CO_2 has risen. But this, of course, is a consequence of human clearance of forests rather than any climatic response on the part of the natural vegetation. The future, however, may follow the same track, so that although the elevated levels of atmospheric CO_2 might be expected to stimulate increased production of biomass, the removal of this commodity by human activity may well continue to exceed any such increase. Unless massive global changes in the policy of resource use are implemented, the vegetation of the earth is unlikely to provide a reliable buffer against the greenhouse effect.

Methane

Although CO_2 is the most important of the potential contributors to greenhouse warming, perhaps being responsible for about 50% of the greenhouse effect (Fig. 5.10), there are other carbon compounds that make a significant contribution, namely methane (CH_4—about 18%) and the chlorofluorocarbons (CFCs—about 14%). There are two other major gases, nitrous oxide (N_2O—about 6%) and ozone (O_3—about 12%), that also need to be reckoned with (Ashmore 1990).

Methane is present in the atmosphere in far smaller concentrations than CO_2, currently about 1.7 ppm (compared with CO_2 at 350 ppm), but its very considerable impact is due to its high capacity to absorb infra-red radiation (about 30 times that of CO_2—see Chapter 2). It is generated naturally by a variety of processes in which decomposition of carbon compounds is incomplete, so that some atoms of carbon are not converted to CO_2. This process is particularly associated with certain bacteria that operate in the absence of oxygen. Peatlands and other wetlands, for example,

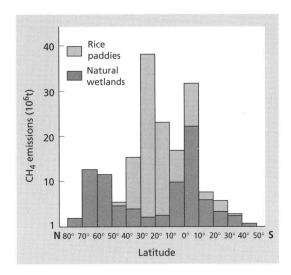

Fig. 5.11 The variation of methane emission from natural wetlands and rice paddies with respect to latitude. Note the concentration of natural wetland emissions in tropical and boreal latitudes and that of rice paddy methane emission in the tropics and subtropics.

generate methane, and this includes agricultural wetlands, such as rice paddies. The rates of emission from natural and man-made wetlands is shown in Fig. 5.11 and it can be seen that there are two main peaks in natural wetland emissions, near the equator and between 50° and 70° in the northern hemisphere (the cool temperate peatlands of the boreal zone). The rice paddies reach their peak between 20° and 30° north. Probably 40–50% of the total input of methane to the atmosphere comes from wetlands (Whiting & Chanton 1993).

Methane is also produced by termites and ruminant animals (such as cattle), both of which have large populations of anaerobic bacteria and protozoa in their guts. Termites are abundant over much of the world and they may contribute as much as 15% of the methane entering the atmosphere (Rasmussen & Khalil 1983). Since pastoralism accounts for a very large proportion of ruminant animals, human activity is again a major factor in methane production. The accumulation of methane in the atmosphere is further enhanced by the methane produced from the burning of vegetation, natural gas exploitation and leakage, and the disposal of waste by landfill operations (Fig. 5.12).

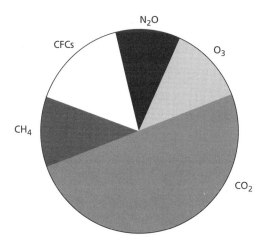

Fig. 5.10 The potential contribution of various greenhouse gases to the greenhouse effect. Water vapour, because of its variability, is excluded from this diagram, but ozone is included.

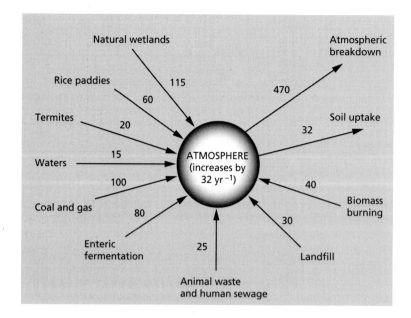

Fig. 5.12 Sources and sinks of atmospheric methane. The figures are in 10^{12} g yr^{-1}. These data, derived from a number of sources, are estimates and the budget does not currently balance.

In Chapter 3 it was shown that methane, like CO_2, has higher concentrations in the atmosphere during interglacials and lower concentrations in the glacial episode, and, as in the case of CO_2 it is difficult to ascertain whether these changes actually drive climate or lag behind it. Ice-core studies have shown that methane has doubled over the past 150 years from a pre-industrial level of about 0.8 ppm (Pearman & Fraser 1988) (Fig. 5.13). It is currently increasing at a rate of about 0.1 ppm every 5 years (Fig. 5.14), but there is some indication that the rate of increase has actually slowed down since 1983 (Steele *et al.* 1992). Even so, the current rate (about 1% per year) is far faster than that of CO_2, but it is not clear which of the possible sources is primarily responsible for the increase in methane concentration in the atmosphere. The possibility of a feedback mechanism operating in which rising global temperature may enhance methane pro-

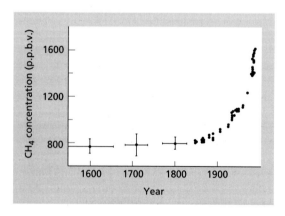

Fig. 5.13 The increase in atmospheric methane over the past 400 years, based on recent measurements and the analysis of air bubbles in ice cores from Antarctica. From Pearman & Fraser (1988).

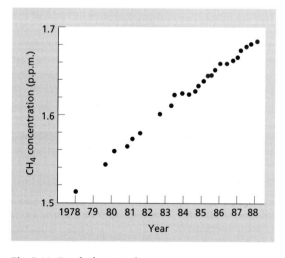

Fig. 5.14 Graph showing the current rise in atmospheric levels of methane. From Ashmore (1990).

duction, for example from wetlands, needs to be considered.

Perhaps as much as 60% of the total methane production results from human activities, so this is clearly a greenhouse gas that could be controlled by changing human land-use policies (Hogan *et al.* 1991). The destruction of methane in the atmosphere depends on the presence of hydroxyl ions that are responsible for its oxidation. But hydroxyl ions are also scavenged by another carbon compound that is currently increasing in the atmosphere, namely carbon monoxide. This gas is itself generated by human activity in the burning of forests (Newell *et al.* 1989) and in motor exhausts. So we are not only producing methane, but also interfering with its natural degradation.

Other greenhouse gases

The chlorofluorocarbons (CFCs) come in a wide range of varieties, the two most important of which, in terms of the greenhouse effect, are $CFCl_3$ (so-called CFC-11) and CF_2Cl_2 (CFC-12). These are man-made compounds used in aerosol propellants, refrigeration and in foam packaging. Although their concentration in the atmosphere is minute (about 0.0003 ppm) they play a significant role in the greenhouse effect because of their strong absorption in the infra-red. They are reckoned to be between 11 000 and 14 000 times as effective as CO_2 as greenhouse gases, so any increase, however small, is likely to be important. In fact, they are increasing by about 4% per year, so they represent a very serious threat to climate, quite apart from their detrimental influence on stratospheric ozone (see Chapter 8).

International agreement has now restricted the use of CFCs-11 and -12 and much effort is being put into alternative compounds that will prove less environmentally damaging. Most attention is being focused on the hydrohalocarbons, which are more easily destroyed in the atmosphere so have a shorter life in which to act as greenhouse gases. They are also about 10 times less efficient as infra-red absorbers. There is a problem with these hydrohalocarbons, however. In the atmosphere they are degraded by the hydroxyl (OH) radical to produce trifluoroacetic acid (TFA) which is virtually non-biodegradable and may accumulate to toxic levels, especially in enclosed water basins, like desert playa lakes. It is still far from certain that these will prove safe alternatives to the CFCs (Schwarzback 1995).

Ozone is recognized to be an important greenhouse gas that is generated in the troposphere by human activity, but it is very difficult to gain precise information about global concentrations and trends because of the wide variability in space and time. In Fig. 5.10 it is suggested that about 12% of the greenhouse warming could be due to this gas.

Another neglected greenhouse gas is water vapour, which is abundant in the atmosphere and also absorbs in the infra-red wavelengths. Again, its variability means that it is difficult to assess its true role and most calculations leave it out of consideration. The wisdom of this attitude may well be questioned as this gas comprises between 1 and 4% of the atmosphere by volume and also because the reaction of water vapour to a rise in temperature could be extremely important in determining the ultimate outcome of climate change. Raised temperature, for example, could lead to higher levels of water vapour in the atmosphere (saturated vapour pressure is raised by higher temperature) and this could enhance the greenhouse effect, leading to a positive feedback. Preliminary data suggest that the water content of the stratosphere is indeed rising. On the other hand, increased water vapour in the atmosphere could lead to denser cloud cover and a lower penetration of solar energy to the earth's surface. This reaction becomes even more complex when one considers the possible interaction of another pollutant gas, sulphur dioxide (SO_2). This is emitted by fossil fuel burning and can stimulate the condensation of water droplets, so enhancing cloud formation. Currently, models in which sulphates are taken into consideration (and which hold back the greenhouse effect) seem to fit the observed data better than when CO_2 is taken on its own (see Fig. 5.18).

In the event of global warming, the earth's hydrological cycle will become even more difficult to predict. Not only cloud cover will alter, but also atmospheric circulation patterns, leading to some areas becoming drier and others wetter than at present. This, coupled with human activity in clearing vegetation, will lead to changes in precipitation and river flow patterns (Rind *et al.* 1992).

The remaining greenhouse gas of any importance is nitrous oxide (N_2O), which has a concentration of about 0.31 ppm in the atmosphere and has risen very little since pre-industrial times but seems to have accelerated recently. It is generated by fossil fuel burning and by biomass fires in general, and is also liberated during the application of fertilizers to land. Forest clearance itself seems to encourage the liberation of this gas from the soil (Keller *et al.* 1993). It is an efficient infra-red absorber (about 200 times as efficient as CO_2), hence is relatively important as a greenhouse gas despite its low concentration.

In the case of all these greenhouse gases, one property that affects their relative importance is their stability in the atmosphere—how long they survive before being broken down. This can be expressed in terms of the expected residence time for a given molecule in the atmosphere. These are rated as in Table 5.1 (Ashmore 1990).

As can be seen, most of these gases are fairly stable, some of them very stable. This means that

even if their emission could be tightly controlled, those molecules already in circulation in the atmosphere would still be with us for decades or even centuries to come.

If we look at the problem from the point of view of human activities that are responsible for generating these greenhouse gases, then the production and use of energy resources certainly tops the list, perhaps accounting for 57% of the potential greenhouse problem (White 1990) (Fig. 5.15). Chlorofluorocarbons rank next in importance, followed by agriculture and land-use changes. These are clearly the areas in which policy changes need to be made in order to reduce the threat of greenhouse warming.

Climate change

Many of the factors described here and in Chapter 2 lead us to expect an increased warmth retention by the earth, leading to increased global temperature. This assumes, of course, that the greenhouse effect is the only operative and that other factors (solar cycles, volcanism, etc.) are not causing significant changes at the moment. We must now face the question of whether such a warming of the earth can be detected.

The main problem is the relatively small change in global temperature one might expect from the greenhouse effect so far, coupled with the high degree of variability in the earth's climate, both spatially and temporally. We can overcome the spatial problem by averaging temperatures from the large number of meteorological stations world-wide and obtaining a global mean temperature. But even then, there is quite a large variation from one year to the next and from one decade to the next. Such factors as sunspot cycles, volcanic activity and the El Niño phenomenon (see Chapter 6) are responsible for much of this variation (Wigley & Raper 1990). These variations may be quite large, leading to fluctuations in the global mean temperature of up to 0.4°C over a period of 30 years or so, with the result that it would be very difficult to detect a greenhouse effect trend of this magnitude within such a time span. Despite this problem, the current upward trend in global mean annual surface air temperature (Fig. 5.16) does point to an increase of about 0.5°C over the past 100 years, and the greenhouse effect is a very likely cause (Wigley

Table 5.1 Residence times of the various greenhouse gases.

Greenhouse gas	Residence time
Carbon dioxide	50–230 years
Methane	10 years
CFCs	65 years (CFC-11)
	130 years (CFC-12)
Nitrous oxide	150 years

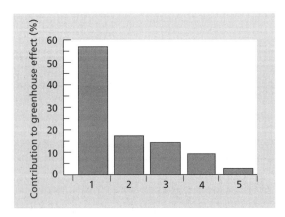

Fig. 5.15 The relative contribution of various human activities to the greenhouse effect. These are (in order of importance): **1** energy production and consumption; **2** CFCs; **3** agriculture; **4** land-use changes; and **5** other industrial processes.

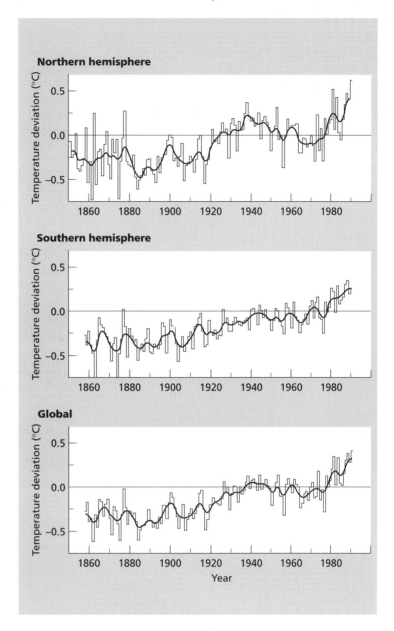

Fig. 5.16 Changes in global mean annual temperature (land-based measurements) since 1860. Temperature is expressed in terms of variation from the mean value between 1950–1979. From UNEP (1991).

1989). The rise has not been a steady one. There was a steep increase between 1910 and 1940, followed by a plateau for 35 years, then a recent steep rise since 1975. More detailed analysis over the surface of the globe shows that this increase is far from even, with some areas, such as Antarctica, central South America, the North Pacific and the North Atlantic actually cooler in 1986 than they were 20 years earlier (Jones & Wigley 1990).

Plate 5.1 illustrates how the temperature changed between 1986–1995 in relation to the period 1961–1980. It is most apparent that the high latitudes have become warmer (especially the land masses) while the tropics have hardly changed. Such spatial variability makes the future climate very difficult to predict in detail.

There can be little doubt, therefore, that climate has changed over the last 150 years and even over the last 10 years, but it would be unwise to assert that the greenhouse effect has therefore been demonstrated. Given the background variability of climate in the longer term, one cannot be sure that the current trends are being determined by human activity. The evidence, however, certainly does not contradict such a view.

Modelling climate

The prediction of the future climate is based upon the construction of mathematical models that account as closely as possible for past climatic observations. In other words, the best test of a proposed model is to test it against past changes.

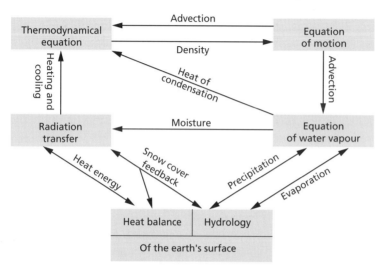

Fig. 5.17 The major components of a global circulation model and their complex interactions. The model is based on the physics of energy transfer between the earth's surface and the atmosphere. From Wetherald (1991).

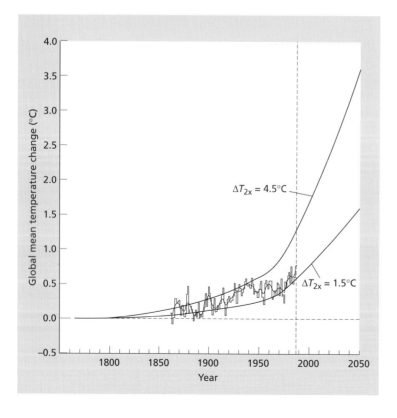

Fig. 5.18 Predictions of global warming from two models compared with the observed temperature since 1860. Both assume a doubling of atmospheric CO_2 levels. The upper line is a model based upon a high-sensitivity system in which global temperature will eventually stabilize at 4.5°C above the 1765 value. The lower line is based on a low-sensitivity system in which stability is achieved at 1.5°C above the 1765 level. Up to 1940, the upper curve provides the best fit, but subsequent records approximate more closely to the lower curve. From Wigley (1989).

Some complex models have been built that are able to account for past records over longer periods of time (such as the 10 000 years of the Holocene). In this case the 'record' of former climate is interpreted from past reconstructions based on geological and palaeoecological studies (COHMAP 1988). The models involved in this type of exercise are termed 'general circulation models' (GCMs) and are designed on the basis of energy balance and atmospheric and oceanic circulation patterns and their sensitivity to change. The essential components of such a model are shown in Fig. 5.17 (Wetherald 1991). Such models are based on the laws of physics and the ways in which they apply to processes such as cloud formation and oceanic deep water movements. All of these interact to produce a very complex system which can be modelled on a computer. As more information becomes available, the model can be altered and tuned so that its predictions on the basis of a given set of past data can come closer and closer to reality. It may then be in a position to predict the consequences of particular circumstances (such as a proposed doubling of CO_2 within a given period).

Using such a refined model, for example, it should be possible to extrapolate the current global mean temperature curve over the past 200 years or so and predict future global temperature building in a doubling of atmospheric CO_2. Figure 5.18 (Wigley 1989) shows such a model prediction allowing for a range of possible climate sensitivities to such a change. The upper curve predicts the outcome assuming a high degree of sensitivity to greenhouse gas changes, while the lower curve assumes low sensitivity. The truth probably lies somewhere in the middle, but the case for a future climate in which human activities play a dominant role seems well established.

Further reading

Butcher, S.S., Charlson, R.J., Orians, G.H. & Wolfe, G.V. (eds) (1992) *Global Biogeochemical Cycles.* Academic Press, London.

Schimel, D.S. (1995) Terrestrial ecosystems and the carbon cycle. *Global Change Biology* **1**, 77–91.

Whyman, R.L. (ed.) (1991) *Global Climate Change and Life on Earth.* Chapman & Hall, London.

Woodwell, G.M. (ed.) (1990) *The Earth in Transition; Patterns and Processes of Biotic Impoverishment.* Cambridge University Press, Cambridge.

Chapter 6

The Marine Realm

The hydrosphere

One of the many unique features of Earth as a planet is that it has two layers of fluid material overlying its surface—the oceans and the atmosphere. The atmosphere is of course continuous over the whole surface, whereas the oceans cover only about 70% of the earth. Some people like to refer to these two fluid 'envelopes' as the atmosphere and the hydrosphere. The term hydrosphere lumps all global water into a single category, whether it is in the oceans, in lakes, in ice or in porous rock under the ground. No-one nowadays needs reminding of the total dependence of life on these two fluid layers: the water of the hydrosphere is the major component of most living things, including ourselves, while all living things depend ultimately on the oxygen and carbon dioxide (CO_2) in the air. All of these vital components of our environment move between the atmosphere, the hydrosphere and the geosphere.

The continuity of the atmosphere over the whole globe, coupled with its low density and viscosity, means that it moves and mixes freely, and so retains broadly the same composition all over the world. There are of course exceptions, derived from the local production of pollutant gases, varying from place to place, and in the amount of water vapour in the air, which varies greatly with temperature and access to water bodies. In contrast, the continuity of the 'global ocean' is interrupted by the major continents, and its higher density and viscosity mean that it moves much more slowly in a number of complex ocean circulation systems.

Water is a remarkable substance in various respects; firstly, it is the only naturally occurring inorganic liquid at the earth's surface. Of the huge range of elements and compounds making up the outer part of the crust, all the others are either solids (as in the mineral matter of the crust) or

gases, as components of the atmosphere. It is further unique in that within the range of temperatures at the earth's surface, it occurs in all three of its physical states of gas (water vapour), liquid (water) and solid (ice). Indeed, its presence as a major component of the earth system may well have had some role in maintaining the temperature within that range.

The hydrological cycle

Evaporation, precipitation and the drainage of water via rivers off the land surface into the oceans drives the circulation of the hydrosphere. Some aspects of this atmosphere land ocean system have already been discussed in Chapter 2. Figure 2.2 is a diagram of these processes, referred to as the hydrological cycle. It illustrates some of the major sinks or global 'holdings' of water, and the fluxes that take place between them. The world's oceans contain 97% of the water available on or in the crust. The polar ice-caps and other ice locked up on the land, lakes, groundwater and water vapour in the atmosphere together make up the remaining 3%. Atmospheric water, as clouds and water vapour, together hold only 0.002%, and yet this minute fraction represents the sole pathway from the oceans back into the freshwater bodies on the land, and is the ultimate source of all the water on which land vegetation (and so virtually all terrestrial life) depends. This huge difference in the proportion of water in the oceans and atmosphere is reflected in the rate of 'turnover' of water in these two settings. The water vapour in the air has an average 'residence time' of 10 days before it returns to the land surface or the oceans as rain or snow. The water in the world's oceans, in contrast, has an average residence time well in excess of 10 years.

The proportion of water which resides in the rocks is also a highly significant feature of the hydrological cycle, in terms of the survival of ter-restrial life. This groundwater is calculated to be about 100 times the volume of freshwater in all the world's lakes and rivers, and is about half that of the combined ice-caps and glaciers of the world.

This chapter is concerned with the role of the oceans in relation to environmental change. The huge body of oceanic water acts as a major heat store and heat transport system, carrying heat both horizontally by the movement of ocean cur-rents, and vertically by the process of convection. It also acts as both a store (a 'sink') and a pathway ('flux') for many substances which are important for the survival of living organisms. Some of these roles of the world's oceans have been dealt with in earlier chapters. Here, we try to bring together the ways in which these many different functions interact and the role that these processes may play in global environmental change.

The ocean/climate system

Movements of the atmosphere and oceans com-bine to distribute the solar energy received by the earth in the warmer equatorial latitudes to the higher, cooler ones. Figure 6.1 shows how the radiation (as watts per square metre) absorbed at the equator diminishes rapidly (with increasing obliquity of the angle of the sun's rays) as you move into higher latitudes (curve A). The outgo-ing, long-wave radiation from the heated earth's surface is much more evenly spread between high and low latitudes (curve B). This means that the earth's surface in the latitudes above 30° north and south is radiating much more energy than it absorbs, while within the tropical latitudes it is absorbing more energy than it radiates.

The circulation of both oceans and the atmos-phere evens out this energy budget to some extent by moving warm water and air from low to high latitudes (right-hand side of Fig. 6.1). The low

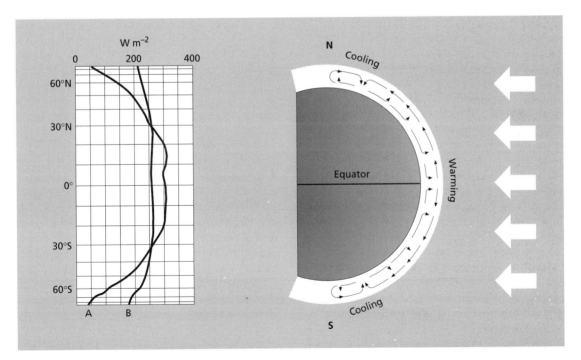

Fig. 6.1 The earth's radiation balance. Curve A shows the amount of radiation absorbed at the earth's surface, while curve B is the outgoing long-wave radiation. At latitudes above 30° the amount of outgoing radiation exceeds that absorbed, while the opposite is the case in lower latitudes. The circulation poleward of the atmosphere and oceans brings about a global balance of heat absorption and radiation. From WOCE Report (1993).

viscosity of air means that winds will generally move at much higher velocities than ocean currents, and so effect more rapid heat flow; however, the specific heat (heat capacity) of water is far greater than that of air. The importance of ocean currents in controlling, for example, the climates of the land around the North Atlantic has already been discussed in Chapter 3. Figure 3.22 shows the oceanic 'conveyor belt' which is one of the several elements of the complex system of currents that redistribute the solar heat energy. We need to look at these in greater detail.

Global movement of ocean waters

The movement of ocean water can be seen on a number of different scales. The surface water is affected most by wind, which in addition to producing wave movement drives a series of surface currents which also help to mix water in the upper part of the ocean. As the surface water moves, it may gain or lose heat. Radiation from the sun and contact with warm air will raise its temperature. Evaporation, contact with cooler air and the proximity of an ice sheet margin, or the icebergs that break from it, can all contribute to cooling the ocean water. Where the warming of the surface water has a significant effect, a vertical gradient of water temperature will be produced (a thermocline), with the lighter warm water overlying the heavier, cooler water beneath. If there is a strong development of the thermocline, the upper layers of warmer water will circulate independently of the underlying cooler water, with relatively little mixing of the two. Since the bulk of marine life is in the upper layers where the light is available for photosynthesis, the uptake of mineral nutrients from the water is concentrated in that part of the ocean. The dead bodies of the floating organisms, both plants and animals, together with the animal excreta will ultimately fall out of the upper layers into deeper water (unless recycled *en route* by micro-organisms). Accordingly, the water above the thermocline will usually become depleted in nutrients, while the deeper water below the thermocline circulates independently and retains most of its available nutrients.

The strength of the thermocline is commonly seasonal in higher latitudes, building up through solar warming in the summer months, and then being turned over as the air temperature falls in the winter, and wave action breaks up the stratification of the water. The build-up of such a thermocline in the north Pacific occurs between June and November when the thermocline develops at about 50 m depth, but this breaks down abruptly with the winter fall-off of radiant energy and the stirring action caused by increased surface water movement and convection. The following summer it gradually builds up again. This seasonal behaviour of the surface water can have a profound effect, both on weather systems and on the availability of nutrients on which the growth of phytoplankton (and ultimately all marine life) is dependent. These aspects are considered further later in this chapter.

In very high latitudes, convection caused by air-cooling of the surface water will carry cold water down to considerable depths. In some circumstances this circulation may extend to the ocean floor, although it is believed that such deep oceanic convection currents may be limited to local 'chimneys' of vertical flow, only a few kilometres across. Such turnover of ocean water is an important process in breaking the almost complete isolation of deep ocean water, below about 2000 m, from contact with the atmosphere. Since the capacity of the ocean water to hold gases (such as CO_2) in solution is very temperature dependent, we need to know more about the processes that control access of deep ocean water to the atmosphere.

The deep ocean currents are driven not by wind but by gravity differences associated with the degree of salinity and the water temperature; cold, salty water is heavier than warm, less salty water. These massive deep water thermohaline circulation currents evidently have a great effect on the fate of dissolved CO_2 and any pollutants introduced into the oceans. Figure 3.22 shows the pathway of what has been described as the 'Great Oceanic Conveyor Belt' (Broecker & Denton 1990) and its role in ice age climatic processes has already been discussed in Chapter 3.

Instability in the ocean/climate system

The unpredictability of the ocean/climate system has become very apparent during the last decade, in connection with the repeated occurrences of so-called El Niño events in the Pacific region.

This recurring instability in the Pacific and South-East Asian area is also sometimes referred to as the El Niño/Southern Oscillation (ENSO) phenomenon.

The term El Niño ('the Christ Child') was based on the observation that off the west coast of tropical South America a periodically recurring warm spell, typically appearing around Christmas time, was correlated with a disruption in the normal pattern of fish abundance. The resulting break in the fishing season would sometimes extend as late as June (well into the southern hemisphere winter). The observed rise in sea temperature at these times was also associated with increased rainfall in that region. The most extreme ENSO event this century appears to have been that of 1982–83, which has received much attention as it was accompanied by very widespread abnormal weather conditions in other parts of the world.

The ENSO system appears to be a function of the interplay between the pattern of the equatorial wind belt across the Pacific, and the effect that this has on the temperature stratification within the Pacific ocean water. The wind-driven surface currents affect the process known as upwelling, which breaks the thermocline stratification and brings cool nutrient-rich water from the ocean depths up into the photic zone, where it becomes available to phytoplankton. Off the Pacific coast of tropical South America, the prevailing easterly winds generally drive far across the Pacific, carrying the surface water westwards, and causing an upwelling of deep nutrient-rich water close to the South American coast. It is this upwelling which gives the nutrient boost which drives the enormous planktonic bioproductivity, and hence some of the richest fisheries in the world, off the coasts of Equador and Peru. A change in the wind pattern disturbs this upwelling, and results in the main impact which an El Niño makes on the fisheries.

Figure 6.2a shows the relationship between the easterly wind, the surface ocean current which it drives, and the upwelling at the eastern margin of the Pacific. A further product of this movement is that the warm surface water driven by the wind is lighter than the general body of the Pacific water, and as a result the sea level is about 60 cm higher on the warm-water (lighter) Asian side than on the cold-water South American side of the Pacific. The thermocline is at this stage strongly tilted to the west, getting close to the surface on the eastern shore (western coast of South America) where the deep ocean water upwells, bringing nutrients into the photic zone. This appears to be the 'normal' pattern between El Niño events. The strongly tilted condition of the thermocline in the 'southern oscillation' is referred to as being in the 'high index state', while the levelling out which accompanies an El Niño (Fig. 6.2b) is called the 'low index state'.

At the onset of an El Niño episode, the easterly winds weaken, so that the surface current being driven by it slackens and the upwelling is greatly diminished (Fig. 6.2b). As the piling up effect of the easterly wind drops off, the sea level drops on the Asian side of the Pacific, and a westerly current (from Asia towards America) replaces the easterly surface flow. With the diminished upwelling, the thermocline levels out across the Pacific, dropping in the east and rising in the west.

Other atmospheric and climatic phenomena accompany the oceanic change of pattern. When the normal (high index state) pattern of easterly winds is running, as in Fig. 6.2a, the cool dense air above the oceanic upwelling remains at low altitude, and does not rise to form clouds with ensuing precipitation. As a result there is relatively little rainfall in the tropical eastern Pacific during the winter months. During an El Niño episode (low index state), the weakened easterlies do not extend so far and rain-bearing clouds extend from Indonesia about two-thirds of the way across the Pacific, bringing rainfall to areas where there would normally be none at that time (Fig. 6.3b).

There are further climatic and biological effects associated with the kind of strong El Niño that occurred in 1982–83. In that year, the reduced availability of fish off the west coast of South America had serious results both for sea birds and for seals, representing higher levels up the food chain. It is estimated that there was a fall of 25% of the fur seal and sealion population along the Peruvian coast. Inland, the increased rainfall in Peru and Equador turned the coastal deserts into temporary wetlands, with the normally arid grasslands now dotted with lakes. The resulting disturbance of habitat favoured an explosion of the toad population, and this in turn cascaded to give a bird population explosion. On the other

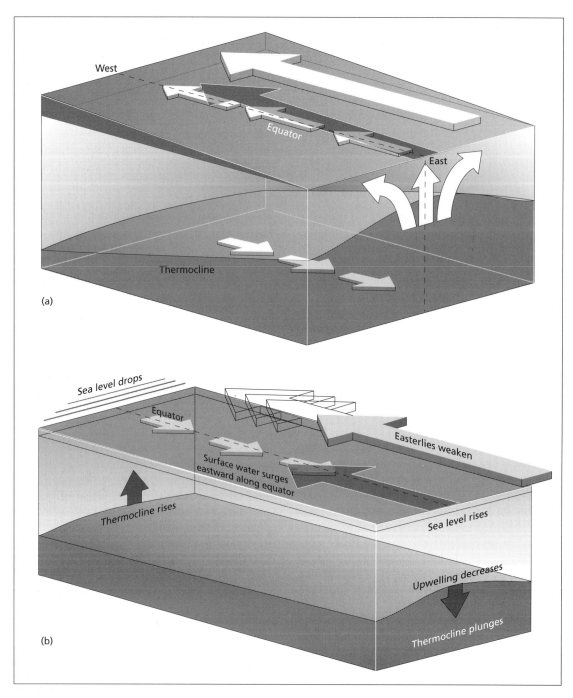

West

Equator

East

Thermocline

(a)

Sea level drops

Equator

Easterlies weaken

Surface water surges
eastward along equator

Thermocline rises

Sea level rises

Upwelling decreases

Thermocline plunges

(b)

Fig. 6.2 An equatorial transect of the Pacific Ocean (a) in its normal condition ('high index state') between El Niño episodes. The easterly winds draw the surface water westward, bringing up nutrient-rich cool water from below, along the Pacific coast of South America (rising arrows at right). As a result, the thermocline is tilted to the west. As the warm water is lighter, the sea level is some 60 cm higher on the western side of the Pacific than on the eastern side. During an El Niño event ('low index state'), (b) the easterly winds weaken from the west, the upwelling diminishes, and both the sea level and the thermocline flatten out. From Wallace & Vogel (1994).

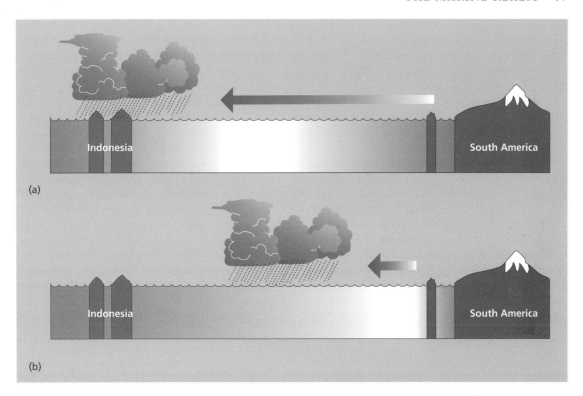

Fig. 6.3 Rainfall distribution in the equatorial Pacific. During normal years ('high index state'), (a) the strong easterly surface winds keep the surface water of the eastern and central parts cool; heavy rainfall is confined to the western (Indonesian) side of the Pacific. During an El Niño ('low index state'), (b) the easterly surface winds weaken, the mid-Pacific warms and the rain area moves eastwards. From Wallace & Vogel (1994).

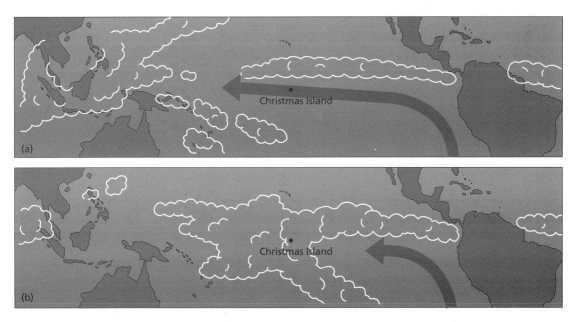

Fig. 6.4 Information derived from satellite images of the Pacific with rainfall represented by cloud cover, (a) during a normal phase in November 1988, and (b) during the El Niño of November 1982, when the western Pacific region was deprived of its normal high rainfall. From Wallace & Vogel (1994).

side of the Pacific, the eastward shift of the main monsoon rain belt (shown as cloud cover in Fig. 6.4a,b) led to a drought in parts of Australia and Indonesia, with outbreaks of serious forest fires. The most notable of these was the so-called Great Fire of Borneo which burned through a wide area of what had been humid tropical forest, a biome not normally associated with a 'fire ecology' (Johnson 1984).

Although the ENSO climatic phenomenon is in a loose sense a periodic event, the strength and periodicity both appear to be erratic, and considerable effort has been directed to modelling the conditions which might favour—and precede—its onset. This modelling of the ocean/climate system probably holds out the best prospect of our being able to predict the arrival and impact of an El Niño episode. But there is a continuing problem in distinguishing what may be part of the recurring cyclicity of El Niño from the anticipated impact of a longer term global warming. As Michael MacCracken (1995) of the Office of the US Global Change Research Program writes, 'Does this ... mean that the recent warming is due primarily to an increased (but not understood) frequency of El Niño Events, or is this the pattern by which greenhouse-gas-induced warming is becoming evident? We don't know'.

Apart from the enormous consequences for fisheries, agriculture and wildlife of this phenomenon, the understanding and prediction of the ENSO process is of great importance for the science of global climatic change. Not least, because it is a phenomenon in which some of the interactions of atmosphere, ocean and biota are seen to be causally linked and at least partially understood.

The ocean and the carbon cycle

Some aspects of the carbon cycle have been dealt with in Chapter 5, and the global ocean is simply one part of this system. But the enormous role that the oceans play in moving and storing carbon makes it necessary to return to some features of the cycle in the context of this chapter.

Firstly, the oceans contain in the combined living and dead organisms, plus various forms of dissolved material, more than 20 times as much carbon as the total terrestrial biomass. Similarly, the atmospheric holding of CO_2 represents only about 2% of the total oceanic carbon. The oceans also represent one of the most important pathways by which carbon is taken out of circulation in the atmosphere and is drawn down into the ocean floor sediments. Despite the much greater area of ocean compared with the land, this total photosynthetic drawdown of atmospheric CO_2 into the ocean is about four-fifths that of the combined terrestrial biomes (estimated at 92 and 121 Gt yr^{-1} respectively; see Fig. 5.1). These figures are at best only estimates, and small shifts in their values could have a very significant cumulative effect on the atmospheric level of CO_2.

In terms of climate change, perhaps the most significant aspect of the ocean's role in the carbon cycle is its capacity to soak up some of the increase in atmospheric CO_2 being produced by fossil fuel burning. We have seen in Chapter 5 how the ocean's phytoplankton take up CO_2 by photosynthesis, and that some of this then falls to the ocean floor to become buried in deep ocean sediments and so taken out of circulation in the carbon cycle for millions of years, at least. The fact that the atmospheric CO_2 is not rising as fast as we would expect from our rate of fossil fuel burning suggests that the oceans may be taking up some of the difference. That discrepancy between increased CO_2 release and the rate of atmospheric CO_2 increase has been referred to as the 'missing sink'. An underestimate of the oceanic drawdown of CO_2 is only one of several possible explanations of the apparent missing sink, as we have noted in Chapter 5. It is of course extremely difficult to investigate the rate of carbon uptake at the ocean surface in a way which gives a full picture, even for one site, over the span of a year. It is even more difficult to extrapolate from such limited studies to assess a global figure. There is a wide range of opinion as to which factors (mineral nutrients, light intensity, temperature, predation) are likely to be the most significant in limiting planktonic drawdown of carbon in any particular location. The simple supposition that mineral nutrients (and especially iron) are the most important limiting factor has been seriously questioned as a result of observations on natural planktonic communities (Moore 1994). None the less, if we want to try to increase the oceanic drawdown of carbon this seems to be the most promising aspect to explore.

Increasing the carbon drawdown in the world's oceans

In tropical waters, where the temperature and light availability are particularly favourable for marine phytoplankton the potential for increasing their bioproductivity by increasing the availability of mineral nutrients looks most promising. In the photic zone, where the phytoplankton are active, iron is believed to be the main mineral nutrient which limits their growth. Morel and Chisholm (1991) suggested that by supplementing the iron available in regions of the ocean where other nutrients do not appear to be at a limiting level, the carbon drawdown by the phytoplankton could be increased. This offered the possibility of compensating for—and even counteracting—the anthropogenic build-up of CO_2 generated by fossil fuel burning. To many scientists this seemed an attractive and environmentally harmless way of delaying the onset of global warming caused by the enhanced greenhouse effect of the rising CO_2 level. Others were concerned at the prospect of interfering with the ocean ecosystem without our having an adequate understanding of what the outcome might be.

Despite these concerns, several attempts have been made to explore the effect of this 'iron fertilization' of tropical ocean water. One of these was carried out by a joint British/American team, which released iron into the equatorial Pacific Ocean about 500 miles south of the Galapagos Islands (Watson *et al.* 1994). The iron was accompanied by a 'tracer' of sulphur hexafluoride (in minute concentration) so that the body of iron-enriched water could be followed as it drifted with the surface movement of the ocean. The site used was chosen as one of the areas in which other algal nutrients (especially nitrate and phosphate) are not completely depleted in the surface water by the planktonic algae, as they are in many areas of the tropics. Such 'high nutrient, low chlorophyll' (HNLC) regions occur also in the oceans around Antarctica; they are so labelled because, despite the seeming availability of some key nutrients, the chlorophyll level (a simple means of assaying the abundance of phytoplankton) remains low.

The outcome of this enterprising experiment was rather discouraging. The immediate and clear-cut result was that within 48 hours of the iron 'seeding' of the ocean water, the CO_2 level in the water fell, and this was attributed to an increase in the photosynthetic activity associated with the enhanced algal nutrition. This fall was sustained, and yet it was only at a level of about 10% of what would have been expected if all the available nitrate and phosphate had been exploited by an expansion of the phytoplankton. In other words, the iron enhancement produced only one-tenth of the increased carbon drawdown that had been anticipated on the hypothesis that iron was the element limiting the exploitation of the other mineral nutrients by the phytoplankton.

Just why the response was so limited is not clear at present. It may be that a single pulse of iron, as in this experiment, did not give the plankton scope to fully utilize the other nutrients, so as to achieve the theoretical maximum productivity. It is also possible that a limiting supply of some other micronutrient in the water is in some way limiting the phytoplankton growth. In any event, the results of this experiment do not rule out the potential of iron seeding to stimulate oceanic drawdown of CO_2. Perhaps the phytoplankton communities in the cooler HNLC waters around Antarctica would give a different response. For whatever reason, it looks as though iron seeding does not offer a quick, simple route to increasing oceanic removal of excess CO_2 from the atmosphere, but further experiments are obviously needed.

The role of calcium in carbon burial

In the discussion of carbon drawdown ensuing from planktonic photosynthesis, we have emphasized the role of the organic carbon compounds–carbohydrate, lipids and protein–which make up the bulk of the substance of the marine biota. But a different form of carbon burial also occurs in the ocean, which involves the calcium carbonate coatings which are formed around the cells of some phytoplankton (for example, coccolithophores; Fig. 6.5) and zooplankton (for example, foraminifera; Fig. 6.6). These calcium carbonate shells are derived in part from calcium ions carried into the ocean from the land surface, and partly from carbonate and bicarbonate ions derived from CO_2 dissolved in the ocean water, and ultimately derived from the atmosphere.

This precipitation of calcium carbonate by both

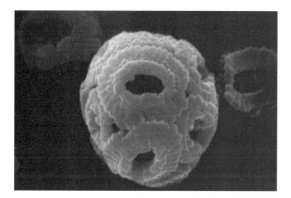

Fig. 6.5 A scanning electron micrograph of a 'coccosphere', the calcium carbonate armour made up of small life-belt-like discs, covering a single-celled planktonic alga which is a common constituent of the present-day ocean phytoplankton. From the Pliocene of Cyprus. Greatly enlarged. By courtesy of A. Lord, University College, London.

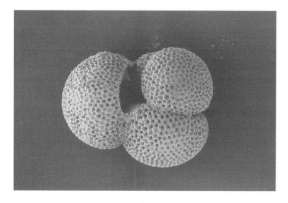

Fig. 6.6 A scanning electron micrograph of the calcium carbonate shell of a planktonic foraminiferan. Such single-shelled animals are common in the present-day ocean zooplankton. From the Pliocene of Cyprus. Greatly enlarged. By courtesy of A. Lord, University College London.

plants and animals is quite separate from the process of photosynthesis, and is not directly connected with it. It is, however, made possible (and so is controlled) by the continued supply of calcium to the ocean reservoir from the land surface. Its ultimate source is the calcium minerals in rocks, most notably in some feldspars (aluminium silicates of calcium, sodium and potassium) which are present in most igneous rocks. Some of it also comes from the weathering of limestones (calcium carbonate) once formed under the sea, from animal shells and the remains of calcareous

phytoplankton, and subsequently exposed to weathering on the land surface. The English Chalk, which forms the 'White Cliffs of Dover', is largely made from the calcium carbonate plates secreted by coccolithophores living in the sea some 70 million years ago. But the initial source of the calcium in all marine carbonates, like Chalk, is the calcium-containing minerals in igneous rocks exposed at the surface of the earth's crust.

Even before there were marine biota as we now know them, the ocean must have been involved in carbon cycling by the purely chemical precipitation of calcium carbonate in warm tropical waters. This would have taken CO_2 from the atmosphere, dissolved in sea water, burying it as carbonate on the sea floor, and so driving a pre-biotic carbon drawdown pathway. This would have been an important process of carbon movement in the pre-photosynthetic world described in Chapter 2.

Time scales of carbon burial

Once the organic matter of carbonate shells generated by marine biota have sunk to the ocean floor, the carbon in them may be regarded as having been 'taken out of circulation', so far as atmospheric CO_2 is concerned. However, on a longer geological time scale, that carbon has the potential to be brought back into circulation again.

As we have seen in Chapter 2, through the course of geological history the plates making up the earth's crust have moved in response to convection currents within the mantle. Where new crust is being formed by the upwelling of molten mantle rock, as is happening along the mid-Atlantic Ridge, ocean-floor spreading is occurring with the basaltic rock closest to the ridge being the youngest, and becoming progressively older as one moves away from its source (Fig. 2.6). In other parts of the ocean floor, oceanic plate rocks are being overridden by other plates, either continental or oceanic in nature, and the old ocean floor is being drawn down or 'subducted' into the mantle (Fig. 2.7). This process of subduction is occurring at the present time off the Pacific coast of North America. It has, in fact, been occurring for some millions of years, and this subduction of the oceanic crustal plate beneath that of the continent has had a profound effect on the geography of the west side of North America. When the convectional down current becomes weaker, the lighter rock drawn down into the mantle then rises isostatically to produce the geological 'uplift' which produces mountains. This process is comparable to the rising of continental plates when the load of the great Pleistocene ice-sheets was removed by melting, as described in Chapter 4.

As a result of the subduction of the old ocean floor the carbon-rich sediments come under the influence of the heat and pressure associated with deep burial in the mantle. Eventually they become in varying degree incorporated in rock magma deep below the surface. Bodies of such magma may then move up into the shallower layers of the crust (right-hand side of Fig. 2.7) eventually to emerge as molten material in the eruption of volcanoes. The line of volcanic vents which lies down the western side of North America is the product of the subduction of oceanic plate rock over the last few tens of millions of years, beneath that side of the continent. Volcanoes such as Mount St Helens in the state of Oregon, which is one of the most recently active members of that chain, are known to release great volumes of CO_2 and other gases (see Fig. 5.1) as well as explosive discharge of volcanic dust into the atmosphere. Such volcanic CO_2 is in large part merely the return of carbon buried on the floor of the ancient Pacific, before it was subducted beneath the western margin of North America. There is in this sense a carbon cycle on a massive scale which involves the subduction of carbon taken out of the atmosphere by marine biota and by mineral weathering, which is released millions of years later as volcanic gases. This cycle obviously moves at a far slower rate than that driven by biological processes reviewed in Chapter 5.

Sea level change

In Chapter 4 we looked at the evidence that during the series of glacial phases that have characterized the last million years, there have been changes in sea level resulting from the accumulation and melting of the polar ice-caps. It was explained in that chapter that we can distinguish, at least theoretically, between two ways in which glacial phenomena can affect sea level. Firstly there is a global process, in that as water is removed from the ocean by evaporation, and some of this is then deposited on the land as ice, above sea level, the level of the oceans will fall all over the world. This is the eustatic movement of sea level. Independently of this, the weight of the great ice-sheets, to be measured in hundreds, and in places even thousands of metres in thickness, depressed the lighter continental crust down into the heavier less-rigid layers of the crust below. This is the process of isostatic movement, which affects sea level on a much more local basis; indeed, its effects can only be seen in the immediate area where the ice-load effect was felt by the crustal rocks. Isostasy is the term applied to the balance that exists between the weight of the upper, lighter parts of the crust and the heavier rock below, in which they 'float'.

When the ice melted on those parts of the crust which had been depressed by the effect of such ice load, then the crust in that area rises again to its former level, restoring the equilibrium with the rocks below. The movement of the blocks of crustal rock under ice load is rather like that of a ship being loaded with cargo. If more weight is placed in the bow of the ship, it will tilt and the bow will lie lower in the water than the stern. This was in effect the condition of the British Isles during the several Pleistocene glaciations. Scotland was more heavily laden with ice than the south of England. As a result, parts of the Scottish coast were depressed to a level well below that of the present day, and wave action cut beaches into solid-rock shorelines, which are now elevated as 'raised beaches' far above present-day sea level (see Fig. 3.7).

It is worth observing that these raised beaches were cut at a time when the global (eustatic) sea level was lower than it is now, during phases of maximum ice-sheet expansion. This emphasizes the independent action of the eustatic and isostatic control of sea level.

This isostatic recovery of load from the ice-sheets of the last glaciation is still continuing at the present time. Scotland is still rising after removal of the ice some 10 000 years ago, and as that end of the crustal block of the British Isles rises, the south-east is sinking (Shennan 1989). Figure 6.7 shows the estimated current rates of this crustal movement in Britain. The positive figures in Scotland indicate crustal rising, between 0.5 and 2 mm yr^{-1}, while in the south-east there is subsidence of the same order of

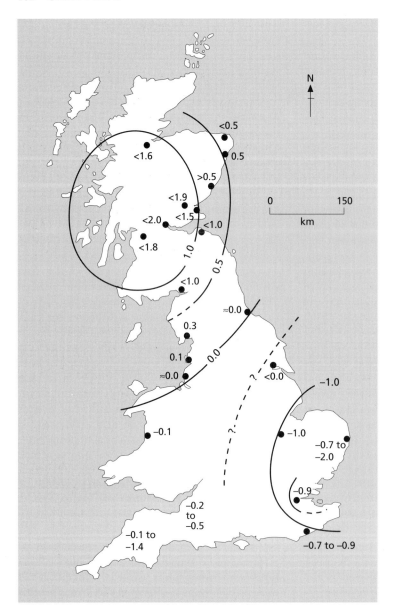

Fig. 6.7 Map of the current rates of vertical crustal movement, expressed in mm per year, in Britain, resulting from the recovery of ice load during the last glaciation. The positive figures (in the north) represent isostatic rise while the negative figures (in the south) are subsidence. After Shennan (1989).

magnitude. The axis of tilt, with zero crustal movement, lies across Britain from North Wales to Yorkshire.

With the global rise of temperature predicted on the basis of enhanced greenhouse effect of rising CO_2, two processes affecting sea level can be anticipated. Firstly, as the major ice-caps of the two hemispheres (Antarctica and Greenland) melt, the volume of water added to the global oceans will cause a eustatic sea level rise. This is not simply a prediction, for we know that this has

happened during each of the warm phases between the glacial maxima of the Pleistocene. In addition, if as would be expected the mean sea temperature rises, expansion of the ocean water (global oceanic thermal expansion) will add significantly to this eustatic rise. Unfortunately it appears to be very difficult to separate the contributions made by these two processes (ice-melt and thermal expansion) in the course of past changes, and hence in predicting future ones. The ensuing difficulty in forecasting changes in global sea levels are considered further later in this chapter.

There is yet another factor in the sea level equation, connected directly with human activity. Many of the manipulations of freshwater for which humankind is responsible have repercussions in terms of sea level. One of the most obvious is in the construction of dams designed to hold large bodies of freshwater on the land surface for purposes such as hydroelectric power or domestic water supply. Such water held above sea level (even though it is being constantly recycled) will of course contribute to a eustatic lowering of sea level, like the frozen water held in the icecaps.

Other human activities work in the opposite direction; the water withdrawn from ancient aquifers (groundwater which has accumulated slowly, over perhaps thousands of years) has been aptly referred to as 'mined water'. If this is then used to irrigate arid or near-arid land, and most of the water is 'lost' by evaporation or evapotranspiration, that water will be recycled back into the oceans by rainfall. The draining of wetlands has a similar effect, in removing water from the land surface, eventually returning it to the ocean. The net effect of water mining and wetland 'reclamation' is to remove water held on or within the continent, and transfer it to the ocean 'sink'. The result of this is then to raise the global sea level.

These are just two instances of the small but complex contributions that human activity can make to the sum of factors influencing global sea level. Sahagian *et al.* (1994) attempt to assess the contribution of all these anthropogenic causes of sea level change to the 1.75 mm yr^{-1} rise in global sea level which they estimate to have occurred this century. They conclude that 'human activities directly contribute at least 30% of the twentieth century sea level rise. Consequently the contributions of glacial melting and ocean thermal expansion are smaller than previously thought'.

Naturally, much research has been directed to answering the several key questions relating to global sea level change. What has been the sea level rise over the last 100 years? What have been the principal causal factors? Can we predict what the effect of global warming on sea level will be over the next 50–100 years? As you might imagine, there are divergent opinions on the answers.

Observed rate of sea-level rise

Rather surprisingly, it is very difficult to get reliable measurement of an absolute rise in global sea level. Obviously, the daily rise and fall of the tides makes the mean tide level in any location an average value of a constantly changing datum, recorded over as long a period as possible. Many northern hemisphere ports (where long-term observations of sea level have been made) are within areas which are affected by recovery from the effect of the Pleistocene ice load, as we have seen for Scotland. This means that they have been

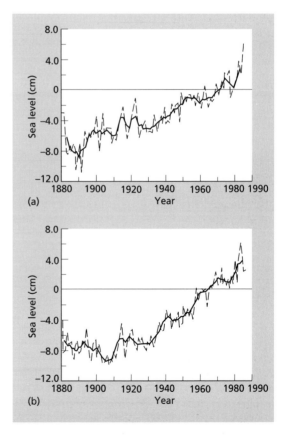

Fig. 6.8 Two composite global mean sea-level curves, derived from two different data sets derived from a large number of tidal stations. The baseline is derived from taking the values from 1951 to 1970 as zero. The dashed lines represent annual means, the solid line a 5-year running mean. From Warrick and Oerlemans (1990).

under the influence of isostatic (local) sea level change, which can of course mask the eustatic (global) sea level rise associated with post-glacial global warming. In addition to this, many factors, as well as the state of the tide, such as local meteorological conditions (especially atmospheric pressure and wind direction and strength), can affect the sea level at any one station from day to day. These processes make it very difficult to arrive at convincing figures of global changes in sea level, even when results are averaged out for a large number of tidal stations world-wide.

Accepting the frailty of such data, the collated evidence suggests that there has been an average rise of mean sea level globally over the last 100 years of between 1 and 2 mm yr^{-1} (Warrick & Oerlemans 1990). The two graphs shown in Fig. 6.8 indicate changes in the annual mean sea level over the last century, based on two separate assessments by different workers combining a large number of observations. Each shows a curve based on the annual means, together with a 5-year running ('smoothed') mean. These two data sets agree in suggesting a total rise of some 120 mm over 100 years, which lies rather low within the range of 1.0–2.0 mm yr^{-1} offered as a global figure by Warrick and Oerlemans (1990).

Prediction of sea-level rise

Any model of global sea level change depends on the prediction of global temperature rise, and what effect this will have, both on ocean-water thermal expansion and on the melting of ice in the ice-caps and montain glaciers. This in turn depends on the contribution to the enhanced greenhouse effect of the rising CO_2 level. Because any future projection of future fossil fuel burning is governed by a series of economic factors modified by legislative action by individual governments, this is perhaps the hardest element in the equation to forecast (Warrick & Oerlemans 1990). A further complication is that the impact of temperature rise alone would be tempered by any change in the pattern of precipitation. If, as some have suggested, global warming produced increased evaporation over the ocean in low latitudes, this could well result in increased precipitation (as snow) over the ice-caps.

If there is no significant change in the current world-wide trend of increase in fossil fuel burning

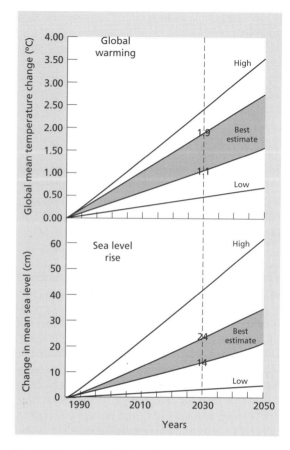

Fig. 6.9 Projected global warming and ensuing sea-level rises. Both show corresponding high, low and 'best estimate' ranges. After Warrick & Barrow (1990).

(the 'business as usual scenario' of the IPCC) it is suggested that a rise in sea level 'best estimate' for the year 2030 is 18 mm (Warrick & Barrow 1990). This projection is shown in Fig. 6.9, together with the global warming forecast on which it is based. Most of this rise is attributed to thermal expansion of the oceans rather than ice melt. Indeed the increased precipitation over the polar ice caps is predicted to have a negative effect due to increased snow accumulation. Ice melt on the smaller (non-polar) ice-caps and mountain glaciers would, however, make some contribution to global sea level rise. Weighing this evidence has given a very wide range of assessments of the relative contributions which may be made by ice-melt versus thermal expansion in future global sea level changes. An estimate of the mean value of some very disparate figures suggests that the thermal expansion will contribute something between 50% and 100% of the predicted impact of global warming on global sea level rise.

Impact of sea-level change on ecosystems and human settlement

In trying to forecast the effect of global sea level change, we must discriminate between the impact on natural or partially managed ecosystems at or close to sea level (mangroves, salt marshes) and those low-lying areas which have been settled by human communities. These include highly populated urban settlements, including many of the great ports of the world, as well as the highly productive agricultural land that has been exploited on many of the great river delta systems such as the Nile and the Ganges.

Considering first the impact on wetland ecosystems, it is important to recognize that all such habitats have experienced major changes in sea level during the course of the last million years of glacial/interglacial climatic and sea level oscillation. Where the sea level is predicted to rise on a low-lying coastal plane—as in many parts of north-west Europe and the eastern seaboard of North America—the normal course of events, without any human intervention, would be a migration of the wetland communities keeping pace with the moving coast line. As with other global change phenomena, the capacity of the biosphere to adjust will be dependent on the rate of the change.

In reality, most of such natural wetland ecosystems in Europe and North America have tended to become involved in some kind of agricultural use. This may range from the use of salt marshes for rough grazing, to the building of sea walls around tidally influenced wetlands. In the latter case, the eventual fate is usually their 'improvement' by drainage and reduction in salinity, and their eventual exploitation as arable land. Similar manipulation of the natural ecosystem has been carried out extensively in many areas which would, if left undisturbed, have sustained mangrove swamp vegetation.

We have little evidence that the migrations of maritime wetlands through the sequence of glacial and interglacial change with successive rises and falls of sea level has caused the extinction of any type of natural community, or indeed of any individual species within them. It is a reasonable supposition that ecosystem movements were able to keep pace with sea level change throughout those global vicissitudes. The principal concern arising from predictions of future change lies in the possibility that this would be much faster than any that has taken place hitherto, and that maritime wetland ecosystems would be destroyed faster than they could migrate to accommodate the change. This would be most likely to occur where incursion by the sea over arable land was fended off by raising the level of sea walls and similar defenses, so that natural communities on the seaward side would become inundated without the opportunity of migrating landwards. The loss of habitat that this would represent would really be no different in kind from that occurring in many settings where human interference has impacted on salt-water wetlands.

In areas where land lying close to mean sea level has been extensively modified for agriculture or urban development, concerns over sea level rise take a different form. Where the value of the land is considered to justify it, and the financial resources are available, the means of protection from gradual sea level rise are well established. Much of Holland, for example, lies below sea level, the land protected from flooding by sea walls. The main factors governing our capacity to survive global sea level rise in those areas of dense human settlement are economic and social.

Further reading

Franks, F. (1983) *Water*. Royal Society of Chemistry, London.

Wallace, J.M. & Vogel, S. (1994) El Niño and climate prediction. *Reports to the Nation on Our Changing Planet* **3**, 1–24. Office of Global Programs, National Oceanic and Atmospheric Administration: US, Department of Commerce, Washington DC, USA.

Houghton, J.T., Jenkins, G.J. and Ephraums, J.J. (1990) *Climate Change—the IPCC Scientific Assessment*. Cambridge University Press, Cambridge.

Chapter 7

Dynamic Responses to Climate Change

One of the major questions raised by the prospect of climate change is how will this affect the geographical distributions of different species of animals and plants. Can we look forward to major extinctions and a loss in global biodiversity? Will special conservation measures be required to avoid such losses? Do we need to assist certain species in spreading to new areas where they can survive? Will the changes affect our domesticated species, particularly plant crops, thus changing global patterns of agriculture?

The geographical range of species

In order to begin facing these questions we need to be clear about what environmental factors determine the natural ranges of plants and animals, and this is a complex subject. Even in a world without human beings there would still be a complex interaction of physical and biological factors that would determine the limits of distribution of any given species, and in a world with humankind in a dominant role the situation is even more difficult to unravel. There is also the question of scale. The answer to the question, why is a certain organism found on one continent but not another, may have a very different answer from the question why it is found above a given altitude and not below, or why it is found on limestone rocks but not on granite.

Distribution patterns on a continental scale can usually be explained by reference to evolutionary processes coupled with the tectonic movements of the earth's plates (see Chapter 2) and the dispersal capabilities of the particular organism (see Cox & Moore 1993). At a lower scale, one is often faced with questions of climatic requirements, the temperature and moisture availability optima and limits for particular species or, in the case of some animals, the limits of their food species. At lower levels again, we

may be looking at the microclimates of habitats, or differences in environmental chemistry between different soil types. At all these scales we must also bear in mind that the biological world is intensely competitive and a species can only survive in a location if it can cope not only with the physical conditions but also obtain the resources it needs (food, sunlight, water, etc.) under the pressure of competition from other species, and at the same time survive the depredations of the predators, parasites and pathogens that it may encounter there.

When we consider the possible effects of climate change on plant and animal distribution patterns, therefore, we cannot think of any one species in isolation, but we must also consider the possible influence of the new assemblages of species in which it will find itself. The predictive process, however, can reasonably begin with a consideration of the physical requirements of individual species. Figure 7.1 shows changes in the photosynthetic rate in the alpine plant, the glacier crowfoot (*Ranunculus glacialis*), in response to changing temperature and light intensity (Ellenberg 1988). At high light intensities the optimum temperature for photosynthesis in this plant is between 20 and 24°C. From such information, gained by laboratory experimentation, it may be presumed that a general rise in temperature as a result of climate change would enable this species to grow more efficiently in sites that may now be too cold for optimum growth, such as those closer to the mountain summits. One can anticipate, therefore, that an alpine plant of this type might extend its range higher up the mountainside above the tree line.

There are complications, however. At lower light intensities the species has its optimum at slightly lower temperatures, between 12 and 20°C, so if other species also respond well to warmer conditions and result in increased shad-

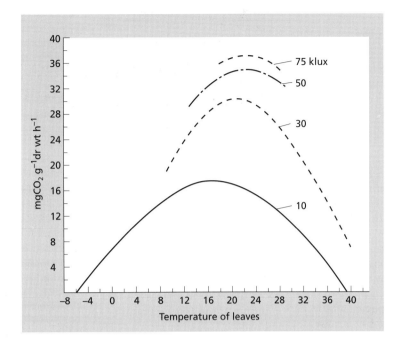

Fig. 7.1 Changes in the photosynthetic rate of the glacier crowfoot (*Ranunculus glacialis*) in response to temperature at different light intensities. At low light intensity its optimum lies between 12 and 20°C, while at higher light intensity optimum photosynthesis takes place between 20 and 24°C. After Ellenberg (1988).

ing for the crowfoot, or if cloud cover is substantially increased in the new climatic regime, then the effects of the warming may not be those anticipated by simple extrapolation from these experimental data. The former possibility, that other plants will also respond positively and may create more shade, is a particularly important one when considering the lower altitudinal limits of the glacier crowfoot because one may well find that there the tree line itself is now situated at a higher altitude, so the plant is gradually forced towards the alpine summits where it will occupy progressively smaller available areas and may be placed in danger of local extinction (see Chapter 1).

We are also assuming here that altitudinal distribution of the glacier crowfoot on the mountain is determined by the plant's ability to photosynthesize at the prevailing temperature. There are, of course, many other aspects of plant growth and survival that could be influential, such as frost resistance, seed ripening, root development, and so on, and there are very few studies of plants that are sufficiently detailed or comprehensive to allow the physiological limitations and geographical distribution pattern to be elucidated. Perhaps the most elegant study of a single species is that of the European lime tree, *Tilia cordata* (Pigott & Huntley 1981). This is a warmth-loving, decidu-

ous tree that is common in central Europe and becomes scarcer towards the north and north-west, having its northern limit in southern Scandinavia and in northern England. It is apparent that this limit is temperature-determined, since it corresponds quite closely to the July 19°C isotherm, suggesting that summer warmth is necessary for sustained survival. The emphasis, however, is on the 'sustained', because geographical limitation is evidently not caused by the temperature requirements of photosynthesis, or even the frost hardiness of the tree, but rather by its ability to set viable seed. North of its natural limit the tree may be planted and will grow, but only in exceptional years will seed be set, so the influence of climate acts through the population dynamics of the species.

Detailed studies on the flowering of the lime tree have shown that the germination of the pollen grain when it lands upon the stigma is closely related to temperature, as is the growth of the pollen tube down the style and into the ovary where it will eventually effect fertilization of the egg cell. The optimum temperature for the growth of this pollen tube is at 19°C, which may well explain the plant's relationship to the 19°C July isotherm. But there are additional considerations, such as the length of time needed for the embryo to mature in the seed, and many plants are more closely related in their distribution limits to some measure of growing season lengths than they are to specific isotherms. A useful measure of growing season is 'growing degree day', defined by Woodward (1987b) as the number of days on which the temperature exceeds 0°C multiplied by the mean temperature over this period. The 'growing degree day' requirement of the lime tree, for example, is 2000.

Studies of this sort permit reasonable predictions about the outcome of climatic changes in terms of the poleward movement of plant species, on the assumption that the possibility of new biological combinations of species will not prove too serious a modifying factor. But this is a big assumption. Consider, for example, the birch trees of North America. There are many different species, each with its own distribution pattern. But species and populations that grow further north have been found to contain higher concentrations of chemical compounds in their leaves that deter grazing animals, such as the snowshoe hare (Bryant *et al.* 1994). The reason for this is not fully understood, but it probably relates to the evolutionary and grazing history of the birches over the past 12 000 years since the last glaciation, when grazing intensity has been greater in the higher latitudes. Climate change could well bring unprotected species and races of birch from the south into the north that would be subject to intensive herbivory and thus might suffer as a result. In this respect the snowshoe hare may not be too serious a pest as it could only graze on seedlings or the lower branches of adults, but if insect pests, particularly certain moths, showed similar preferences for the southern varieties then the selective pressures could be seriously stacked against the success and perhaps even the survival of the immigrant trees.

One group of plants that has featured highly in discussions about the future of vegetation in a greenhouse world has been the so-called C4 plants. In the 'normal' type of photosynthesis found in the majority of high-latitude plants, termed C3 plants, CO_2 is first fixed to produce 3-carbon chains that can subsequently be constructed into larger molecules. Another type of fixation method was described first in sugar cane, a tropical grass species, in which CO_2 is first

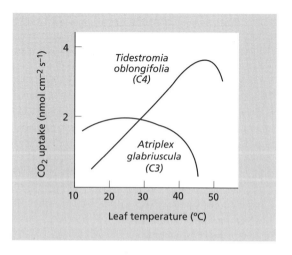

Fig. 7.2 Variation in photosynthetic optima for C3 and C4 plant species. *Tidestromia oblongifolia* is a hot desert C4 species that thrives in Death Valley, while *Atriplex glabriuscula* is a C3 plant of cooler coastal habitats in many parts of the world, including the California coast. From Bjorkman (1975).

attached to a 3-carbon unit to make a 4-carbon molecule, hence the term C4 plants. These 4-C molecules, however, are relatively short lived, and are transported to specialized cells around the vascular bundles where they are broken down and the normal C3 type of fixation takes place once more. The advantage of the C4 mechanism centres upon the efficiency of the enzyme involved in this primary fixation, phosphoenolpyruvate (PEP) carboxylase. This has a much higher affinity for CO_2 than the C3 enzyme, rubisco, so that it can accumulate CO_2 more effectively from low concentrations. Also, it does not suffer from oxygen interference (photoinhibition) as does rubisco and this means that less CO_2 is wasted. The system has, in addition, a higher temperature optimum than the C3 mechanism, so it is particularly effective under conditions of high temperature and light availability (see Fig. 7.2). Its efficiency means that when water is in short supply the plant can be more economical in the time it spends with its stomatal pores open for gaseous exchange, so it loses less water vapour in transpiration. Put all of these things together and you have a plant that is well suited to life in hot, dry conditions and this is precisely where C4 plants are most abundant. Here they rank among the most productive of plant species. Eighty four per cent of the grasses of California are C4, while only 6% of the grasses of the State of Washington have this mechanism (Teeri & Stowe 1976). Not only grasses, but many other groups of plants contain C4 members, from goosefoots to orchids (Moore 1981).

One of the most important aspects of this particular biochemical reaction as far as humanity is concerned is that a number of our important crop species are C4, including maize and sugar cane. The question of the future of C4 species in a greenhouse world is therefore of both ecological and economic importance. At present, in North America, the advantages of the C4 system are most apparent at latitudes below 45°N (Fig. 7.3) (Ehleringer 1978) because of their ability to grow fast in the hotter climate. In a greenhouse world, one might expect such species to flourish, but one of their main advantages, the ability to scavenge for CO_2 under low concentrations is lost, and this may well tip the balance the other way. Figure 7.4 shows the response of wheat and maize to elevated CO_2 levels in terms of their photosynthetic rates

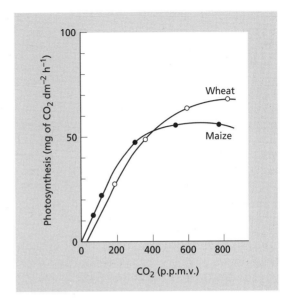

Fig. 7.4 The photosynthetic response of wheat and maize to increasing CO_2 concentration at constant light levels (0.4 calories cm^{-2} min^{-1}). Below about 400 ppm CO_2 maize (C4) has the growth advantage, but it becomes saturated at about 450 ppm while wheat (C3) is still CO_2 limited and continues to increase its photosynthesis at higher CO_2 concentrations. A greenhouse world could therefore favour wheat, depending upon the effects of temperature, light and humidity.

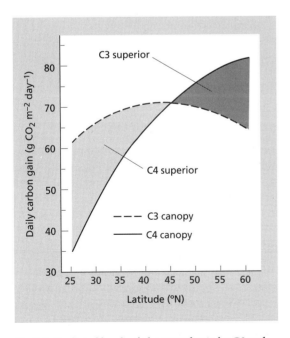

Fig. 7.3 Predicted levels of photosynthesis for C3 and C4 species over a range of latitudes in the Great Plains of North America during the month of July. The advantages of the C4 mechanism are evident only in latitudes south of 45°N. From Ehleringer (1978).

while the light intensity and temperature are maintained steady. Maize has an advantage over wheat at low CO_2 levels, up to about 400 ppm, but once this threshold is exceeded maize photosynthesis levels off while wheat overtake it. In a greenhouse world, therefore, the competitive balance between C3 and C4 plants is quite difficult to predict. Raised temperature will permit the growth of C4 plants in higher latitudes, so the opportunity will arise to grow some of our C4 crop species at higher latitudes, as shown in Fig. 7.5. This diagram (Parry 1990) predicts the future of maize production in the British Isles where it is currently used almost entirely as a silage crop because summers are not warm or prolonged enough for grain ripening (the unusually hot summer of 1976 excepted, as shown in Fig. 7.5). So warmer conditions will result in agricultural changes because C4 species can be grown further north.

The crop plant, however, is in a privileged and protected position compared with its wild rela-

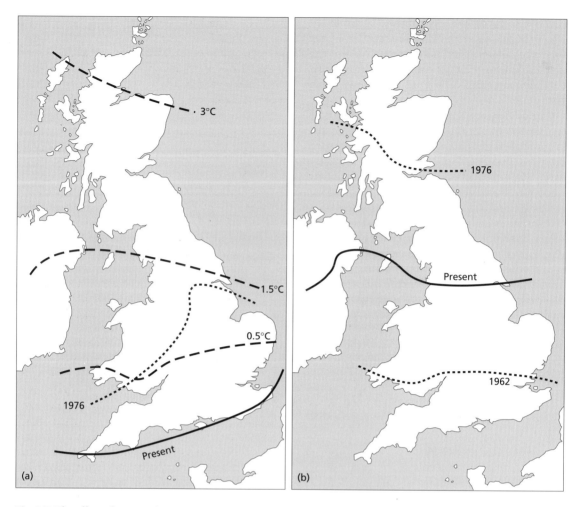

Fig. 7.5 The effect of projected rises in mean temperature on the growth of maize as a crop plant in the British Isles. (a) As a grain crop maize is currently unsuited to the British climate, being capable of ripening its fruit only in exceptional summers (such as that of 1976). At elevated temperatures it becomes possible to grow maize for grain throughout the islands. (b) Maize is currently employed as a silage crop (using leaf and stem tissues) mainly in the southern half of the British Isles, but this varies from year to year (1976 was a long hot summer, while 1962 was cool). After Parry (1990).

tives. The productive advantages conferred upon C4 plants by higher temperatures will be largely lost because of the associated high CO_2 levels. So, in natural vegetation and in weed communities the competitive edge may well move towards the C3 species and we may observe an overall decline in wild C4 plants at higher latitudes.

Carbon-dioxide fertilization

In addition to its role as a greenhouse gas, affecting the mean global temperature, an elevated level of atmospheric CO_2 has a potentially much more immediate effect on plant life. The possibility of increasing the productivity of C3 plants by raising the CO_2 level has been acknowledged for almost as long as CO_2 has been recognized as the gaseous substrate of photosynthesis (Wittwer 1985). It seems self-evident that if one increases the supply source of carbon, this will automatically increase the photosynthetic rate, unless some other limiting factor intervenes. Early experiments in which CO_2 was supplied to greenhouses in which crops were being grown gave yield increases in lettuce, tomatoes and cucumbers; even if all other factors were held constant, yield increases of the order of 30% could be achieved (Parry & Swaminathan 1992). However, even then the process of photorespiration, the uptake of oxygen and ensuing release of CO_2, means that nearly half of the 'extra' carbon fixation from the elevated CO_2 available is lost by that process in C3 plants. But although some increase in biomass production invariably occurs with elevated CO_2, its cause cannot always be simply attributed to increased carbon availability since the CO_2 has the added effect of causing partial stomatal closure, resulting in increased water use efficiency and this is considered further below.

Although it is difficult to project the quantitative results of a given level of CO_2 enhancement on bioproductivity in a crop plant, never mind a complex natural community, it seems likely that the post-industrial rise in CO_2 has been partially counteracted by increased photosynthesis. A simple measure of this effect may be obtained by comparing recent growth increments (thickness of the 'annual rings') in forest trees with those from earlier years. Kauppi *et al.* (1992) show from their observations of forest tree increments from many European sites over the period 1950–1980 that there has been an increase in the average annual stem increment of the trees over that period. If this has occurred on a global scale, as seems likely, it means that this increased holding of carbon in the form of forest biomass may be part of the explanation as to why the atmospheric CO_2 rise has not been as great as the rate of fossil fuel burning would lead us to expect. In other words, enhanced tree productivity resulting from CO_2 fertilization may form a significant part of the 'missing sink' of carbon (see Chapter 5). While on the one hand a human waste product is enhancing forest growth, we also face a problem of trees dying from the effects of acid rain. Kauppi *et al.* (1992) suggest that the industrial pollution across Europe is in fact offering a fertilizing effect, particularly perhaps in the form of nitrates (up to 4 g m^{-1} yr^{-1} in parts of Germany), and that this allows faster tree growth and consequent increased carbon fixation. The unhealthy appearance of trees may then ensue as other elements, such as magnesium, become limiting (see Chapter 9). This is a further illustration of the complex interactions involved in global change.

Further confirmation of global biomass growth has come from measurements of oxygen in the atmosphere. Oxygen should decline as CO_2 is produced, but it has done so less quickly than expected. Additional oxygen must have been produced from enhanced photosynthesis (Keeling *et al.* 1996).

Stomatal responses to environmental change

Stomata represent the most important pathway between the internal atmosphere of plants and that of the outside air. They are the route by which plants take up CO_2 and, while photosynthesis is taking place, allow oxygen to diffuse out. They are also, of course, the route by which plants lose water vapour in the process of transpiration. Early botanists believed that transpiration was the main means by which the stream of water was drawn up from the soil, carrying mineral nutrients into the plant. We now see transpiration in a rather different light, as being an inevitable and not always welcome accompaniment to the process of drawing in CO_2. For the great majority of land plants, it is a struggle to maintain an adequate water supply for their metabolic functions, and to sustain turgidity in their living tissues. Transpiration loss of water is a constant threat to this. If a plant has a high density of stomata on its leaves, and opens them all to maximize the photosynthetic rate, then the transpirational loss may exceed what the plant is capable of supplying with uptake by its roots. The density of stomata on the leaf surface, and the proportion of time that they are held open, has to be a compromise between maximizing carbon uptake and minimizing water loss. This compromise is referred to as the 'water use efficiency' which is defined as the number of molecules of CO_2 fixed by photosynthesis divided by the number of water molecules lost by the plant over the same period.

In a very humid atmosphere, saturated with water vapour, the plant can open its stomata to maximize photosynthetic uptake, with minimal transpirational loss. However, if the atmosphere has a very low relative humidity, the transpirational loss will be far greater for a given carbon uptake, and the water use efficiency will be proportionately lower.

The capacity of a plant to exploit available

atmospheric CO_2 will depend ultimately on the density of stomata on the leaves. The hour-to-hour, day-to-day adjustment to water availability can be made by opening and closing the stomata in response to other environmental factors, above all to light intensity. But the stomatal density becomes fixed during leaf development, and appears to be fairly tightly constrained for a given species of plant. However, within the constraints of the plant's genetic make-up, the stomatal density responds during development to the ambient CO_2 level (Woodward & Bazzaz 1988).

Stomatal density of the leaf therefore represents a 'permanent' record of the response of the plant to the ambient CO_2 at the time of its formation. Ian Woodward (1987a) has made the interesting observation that the stomatal density shown by the leaves of a number of native British trees collected and preserved as herbarium specimens over the last 200 years gives a 'signal' of the CO_2 level at the time of their formation. The absolute stomatal density for the leaves was different for the several species concerned, but he normalized the comparison by treating the present-day stomatal density as 100%, and then expressing the

earlier (higher) stomatal densities as a percentage of the present-day level (Fig. 7.6). As the CO_2 has increased, largely in consequence of fossil fuel burning (see Chapter 5), the stomatal density of the eight tree species studied has fallen over the last 200 years; a regression on the data from these species gives a mean stomatal density 200 years ago equivalent to 140% of their present values. This suggests that selective pressure has favoured plants increasing their water use efficiency under the rising CO_2 as the CO_2 fertilization effect has enabled them to reduce the transpirational loss without a corresponding reduction in carbon uptake. It also means that one of the plus effects of rising CO_2 is that plants may be able to tolerate water stress to a higher threshold than under the lower CO_2 levels of the pre-industrial world. Woodward confirmed the pattern of these observations by experiments on a selection of the species that he had observed in his herbarium material. He showed that young trees grown under reduced partial pressures of CO_2 showed the same pattern of increased stomatal density as in the historical material. The level of ambient CO_2 evidently functions as a plant growth regula-

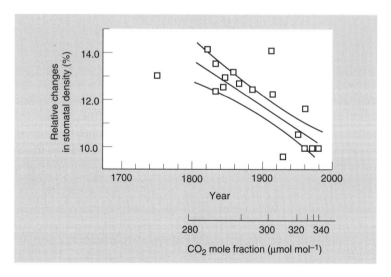

Fig. 7.6 Stomatal densities of modern leaves and specimens preserved in herbaria over 200 years, from eight British native or naturalized tree species. All of these species show a comparable reduction in stomatal density as CO_2 has risen. It is known from ice-core studies (see p. 77) that CO_2 has risen from less than 280 to over 340 ppm (equivalent to μmol mol^{-1}) during this period, and these CO_2 values are shown below the date scale. The stomatal density values are expressed as a percentage of the value for each species from 1981 (taken as 100%). Although the actual stomatal density varied between the eight species by a factor of two, the regression line, with 95% confidence limits, shows a mean reduction of 40% of the 1981 value through the last 200 years. From Woodward (1987a).

tor in its own right, since it is actually affecting the proportion of epidermal cell initials which become stomatal guard cells.

The demonstration that stomatal density can carry a signal of past CO_2 levels has encouraged the investigation of fossil leaves from the last glacial cycle, to test whether they are consistent with the levels of global CO_2 reported from arctic and antarctic ice cores. Beerling & Chaloner (1994) show that for the dwarf willow, *Salix herbacea*, which has a good fossil record through the late glacial and early post-glacial, the stomatal density displays an interesting fluctuation corresponding to that of the CO_2, linked to the Allerød climatic fluctuation of the late glacial (see Chapter 4).

Migration dispersal and spread

From these examples we can see that even if we were able to predict with accuracy the future changes in local or global temperature, and even if we knew precisely how the distribution of a species was actually or potentially limited by climate (neither of which is the case!), it would still be very difficult to predict changes in plant and animal patterns because of the way they interact with one another by competition, predation and parasitism. There are two further questions that need to be addressed, just to complicate issues further: one is the capacity of organisms to change their ranges in the modern world, and the other is how quickly they could adjust.

Range change

Climatic changes, as we saw in Chapters 3 and 4, are not new to the earth and the very considerable climatic changes of the past 20 000 years do not appear to have resulted in cataclysmic episodes of extinction. Many of the more spectacular losses, such as the megafauna (cave bears, mammoths, giant sloths, etc.), may well have been lost as a result of direct human activity rather than as a consequence of climate change. The progressive warming of the climate between about 10 000 and 5000 years ago saw a systematic process of recolonization of the higher latitudes by plants and animals that had been forced to warmer climes by the advancing ice. We have, therefore, historical evidence of the ability of vegetation and fauna to respond successfully to changing climate by extending their ranges into new territory as it becomes habitable and retreating from unsuitable areas as physical or biological factors become too severe. Is it not reasonable to suppose, therefore, that the earth's resilient biota will once again meet the test and cope with the new climate change? The main flaw in this argument is that our own species has now changed the face of the earth beyond recognition. The gradually changing gradient of vegetation and habitats that once spread over our land surface, linking biome with biome, has now become broken up into a mosaic of different land-uses with only isolated patches of the original habitats. And this picture, long familiar to the inhabitants of the ancient civilized worlds of Europe, Asia and Central America, more recently established in North America and Australia, is rapidly becoming established in the tropical regions of the world. Can the old game of biogeographic chess be played on the new board?

One of the most difficult problems facing wildlife conservationists is that of habitat fragmentation. It is this, above all other current global changes, that is the cause of immediate extinction losses of species. Besides its direct influence on species loss, however, it could pose an even greater threat in a world where climate is changing and species need to alter their ranges. In such circumstances, isolated, fragmented habitats separated by inhospitable regions (of agricultural or urban development), could prove inappropriate for the movements of some species. Those plants and animals that have poor powers of dispersal could find themselves trapped between a rock and a hard place—incapable of moving into new, hospitable land and yet unable to breed, feed or compete in their current locations where they will suffer from both physical change and the immigration of new, competitive species.

We already have some evidence concerning the impact of habitat fragmentation and shrinkage upon the mobility of organisms in the studies of migrant birds, and these can act as useful response models, as long as one bears in mind that here we are dealing with organisms that are likely to be least affected since they are among the most mobile.

The temperate latitudes are populated by many species of migrant birds through the summer season, when these highly mobile organisms are attracted by the abundance of available food resources and the long days in which they are able to feed their offspring and maximize reproductive success. The harsh physical conditions of winter and the lack of food supply then drives them back into lower latitudes, often into the tropics or subtropics. Such seasonal movements obviously place stress upon migrant birds, especially those that are small and have little stored fat as an energy supply for the journey, and those that need to cross inhospitable country such as oceans or deserts.

Examples typical of the journeys involved are shown in Fig. 7.7, giving the breeding range and wintering range of two American thrushes, the Swainson's thrush and the northern waterthrush (Winker *et al.* 1992). Both of these summer in Canada and the North-West and then migrate southwards to Central and South America in the fall. The journey length is of the order of 2500 km

and these birds are too small to be able to store enough fat reserve to complete the journey in one step. Instead, they need to make stop-over breaks in which to recover physically and build up their reserves once more. Often these stop-overs last for only a single day and the bird is on the move by nightfall once more, but the availability and the reliability of the location of such sites is clearly a matter of life and death to the migrating bird. The fragmentation or the loss of such sites can easily result in the failure of birds to complete their journey.

It is a matter of great concern that many migratory birds both in the New and the Old World are in decline (see Table 7.1) and the most likely cause of this is the continued destruction, modification and pollution of former stop-over sites, together, of course, with similar changes in their wintering and breeding sites. In the case of the wood thrush, for example, as its name implies, it

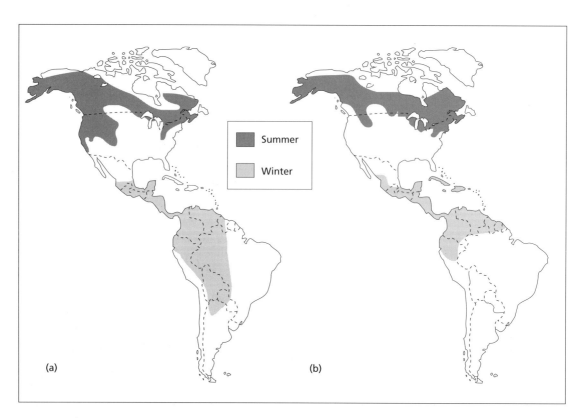

(a)   (b)

Fig. 7.7 The summer (northern) and winter (southern) distributions of two American song birds: (a) Swainson's thrush (*Catharus ustulatus*) and (b) Northern waterthrush (*Seiurus noveboracensis*). The success of

migration between the two areas (about 2500 km apart) depends on the existence of suitable stop-over sites. From Winker *et al.* (1992).

requires forested habitat both in transit and in its destination. Studies in the Yucatan Peninsula of Central America (Lynch 1992) have indicated that this type of habitat is rapidly being lost as timber is being harvested; the mahogany trees of this area have now practically vanished. This has undoubtedly contributed to the decline in overwintering, forest-loving birds such as the wood thrush (see Fig. 7.8). But a recent change in agriculture has involved the cultivation of the zapote tree

Table 7.1 Declines in some New World migrant birds between 1978 and 1987. From Robbins *et al.* (1989).

Species	Annual decline (%)
Yellow-billed cuckoo	5
Wood thrush	4
Northern oriole	2.9
Canada warbler	2.7
Ovenbird	1

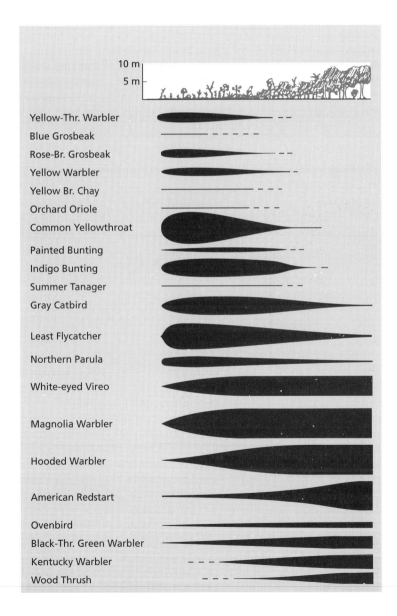

Fig. 7.8 Habitat preferences for a range of migratory American birds during their overwintering period in the Yucatan Peninsula, Mexico. The width of the black band indicates their relative abundance within the habitat gradient shown at the top of the diagram. Recent destruction of forest has put particular pressure upon those species reliant on the woodland habitat. From Lynch (1992).

(*Manilkara zapota*) that produces a milky sap used in the chewing gum industry, and this seems to provide an alternative to the natural forest for the shade-requiring birds.

The lesson that emerges from these studies of migratory birds, however, that is relevant to considerations of climatic change is that habitat fragmentation and loss can be critical to the survival of even the most mobile of animals, the birds. If this is the case, then less mobile, flightless animals, and plants that have poor systems for dispersal or long generation times, are likely to fare badly when new demands are placed upon them. Migration, dispersal or range spread in a world so heavily modified by our agricultural activities could prove a severe challenge to many species and we may face the question of whether we will need to lend them a hand by transplantation and introduction, with all the problems and potential dangers that this may imply. Such are the responsibilities of a 'global gardener'.

Extent and speed of change

The second major consideration in predicting species response to climate change is the degree and the rapidity of the change. Estimations of climate change, however, are themselves tentative and are dependent on a series of assumptions. The approach to such prediction has been based on the construction of models of the way in which the earth and its climate is known to behave, and the various roles of atmospheric and oceanic circulations, coupled with the natural geography of the earth. A number of different models (called 'Global Circulation Models', or GCMs, see Chapter 4) currently exist and permit meteorologists to tinker with various components, such as the CO_2 concentration of the atmosphere, and to observe the outcome as predicted by the computerized model. The results vary with the assumptions that are made about future events, but there is broad general agreement between GCMs that a doubling of atmospheric CO_2 would result in an overall rise in global temperature of between 2 and 4°C. It is also generally agreed that this warming effect will not be uniform over the surface of the earth, but will be most apparent in the higher latitudes, especially the Arctic and Antarctic (see Plate 5.1).

The Intergovernmental Panel on Climate Change (IPCC), which has tried to collate the views expressed by the various attempts to operate GCMs for predictive purposes, has concluded that if we assume no major changes in human activities (the 'business as usual' scenario), global mean temperatures may be expected to rise above present levels by about 1.1°C by 2030 and 3.3°C by 2090. But there are considerable ranges of uncertainty either side of these figures. These are still only global averages and if the study concentrates on specific parts of the world, such as central North America, then the prediction for 2030 would be warmer winters (by 2–4°C), perhaps slightly moister and warmer summers (by 2–3°C) with less precipitation (down 5–10%). Overall, soil moisture would decrease by perhaps 15–20% for this area. In the Arctic we could possibly see an increase of summer temperature by as much as 10°C over the next century.

Most climatologists are still very cautious about the value of such predictions, especially the precipitation projections, but these figures provide us with some estimates of the extent and speed of change against which to evaluate the likely response of individual species of plants and animals.

Since the Arctic is picked out as an area likely to experience the most extreme change in climate over the next century, this seems a most appropriate area to examine in more detail. It is also a region in which there is increasing evidence for long-term survival of certain plants during the Quaternary. On the basis of fossil peat beds on the islands of Novaya Zemlya and Spitzbergen (Crawford & Abbott 1994) and lake sediments in northern Norway (Alm & Birks 1991) (see Fig. 7.9 for locations) it may be concluded that tundra vegetation of some kind has existed in the unglaciated parts of this area certainly since the height of the last glaciation and very probably for longer. Such vegetation has certainly existed and was evidently fairly widespread during the transition from the last glaciation to Holocene times, around 10 000 years ago when the temperature rose by about 7°C within the space of a very few decades. In the Arctic today, therefore, we have a living reminder that certain species can cope with very considerable and very rapid changes in climate. Indeed, even within recorded history there is evidence of very sudden and profound changes in climate in Spitzbergen. According to Crawford

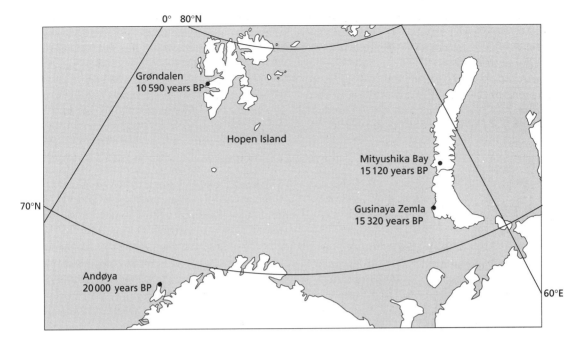

Fig. 7.9 Map of part of the Arctic north of Norway and Russia showing the locations of fossil tundra vegetation and their dates. From Crawford & Abbott (1994), based on Serebryanny & Tishkov (1996) and Skye (1989).

and Abbott (1994), it is the high genetic variability found within the Arctic flora that has provided it with this resilience in the face of change. Many plant species (the mountain avens, *Dryas octopetala*, is a well-studied example) have numerous genetically distinct ecotypes within their populations and this genetic diversity (some of which may be resident dormant in the soil seed bank) allows shifts in the general physiology of a population when the demand occurs.

There is a salutary lesson to be learned from this Arctic vegetation that has been subjected to rapid climatic change for several thousand, perhaps even several million, years. The secret of its success is its genetic diversity. Although its total species diversity is very low (especially when compared with the tropics), each species has an abundance of genetic variability that ensures its ability to adapt to change. If the floras and faunas of other parts of the world are to prove equally resilient the conservationist must be concerned to secure from extinction not only species, but as much as possible of the genetic variation within the species.

The genetic diversity shown by the components of the Arctic tundra vegetation is undoubtedly connected to the diversity of microhabitats, associated particularly with topography and aspect in that particular biome. Aspect, especially, is of increasing significance as an environmental factor the higher one goes in latitude. The difference between a north- and south-facing slope in an equatorial setting is of no great consequence; but at high latitude, such as that of Spitzbergen, the difference in microclimate between north- and south-facing slopes can be enormous, because of the much lower angle of incidence of sunlight. The effect of this genetic heterogeneity means that there are minor races of many of the species, well adjusted to local climates which lie at either side (warmer, cooler) than the mean. These are of course in this sense pre-adapted to future changes in climate, as they have proved to have been in their survival of the viscissitudes of the past series of Pleistocene glacial cycles. To this extent it may be that these Arctic communities, perceived by some as being vulnerable to global warming in particular, may be better able to respond to it than in some lower latitude biomes, where the component communities extend across less heterogeneous environments and have a lower level of genetic diversity.

The extent and speed of the anticipated climate change of the coming century, therefore, is probably not too great for plants and animals, given two provisos. First, they must be permitted to change their distribution patterns without hindrance (an almost impossible requirement given the fragmented nature of the human-modified landscape) and, second, they must have retained sufficient genetic diversity to permit an evolutionary response (a requirement that may well no longer be met by many rarer species). If the IPCC predictions are near the truth, then considerable human involvement will probably be required to avoid a cascade of extinctions. And even if we do interfere, further genetic depletion of many species is almost inevitable.

The impact of climate change on communities

So far we have considered the ways in which individual species may react to a change in climate, but in nature species are assembled together in communities. Is it possible that whole communities will respond *en masse* to alterations in climate? In order to predict this we need to examine the nature of the community.

The nature of the community

The survival of any plant or animal is not simply the outcome of the tussle between it and its physical environment, but a combination of this with the interplay between members of the community of which it is a member. The way in which whole communities have responded to climate change in the past, and are likely to do so in the face of future change, raises fundamental questions about our concept of the community. One of these questions has been with us since the earliest days of the science of ecology. Does a community function as an integrated unit, rather like a complex organism with its component organs? Does damage or removal of one of the component species have a profound or even fatal effect on the working of the whole community? Or is the community simply a mixture of its constituent species, of somewhat varied composition, from which one or two might be removed without seriously disturbing the stability of the whole? If one wanted to seek some analogy between these two

interpretations from the physical world this might be seen in the difference between a compound and a mixture. In the former, the constituents are present in a precisely fixed ratio, and are chemically bound to one another; whereas in the latter, the proportions of the components can be varied indefinitely and there is no bonding between them.

These two contrasted pictures of the community have been labelled the 'organismic concept' and the 'individualistic concept' respectively. Most early plant ecologists favoured the idea of the community as an organism. The American ecologist Frederick Clements (1936) called the community a superorganism, while the British pioneer ecologist Arthur Tansley (1935) referred to the community as a quasi-organism. The first serious challenge to this idea came from another American ecologist Gleason (1926) who regarded the community not as 'an organism, scarcely even a vegetational unit, but merely a coincidence'. This debate might at first sight be regarded merely as a rather rarified quibble among ecological theorists; but it has come to have particular significance at the present day in the context of the response of communities to environmental change. For example, does the integration of the members of a community mean a total interdependence on one another? Is the degree of integration such that the extinction of one or two seemingly trivial elements in a community can cause its total collapse?

At a simple level, the dependence of a flowering plant on a specialized pollination vector illustrates the type of mechanism that is involved. If the vector, perhaps a species of insect, becomes extinct, or is wiped out in a particular area, the plant concerned will fail to set seed. Long term, this will result in the regional or local extinction of the plant—unless some other pollinating organism takes over the role of the original vector. How rapidly this occurs—either the extinction or the adaptation—will depend among other things on the life-span of the plant. A standing forest might continue for many decades before failure of one species to regenerate resulted in its replacement by another community. Even a short-lived (annual) weed might be sustained over several years from its 'seed bank' in the soil where it was living. However, an animal dependent upon the seeds of that plant as its major food source could suffer an immediately negative response from a failure of that seed crop. A predator of that animal might in turn be severely affected by the original pollination failure. The elimination of a keystone species in the complex system has led to extensive repercussions.

Community responses to environmental change

Time scales are clearly important here and it is helpful to examine the rates of environmental change in the past if we are to predict the responses of communities in the future. It may be appropriate to return to the record of events of the Pleistocene at this point, the last two million years in which there have been global climatic changes of a major order (see Chapters 3 and 4). One of the most important sources of evidence of plant responses to climate change has come from the record of pollen in bog and lake sequences. It has generally been the tradition of this science of palynology to interpret the pollen spectra at different horizons in the sequence, in terms of modern plant communities, migrating in response to climatic change. This gives a very adequate general picture of the biological events, but it begs the question of whether the composition of the communities remains more or less constant, or whether the species respond individually and differently, so gradually changing the community composition. In other words, have the plant communities behaved in an 'organismic' or an 'individualistic' way in response to known climatic change? The western half of north America has a remarkable range of habitats, and hence communities, with its wide latitudinal extent, its striking topography and a wide range of bedrock types. There is a record of plant life preserved in a variety of ways from leaves buried in volcanic ash to leaves preserved in the debris dumps of desert packrats (packrat middens), ranging over some 60 million years. In a review of this vast body of evidence, and particularly that of the last 20 000 years, Keeley and Mooney (1993) strongly favour an individualistic interpretation. They claim that species have responded individually, changing

their range and distribution without regard to their community associates; also, that individual species have spread—both in their altitudinal and geographical ranges—at different rates. As a result of this divergence of reaction, community composition has not been static, and Pleistocene plant communities have often differed quite strongly from those of their modern counterparts even when common and dominant species are found in both past and present associations. This individualistic nature in the spread of plant species is illustrated in the fossil pollen data shown in Fig. 4.5.

We conclude, therefore, that the most appropriate way to tackle the question of the impact of climate change on vegetation and on its animal associates is to begin with a consideration of the responses of individual species. It is still necessary, however, to take into consideration the competitive interactions that may face a species on moving into new territory, and the new predators and parasites it may encounter there. New barriers have been erected by the process of fragmentation that may make the process of spread more difficult than formerly, and one must always bear in mind that the outcome of complex interrelationships in the field is often determined by who gets there first. We may well find that our new world offers greatest opportunities to those species with wide tolerances and efficient dispersal techniques. Humans have coined a word for many of the plants and animals that share these characteristics. We call them weeds and pests.

Further reading

Huntley, B. (1991) How plants respond to climate change: Migration rates, individualism and the consequences for plant communities. *Annals of Botany* **67** (Suppl. 1), 15–22.

Mooney, H.A., Fuentes, E.F. & Kronberg, B.I. (1993) *Earth System Responses to Global Change.* Academic Press, San Diego.

Woodward, F.I. (1987) *Climate and Plant Distribution.* Cambridge University Press, Cambridge.

Woodward, F.I. (1987) Stomatal numbers are sensitive to increases in CO_2 from pre-industrial levels. *Nature* **327**, 617–618.

Chapter 8

Ozone: Toxic Gas, UV-Screen

Ozone is probably the most paradoxical item in the whole 'global change' scene. At the level in the atmosphere where we live and breathe, the troposphere, ozone is rightly regarded as an atmospheric trace gas which can be extremely harmful to plant and animal life if it exceeds its normal very low level. But the ozone present in the upper parts of the atmosphere (between 40 and 60 km above the earth, in the stratosphere; Fig. 8.1) has an important function in screening out part of the ultraviolet (UV) radiation coming to us from the sun. In that role it is protecting life on earth from the potentially injurious effect of

the UV-B—the most damaging part of the UV spectrum. So ozone can be either 'good' or 'bad', depending on where it is. The situation is aggravated by the fact that human activities have been damaging the ozone in the stratosphere, where it is 'good', and adding to it in the troposphere where it is 'bad'.

Formation and destruction of ozone

Ozone is a form of oxygen in which each gas molecule is formed from three atoms of that element instead of the two present in a molecule of normal

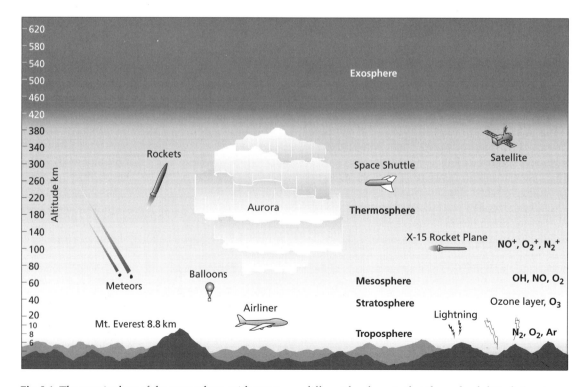

Fig. 8.1 The terminology of the atmosphere, with some of the phenomena and human activities occurring at different levels; note that the scale of altitude is not linear. From Torr M., *New Scientist* 23 Aug, 1984.

gaseous oxygen. While oxygen forms about 21% of the atmosphere, ozone, even at its highest concentration in the stratosphere does not exceed 10 ppm, and as such it is treated as a 'trace gas'. It is estimated that if the total ozone content of the atmosphere was brought down to ground level, and separated as a pure layer of that gas at normal atmospheric pressure, it would only be about 3 mm thick. A rather different way of expressing its role in the atmosphere is the fact that it represents less than a thirtieth of the average carbon dioxide (CO_2) concentration.

Ozone is considerably less stable than oxygen, and is a strong oxidizing agent. It has a pungent smell, and even in quite low concentrations can cause injury to the respiratory system in animals, and to the leaf tissue of plants. Ozone is produced spontaneously high in the atmosphere by the action of UV-radiation (UVR) on oxygen. The effect of the UVR in the waveband 180–240 nm (Part of the UV-C band: see Fig. 8.3) is to break down oxygen molecules to atomic oxygen which then combine with diatomic oxygen gas to form ozone (Fig. 8.2). This formation of ozone is in equilibrium with its natural destruction, which is brought about by UVR of rather longer wavelength, in the UV-B waveband. So UVR from the sun is responsible for this spontaneous formation and destruction of ozone, and this has maintained a very low ozone level in the atmosphere since long before the intervention of humankind.

In addition to this equilibrium between ozone formation and breakdown, the destruction process can be accelerated by the presence of several catalysts. Two strongly oxidizing substances—chlorine and nitric oxide—both act to catalyse the ozone breakdown under the effect of UVR in the atmosphere (Fig. 8.2). The combined effect of UVR in the presence of these simple catalysts at the very low pressures at high altitude results in a series of reactions which even now are not fully understood. The science of photochemistry, which studies the effect of light in chemical processes, is bringing us nearer to the stage at which accurate forecasts can be made of the impact on the ozone in the atmosphere of man-made catalytic substances accelerating its breakdown.

Atmospheric pollution and ozone

Many substances produced by the chemical industry, and which add greatly to the comfort, health and safety of life in the 'developed world', escape into the atmosphere and play some role in the catalysis of ozone destruction. Most notable of these are the hydrocarbon gas methane, nitrous oxide, and the now notorious chlorofluorocarbons (CFCs) once used extensively as aerosol spray propellants, as refrigerants in domestic refrigerators and as an inflating gas in plastic foam products. The gases most strongly implicated in ozone destruction are shown in Table 8.1. In addition to the industrially important CFCs and related organic compounds, these include gases which are produced naturally by the biosphere, such as CO_2 and methane. Table 8.1 shows the important statistics of these trace gases and their residence time in the atmosphere, their average concentration and their estimated annual rates of increase. It is obviously the most stable of these substances (hence with the longest 'residence time'), coupled

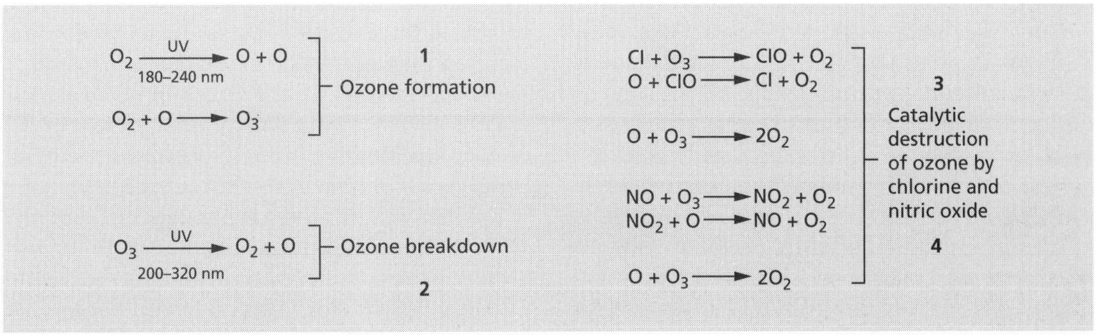

Fig. 8.2 The formation and breakdown of ozone in the stratosphere, and the role of UVR and other tmospheric components. From the Ozone Layer: United Nations Environment Programme (1987b).

Table 8.1 Trace gases in the atmosphere which affect ozone concentrations. Note that only the first four are entirely the products of industrial processes, while the remainder are formed both by natural processes and by human activity. The influence of CO_2 is simply in its affecting climate, and hence the rate of chemical processes in the atmosphere. From the United Nations Environment Programme (1987b).

Gas	Formula	Average lifetime in atmosphere (years)	Average global concentration (ppbv)	Annual rate of increase (%)
CFC-11	$CFCl_3$	75	0.23	5
CFC-12	CF_2Cl_2	110	0.4	5
CFC-113	$C_2F_3Cl_3$	90	0.02	7
Halon-1301	CF_3Br	110	very low	11
Nitrous oxide	N_2O	150	304	0.25
Carbon monoxide	CO	0.4	variable	0–2
Carbon dioxide	CO_2	7	344000	0.4
Methane	CH_4	11	1650	1

with a high rate of annual increase, which are the most serious cause for concern.

Ample coverage has been given in the media to the slow realization of the role of CFCs in the damage that they are believed to have caused to the 'ozone layer'. These halogen-containing compounds produced in the troposphere can slowly diffuse up into the stratosphere where their breakdown under the action of UVR causes release of free chlorine, with its known potential for catalysis of ozone breakdown. Over recent years a number of substitutes for CFCs have been developed, which can be used as refrigerants and in most of the industrial processes for which CFCs had been widely employed. These alternatives, such as hydrofluorocarbons (HFCs), although containing halogens, are believed to be less injurious to the stratospheric ozone. This is because they are readily oxidized while still within the troposphere, largely by the action of hydroxyl radicals, and so have a lifetime of between one and 40 years—significantly less than the 75–110 years cited for CFCs.

Many other products of our industrial world are under suspicion of accelerating ozone breakdown in the stratosphere. Nitrous oxide, which is one of the gases so implicated, is produced by almost any process in which air (consisting largely of nitrogen gas plus oxygen) is raised to a high enough temperature to effect oxidation of the nitrogen. As a result, it is produced by petrol and diesel engines, as well as in many industrial processes. Nitrous oxide is also increasing at least in part as a result of the extensive use in agriculture of man-made nitrogenous fertilizers. The products of incomplete combustion of aviation fuel, released

into the upper part of the atmosphere by jet aircraft, are also under suspicion of catalysing ozone breakdown. Paradoxically, it is now also acknowledged that the result of the burning of hydrocarbon fuel in subsonic aircraft is probably causing some increase in the ozone in the upper part of the troposphere! So far as the UV-screening role of the ozone is concerned, these two processes of ozone breakdown and augmentation may cancel each other out. But it is fair to say that CFCs have caught the brunt of media alarm over damage to the stratospheric ozone. When we have considered the role of ozone as a UV screen, and the action of UV on biological systems, we will return to the effect of human activity on atmospheric ozone and the prediction of such effects in the long term.

Ultraviolet radiation and the ozone screen

Figure 8.3 shows the energy flux of various parts of the solar radiation spectrum, together with a plot of the radiation actually received at the earth's surface under blue sky conditions. The fact that the radiation received is of considerably lower energy at all wavelengths from UV to infrared is a result of absorption by gases present in the atmosphere, many of them (like ozone) being in the category of trace gases. But other components of the atmosphere have a significant effect on the spectrum of radiation reaching the earth's surface. Some of these are produced by natural phenomena such as cloud cover and wind-borne dust and volcanic ash. Others, such as industrial smoke and other, gaseous, emissions are products of human activity. Smoke and gaseous products of

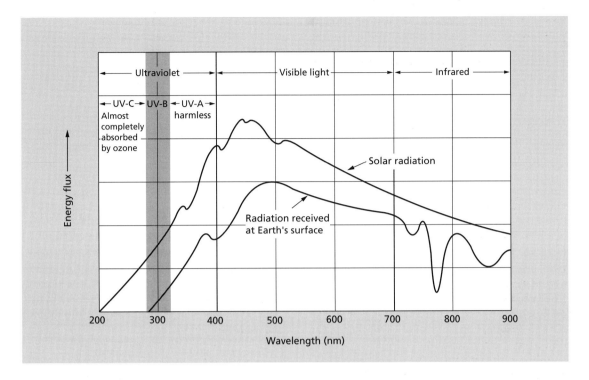

Fig. 8.3 The range of solar radiation, showing the energy flux at different wavelengths. The difference between the two curves represents the absorption by the atmosphere. From the United Nations Environment Programme (1987b).

forest burning are generated naturally by wild-fires, but may be produced in greater quantity as a result of human intervention.

The strength of incident radiation of all wavelengths is also enormously influenced by the angle at which the sun's radiation enters the atmosphere. This is largely because the lower the angle of incidence (that is, the higher the elevation of the sun) the less the area over which a given flux is spread. For example, the radiation flux reaching the ground is far greater in any location at mid-day than in the early morning. This is familiar from the heating effect in these two situations of the infra-red radiation, which makes itself most immediately felt. But the incident angle is important in another aspect; the lower the angle of sunlight, the more atmosphere the flux has had to pass through, and the greater the absorption by any atmospheric constituents. This is of course relevant in relation to high-latitude conditions as well as to changing angles of the sun through the span of a day. In considering the effect

of changes in the UV-screening capacity of atmospheric ozone it is important to bear in mind these other constraints on intensity of solar radiation.

The ultraviolet part of the spectrum between 200 and 400 nm is divided into three bands of UV-A, -B and -C (Fig. 8.3). The UV-A, nearest to the visible end of the UV spectrum, is biologically relatively harmless, compared with the effect of UV–B and –C on living cells. Even so, its energy flux is more or less halved by atmospheric absorption (measured by the gap between the two curves within the UV-A band in Fig. 8.3). The UV-B, between 280 and 320 nm, is biologically the most significant; this part of the UV spectrum is implicated in skin cancers, in the formation of cataracts, potentially a cause of blindness in human beings, and in damage both to planktonic organisms in the oceans and lakes and to plant life on land. It is also implicated in ageing of the skin, and in the development of a sun-tan. But it has a benign role too, in enabling people to synthesize their own vitamin D—a role often overlooked in the contemporary concern with skin cancer. These biological effects of UV-B will be discussed further below.

The shortest wavelength UV, the UV-C nearest to the X-ray end of the spectrum, is almost completely absorbed by ozone and other atmospheric constituents, and virtually none penetrates the atmosphere to the earth's surface. This is shown by the absence of any 'radiation received' within the UV-C band at the lower left in Fig. 8.3.

Distribution of atmospheric ozone

Because of the role of stratospheric ozone in screening out much of the UVR as it first enters the atmosphere, the impression is often created that there is an ozone 'shield' concentrated entirely in the stratosphere—that is, in the uppermost part of the atmosphere only. This picture is accentuated by the tendency to quote ozone concentration in terms of parts per million—that is, relative to the other constituents of the atmosphere. Like percentages, this can be misleading; a percentage figure can rise either because that component is itself rising, or because the other components of the system are falling. Figure 8.4a shows the ozone concentration (as parts per million) relative to other gases in the atmosphere, and Fig. 8.4b its absolute concentration as molecules per cubic centimetre, at different heights above the earth. Because the ozone is actually being generated by the action of UVR in the stratosphere, it is not surprising to find that its relative concentration is

at its highest between 30 and 40 km above the earth's surface. However, owing to its chemical instability discussed above, its residence time in the atmosphere is short and its relative concentration falls off as one moves down the atmospheric column into lower altitudes. In the lowest part of the troposphere, where ozone is not being generated by UVR, the proportion of oxygen and nitrogen rise relative to the short-lived ozone.

The absolute concentration of ozone in the atmosphere expressed as molecules per unit volume (Fig. 8.4b) presents a different picture. This value peaks at between 20 and 30 km altitude, but above that it falls off rapidly, as do all the atmospheric gases. Below this height, there is no rapid fall in ozone, there being actually more atmospheric ozone below 30 km than above. This seemingly rather puzzling distribution pattern is important when we come to consider likely future changes in the ozone distribution later in this chapter.

UV-B and vitamin D

With all the current concern about skin cancer and other adverse effects of UVR it is easy to forget that human beings actually need a low dosage of UVR to enable our bodies to synthesize vitamin

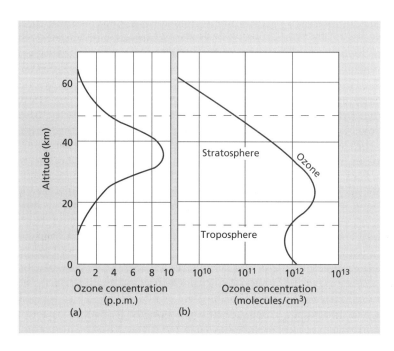

Fig. 8.4 Ozone distribution at different levels in the atmosphere. (a) The relative concentration with respect to other atmospheric components, in parts per million (ppm). (b) The absolute ozone concentration, in molecules per cubic centimetre. Note that the scale is arithmetic in (a) but logarithmic in (b). From the United Nations Environment Programme (1987b).

D. Vitamin D is a fat-soluble vitamin (or vitamins), a group of related steroids, which are required for the uptake of calcium from the diet. A deficiency results in abnormal bone and tooth development, particularly in young children, and in an extreme case can result in the deficiency disease called rickets. Children receiving an adequate diet, and with moderate exposure of their bodies to UVR from sunlight, are able to synthesize vitamin D from precursors in their skin. Lacking this opportunity for synthesis in the body, deficiency symptoms can be avoided by consumption of foodstuffs rich in vitamin D such as most fish liver oils, egg yolk or milk. Only half a century ago, many children living in high-latitude industrial centres were liable to develop rickets as a result of inappropriate diet, coupled with lack of UV irradiation in their environment. The latter combined cloud cover with temperatures unconducive to exposure of the skin to sunlight (both associated with higher latitudes) coupled with UVR screening by excessive smoke and industrial haze. In the early years of this century, one of the ways of ameliorating the tendency to rickets was therefore to offer facilities for UV irradiation to young children, often run on a municipal or communal basis. This makes an interesting contrast with the present concern to avoid UVR exposure on account of perceived problems of another kind.

The dangers of UVR to humankind

Very different from this benign effect of low-dosage UV irradiation are the various forms of damage that higher dosages can cause to animals and, expressly, to people. Pale-skinned Caucasians are particularly liable in the short term to sunburn, causing in an extreme case severe blistering and resulting discomfort and systemic upset. While moderate exposure to UVR causes stimulation of the melanocytes in the skin to produce the pigmentation that causes a sun-tan, excess radiation has several adverse effects. The long-term result of over-exposure to UVR is a thickening of the epidermis, with loss of elasticity, premature wrinkling and over-stimulation of the melanocytes to produce a blotchy skin pigmentation. Many of the visible effects observable as people get older are the result of this 'photo-ageing' rather than 'biological ageing' alone.

Excessive exposure to the UVR of sunlight can also result in various forms of skin malignancy. These may be divided into 'non-melanoma skin cancer', which includes basal cell carcinoma ('rodent ulcers') and squamose cell carcinoma, according to the type of skin cells involved, and the more serious 'malignant melanoma'. The latter is fortunately far less common than the non-melanoma skin cancers, which can usually be dealt with by prompt and early surgical treatment. But at present there is great concern over melanoma because of its steadily increasing incidence, wherever accurate records of its occurrence have been kept. In many of the sunny parts of the world, incidence of malignant melanoma has doubled within the last decade.

A recent study of the death rate from melanoma in the United States shows a clear correlation with latitude (Fig. 8.5). This correlation is due to two closely linked factors. In the southerly, 'sunshine states' of Florida and Texas, the actual flux of UVR is higher for the several reasons discussed earlier in this chapter, linked to the latitude alone. But the human behavioural response to climate is also important. Those southerly states have a higher average summer temperature, so that people are more inclined to spend time out of doors, with more skin exposure than in the cooler northern states. It appears that the incidence of skin cancer is influenced by a complex of factors of latitude, cloud cover pattern and the effect of sunshine and ambient temperature on the behaviour of people in exposing themselves to the sun.

Despite the widespread concern that attenuation of the stratospheric ozone shield is in some way involved in the rising incidence of skin cancer, there appears to be at present no direct evidence that this is having a perceptible effect. R.M. MacKie (1993), of the Department of Dermatology of the University of Glasgow, attributed the increasing incidence of skin cancers to all types of changing habits of sunlight exposure rather than to 'any possible excess UVR as a result of ozone depletion'. By the use of appropriate clothing, sun-glasses and sunscreens we have it within our capacity, in the 'developed world' at least, to control the health hazards of excess UVR, regardless of whether or not human activity is responsible for any increase in radiation received at ground level.

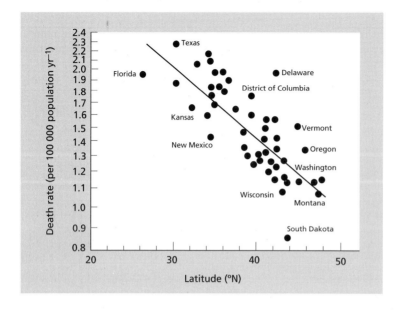

Fig. 8.5 Death rates from the serious skin cancer, melanoma, show variation with latitude in these data collected from the United States. The regression line shows that there is a correlation between higher incidence of this skin cancer in the warmer, sunnier lower-latitude states. From the United Nations Environment Programme (1987b).

UVR damage to other organisms

It is widely accepted that the ionizing effect of UVR can damage DNA in living cells, and that as such it can have a mutagenic, or indeed lethal, effect. This is of course the basis of its being a cause of skin cancers. For the same reason UVR is widely used as a means of suppressing microbial action, and hence for clinical sterilization of a fairly gentle nature. Where micro-organisms are grown in laboratory research on an agar gel surface, UVR is routinely used as a means of producing mutations. The effect of UVR on larger multicellular organisms is generally less dramatic, but we now have well-documented information on the type of damage that different parts of the UV waveband have on plant metabolism (Caldwell *et al.* 1989). Figure 8.6 shows the action spectrum of different parts of the UVR range in terms of their deleterious effect on the Hill reaction (an essential process in all higher plant photosynthesis), on CO_2 uptake by *Rumex* leaves, and damage to DNA (the cause of mutation). Although the gradient and shape of the three curves is different, there is clearly a markedly stronger effect in the UV-B (below 320 nm) than the UV-A part of the spectrum (that above 320 nm). Other experiments on living plants show that different species react very differently to increased levels of UV-B. Work on soybeans, cited in the UNEP (1987b) ozone report, indicate that the estimated increase of UV-B radiation resulting from a 25% lowering of stratospheric ozone would result in a 25% reduction in crop yield.

Despite experimental work of the kind just described, the importance of stratospheric ozone in shielding terrestrial life from UVR is a subject of very divergent opinions. A widely-held view is represented by a statement from M.K. Tolba, Executive Director of the United Nations Environment Programme. He states (UNEP, 1987b): 'Life on Earth depends as much on the presence of ... ozone as it does on the presence of oxygen and water. Without it, lethal levels of ultraviolet radiation would reach the Earth's surface extinguishing life ... as surely as if the atmosphere were removed altogether'. On the other hand, James Lovelock, author of the Gaia hypothesis, writes (Lovelock 1995) that 'the ozone layer certainly exists but it is a flight of fancy to believe that its presence is essential for life'. He goes on to point out the inefficacy of UV irradiation in producing sterilization in the working hospital environment. There is undoubtedly still a great need for experimental work on the effect of UVB on both crops and natural plant communities, and on the actual levels of UVB experienced in real life situations at ground level.

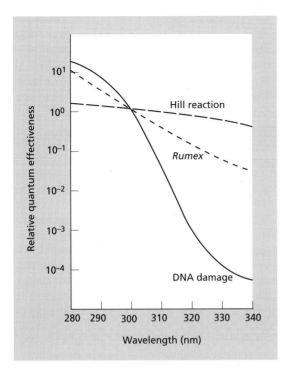

Fig. 8.6 The relative effect on three aspects of a flowering plant, *Rumex*, of ultraviolet light of different wavelengths, ranging from UV-B at left into UV-A at right. The Hill reaction is part of the process of photosynthesis, essential to all plant bioproductivity; the *Rumex* curve is a measure of its growth; the DNA damage is a measure of the potential of the UVR for causing mutation of the genetic material. From Caldwell *et al.* (1989).

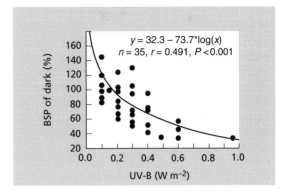

Fig. 8.7 The response of marine bacterial productivity to changing levels of UV-B irradiation. The bacterial secondary production (BSP) was measured as the rate of incorporation of a labelled amino acid into the bacterial protein, compared with the rate observed in the dark, expressed as a percentage. The measure of bacterial activity falls off with increasing UV-B flux. From Herndl *et al.* (1993).

UVR and marine life

The carbon fixation occurring as a result of the photosynthesis of marine phytoplankton constitutes the primary productivity which underlies marine biomass production and, with it, the world's sea fish production. It also plays a long-term role in the cycling of carbon, since that is the main route by which CO_2 may be taken out of the atmosphere to be incorporated in deep-sea sediments and so effectively 'taken out of circulation' (see Chapters 5 and 6). The possible effects of increased UVR on the surface of the open sea, resulting from thinning of the ozone screen, is therefore of great importance to the biology of the oceans.

As with the role of UV-B in relation to life on land, some scientists have questioned whether intense UVR had a severely limiting effect on early metazoan life in the oceans. Again, Lovelock (1995) is sceptical. Writing of the Precambrian environment he says 'the abundant ions of such transition elements as iron, manganese and cobalt are intense absorbers of ultraviolet ...; even if the full unfiltered ultraviolet shone on the surface [of the oceans] it would still not have much hindered life'. However, recent work of Herndl *et al.* (1993) casts some doubt on this view. Studying marine bacterioplankton in the Adriatic, they show that solar radiation suppresses bacterial metabolism by 40% in the top 5 m of inshore waters. They investigated this by comparing the incorporation of labelled leucine into bacterial protein in illuminated and dark conditions within the water column. The results of one of their experiments are given in Fig. 8.7, showing the effect *in vitro* of different levels of solar UVR. As the UV-B radiation increases, the bacterial activity (BSP, 'bacterial secondary production', as measured by leucine incorporation) drops off below the 100% level, based on their activity in the dark. However, it can be seen in Fig. 8.7 that at low UV levels, less than 0.3 W m^{-2}, some of the BSP values are above the 100% (dark) value. This was probably a result of associated phytoplankton releasing readily-metabolized photosynthate by leakage, masking the negative effect of the UV at those low levels. However, as the UV-B flux increases there is a clear fall off in the bacterial activity.

Herndl *et al.* emphasize the role of marine phytoplankton in recycling dissolved organic matter (DOM) in the ocean; if ozone depletion resulted in increased UVR, then we should expect a reduction in that bacterial scavenging of DOM. The full impact of this is hard to assess. If the main photosynthetic productivity was by cyanobacteria ('blue–green algae') and they responded with reduced activity in the face of increased UVR, then the rising UVR would reduce the CO_2 drawdown in the oceans. On the other hand, the increase in absorption by the ocean water following reduced rate of removal of DOM might shade all the micro-organisms from the penetration of the UV-B. This would act as a negative feedback, reducing the impact of the increased incident UVR. Other possible effects should also be borne in mind. Reduced scavenging of DOM would result in more of it accumulating in the surface water instead of its being recycled back into atmospheric CO_2. The effect of this would mean that the ozone depletion would have reduced the build up of the principal greenhouse gas, CO_2, which would be seen by most as a desirable outcome. As with many global change problems, the interaction of several environmental processes makes prediction of the outcome of changing just one of them in isolation very problematical.

Ozone thinning, ozone holes and the UVR problem

A key question in global change research is how far human influence on the atmospheric ozone will actually increase the deleterious effect of UVR reaching the earth's surface. Concern over the thinning of the stratospheric ozone goes back to the International Geophysical Year of 1957, when an international network of 'Dobson Stations' was set up to monitor atmospheric ozone using a technique pioneered by a scientist of that name. One of these stations was set up at Halley Bay by the British Antarctic Survey, which was later to play an important part in the recognition of the Antarctic thinning of the stratospheric ozone layer. By the 1970s CFCs were already being detected in minute amounts in atmospheric samples taken at ground level. It was realized by

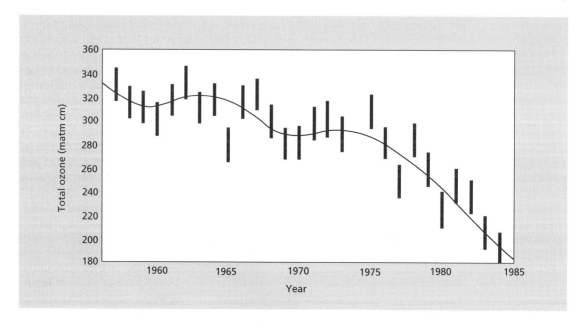

Fig. 8.8 Changing ozone levels over Halley Bay in Antarctica, shown with error bars on the values of the total ozone above the site. This shows the significant fall off observed between 1957 and 1985, which alerted the world to the thinning of the Antarctic stratospheric ozone layer. From the United Nations Environment Programme (1987b).

then that the likely fate of these relatively stable compounds would be to eventually diffuse upwards to altitudes at which the UVR would cause their photodissociation. This would then lead to release of chlorine, which would in turn catalyse destruction of the ozone at that high altitude. There was also concern at that time about the effects that high flying aircraft might have on the stratospheric ozone, although as mentioned earlier in this chapter, that concern now seems to have been exaggerated. By 1985 observations by the British Antarctic Survey showed that there was a seasonal depletion of the springtime ozone layer above Antarctica (Fig. 8.8). This was soon confirmed by satellite observations, and the phenomenon of the stratospheric ozone depletion acquired the evocative label of the 'ozone hole'. Further observations on the stratospheric atmosphere by ground and airborne instruments demonstrated the presence of chlorine and bromine, strengthening the belief that halogenated organic compounds (which include CFCs) were implicated in the ozone thinning.

It is perhaps comforting to consider that if we had to have ozone thinning somewhere over the earth's surface, with its associated danger of increased penetration of UV-B, the Antarctic is, fortuitously, one of the least tiresome parts of the globe for this to occur, at least so far as human life is concerned! Antarctica combines a uniquely low indigenous human population (which is of course zero), with a very low angle of incidence of solar radiation, and a correspondingly low level of radiation flux at all wavelengths, compared with lower latitudes. The prevailing low temperature is also not conducive to sunbathing, and indeed the scientists working there tend to wear protective clothing on grounds of temperature alone! None the less, this thinning of the ozone 'screen' gave good cause for concern about the future threat of atmospheric ozone attenuation, with all that this implied for increased UVR at lower latitudes.

The strength of this threat, coupled perhaps with a feeling that this was one aspect of adverse global change about which 'something could be done' led to the Montreal Protocol. This agreement on an international restriction to the production and use of 'substances that deplete the ozone layer' was signed by the governments of most nations at a meeting in Montreal in 1987, aiming at a 50% reduction in production of CFCs by the year 2000. Tighter strictures were agreed for the developed countries (which had led the way in massive CFC emissions) than for the developing countries, which are still far short of their potential consumer demand for refrigeration and other CFC uses.

Most people would see the signing of the Montreal Convention as a great victory for those concerned with global issues, as indeed it was. None the less, the interplay of politics and economics in this type of international agreement are rarely as simple as they may seem. The environmental danger represented by CFCs had been evident to the chemical industry for some years. Those that had invested significantly in research and development directed to finding alternative propellant and other substitutes for CFCs (such as HFCs and HCFCs) were naturally interested in seeing CFCs put under restriction. The Montreal Convention took care of those interests.

Modelling ozone changes

As soon as the complex reactions between ozone and other atmospheric trace gases (of both natural and industrial origin) became known, efforts were made to model the changes that would occur in the stratospheric ozone. The main objective in any such modelling is, of course, to predict the effects of various strategies which might be designed to minimize the environmental impact of atmospheric pollution.

Attempts to model the effect of changes in ozone distribution have naturally centred on the interaction of CFCs and stratospheric ozone, and the effect that these will have on UVR damage to human health and to plant and animal life on land and in the oceans. Since both ozone and the CFCs are themselves powerful greenhouse gases, changes in their distribution in the atmosphere would also have climatic implications (see Chapter 5). Different strategies of restriction in CFC production and emission produce different scenarios of atmospheric impact. Figure 8.9 shows a prediction set out by the Intergovernmental Panel on Climate Change (Houghton *et al.* 1992), of the effect of the Montreal Protocol on the emissions of various halocarbons. One of the most difficult parameters to incorporate in any

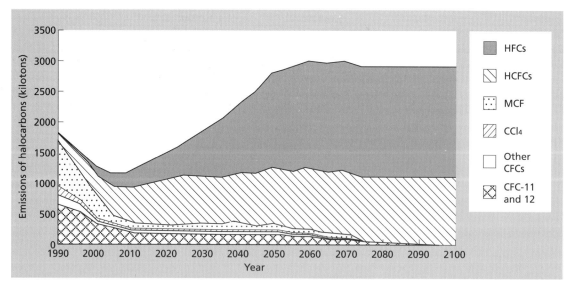

Fig. 8.9 Estimate of the emission of halocarbons assuming a realistic level of compliance with the Montreal Protocol, and a complete end to CFC production by the year 2075. From Houghton *et al.* (1992).

such prediction is the likely degree of compliance with internationally agreed levels, which can only be conjectured. In this projection it is suggested that while the release into the atmosphere of the less harmful halogenated hydrocarbons (HFCs and HCFCs) will level off by 2050, the use of CFCs will be phased out completely by 2075. It is then possible to further project what the actual atmospheric concentration of CFCs might be under different emission scenarios, into the next century. Such a prediction for CFC-12, currently the most abundant of the range of CFCs in the atmosphere, is offered for four different emission models in Fig. 8.10 (from Houghton *et al.* 1990).

Models such as these can be used to predict the changes in ozone at different heights in the atmosphere, and thus the impact on UVR penetration. Figure 8.11 shows such a prediction based on the production of CFCs rising at 1.5% per year, which approximates to a partial implementation of the Montreal Protocol. It is interesting to see how the percentage changes are markedly different at different altitudes in the atmosphere. In the strato-sphere, at around 40 km, where the ozone concentration is greatest (see Fig. 8.4) there is a predicted 40% fall in the ozone, caused by CFCs, within 100 years. In the troposphere, between

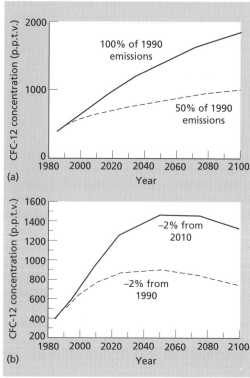

Fig. 8.10 Estimates of the CFC-12 concentration in the atmosphere under four different scenarios: (a) assuming the levels of the 1990 emissions being sustained; (b) it is assumed that emissions increase by 2% per year until 2010, after which there is an annual reduction of 2% of emissions (upper curve); or that emissions fall by 2% per year from 1990 onwards. From Houghton *et al.* (1990).

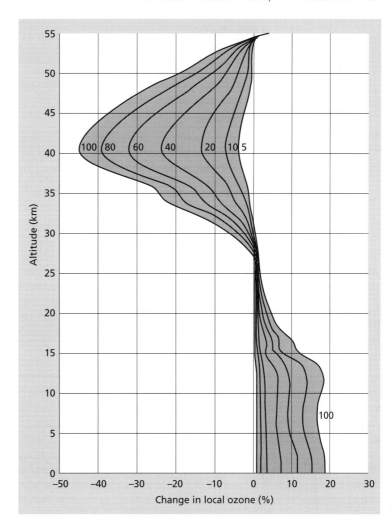

Fig. 8.11 The predicted effect of CFC emissions on the ozone level at various altitudes, assuming an increase of production of 1.5% per year, after 5–100 years. Notice that as the ozone level falls by 40% after 100 years at an altitude of 40 km, at ground level it has risen by nearly 20%. From the United Nations Environment Programme (1987b).

15 km and ground level, a rise in ozone is predicted of some 20% over the same period. This of course enhances the UVR screening effect, significantly offsetting the diminished stratospheric ozone. But in terms of human health, the stratosphere is a far better setting for an ozone screen than having any rise in this irritant, toxic gas in the troposphere where it will be inhaled.

Predicting the biological effects of ozone changes

So far we have tried to concentrate on what we know, both of the present composition of the atmosphere, of atmospheric chemistry so far as it involves ozone, and the role that ozone plays as a screen against UVR. We must now look at the questions which are foremost in the minds of the public, and hence of politicians and other policy-makers. What will be the effect on people of atmospheric changes in ozone, in an everyday situation? Can we put figures on predicted increases of deleterious effects on human health resulting from CFC emissions? Will the impact of increased UVR on a global scale seriously affect natural ecosystems? Or crop yields? Will it affect bioproductivity in the sea—in particular, the marine life exploited by humankind as food? What of the predicted rise in ozone in the troposphere where most of us live and breathe? What effect will that have on people, crops, plant pests and indeed the whole agricultural system on which we are all dependent for food?

If we want to predict future impact in these

terms, it is important first to look dispassionately at what has actually happened already, which scientists are capable of measuring in the present-day environment. The fall in atmospheric ozone is well documented, but what effect has this actually had on UVR received at street level? The measuring of different energy levels across the UV spectrum at ground level is much less exciting as a research exercise than using satellites and balloons to monitor the polar 'ozone holes' high in the atmosphere. Ironically, it is also in some ways more difficult technically. Perhaps both these aspects—psychological and technical—have concentrated attention on ozone monitoring rather than on ground-level UVR measurement. The principal interest in all such measurement is whether we can detect long-term trends in observations, which would assist in risk assessment.

Measuring changes in ground-level UVR

There are a number of different techniques available for measuring UV irradiance with ground-level instruments. The most precise can measure the energy flux at each 1 nm waveband between 280 and 400 nm (spanning the UV-A and -B parts of the spectrum; see Fig. 8.3). Other less precise measurements can give values for broader bands of selected wavelength ranges, or fluxes passing through filters of known spectral characteristics.

In any attempt to collate global data of this type, there are problems in standardizing the instruments used at widely separated sites. Because there are a number of different types of measurement available, the results from each are not immediately comparable. Even the best means of effecting cross-callibration are not generally agreed. These can vary from treating solar radiation 'on a clear day' as standard, to using solar simulators such as a xenon fluorescent light or a tungsten filament in a halogen-filled bulb. The continuity of measurements seasonally, and even within a day, vary greatly between different monitoring systems. Changing weather systems from hour to hour, and above all cloud cover (which constitutes a very effective UV screen) will of course affect measurements. Within the UK, Driscoll (1993) reports that 'clear sky' conditions are observed for less than 15% of the days of the year, and near-total cloud cover for between

50 and 70% of daylight hours, depending on the season and latitude of the site.

Figure 8.12 shows the differences between three stations within the British Isles, monitoring the energy flux (in micro-watts per square metre) within the total UVR band, taken at mid-day between 1988 and 1992. As would be expected, the differences between January and the summer May-to-August period are enormous, reflecting both the effect of changing angle of incidence of mid-day sunlight, day length and cloud cover through the seasons. The effect of the latitude difference is also perceptible, although this is blurred by differences in weather patterns associated with latitude, and in local atmospheric pollution.

Perhaps most surprising is the observation that through the 5-year period of 1988–1992, during which concern has grown over the implications of Antarctic stratospheric ozone depletion, the record of solar UVR levels in the UK have shown no detectable trend of change (Driscoll 1993). In some parts of North America, there has actually been a fall in measured UV-B irradiation at ground level, perhaps as a result of increased emissions of pollutant gases and particulate matter into the troposphere (Ashmore & Bell 1991).

A recent study carried out in Toronto (Kerr & McElroy 1993) monitored the UV-B radiation several times an hour, throughout the daylight hours, for the summer months (May–August) and the winter months (December–March) during the years 1989–93. The comparison between these two periods in the year is of critical interest, since over that period the total ozone levels over Toronto decreased by an average of 4% per year in winter, and by 2% per year in the summer. These Canadian workers measured the UV-B radiation at two narrow wavebands within the UV-B range of 280–320 nm (Fig. 8.13a and b). In the middle of that range (at 300 nm) they record an increase of 35% in the 5-year period of their study, during the (maximum) winter thinning of the ozone layer (the gradient of the regression of the winter values, the lower line across Fig. 8.13b). The corresponding summer increase in UV-B is only 7% (gradient of the summer regression line in Fig. 8.13b). In contrast to this, the level of summer radiation received at the top of the UV-B range, where the ozone absorption is far less, actually shows a slight fall (Fig. 8.13a).

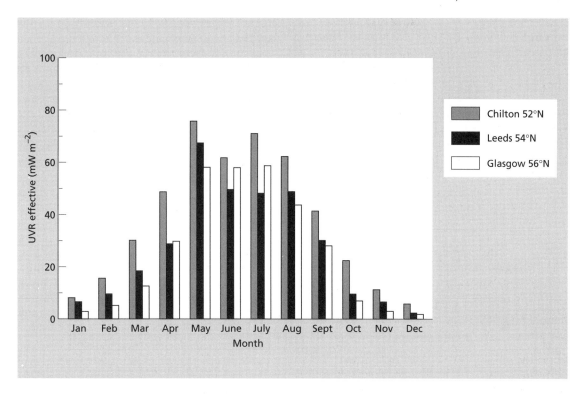

Fig. 8.12 The mean observed ultraviolet radiation at three sites in Britain, taken at mid-day over the period 1988–92. There is a great seasonal difference between summer and winter levels, and a small difference resulting from latitude. From Driscoll (1993).

In assessing the significance of these observations, it has to be borne in mind that the received radiant energy of the UV-B at 300 nm is only about 1% of that at the top of the UV-B range (where there is no significant rise). As those authors themselves note, 'the observed large (35%) increases in UV-B radiation near the 300 nm in winter are large fractional increases in small values. For this reason it may not represent a significant increase in terms of biological impact'. The force of their point is evident when you note that the vertical (energy) scale in Fig. 8.13b is exaggerated 100 times compared with that in Fig. 8.13a.

As with the UVR data considered above, it is important to assess its impact in the context of other biologically relevant aspects of the environment and the living organisms. The total incidence of solar radiation in the latitude of Toronto is of course far lower in winter than in summer, and this applies equally to the UV-B (seen in the low values of both Figs 8.13a and b at the close of each year). This parallels the data given for the three British sites in Fig. 8.12. In addition to the usual seasonal factors reducing winter irradiation discussed earlier in this chapter, the flora (and associated fauna) in the Toronto area would also be protected for several months of the winter by snow cover. As far as human behaviour is concerned, most people out of doors in that part of Canada will be very well covered up in the winter, while sun-bathing enthusiasts will generally only be significantly exposed to solar radiation during the summer months. While the trend of UV-B rise in Toronto is disquieting, its impact on human health does not appear threatening, because it lies in a 7% per year rise in a part of the UV-B spectrum which itself carries only about 1% of the total UV-B energy spectrum.

Predicting health hazards of increasing UVR

A number of attempts have been made to extrapolate from estimates of increased UVR at ground level to predictions of skin cancer and other health effects. One of the most frequently quoted

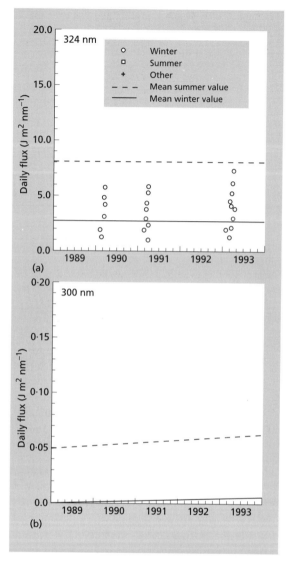

(a)

(b)

Fig. 8.13 The UV-B flux at two narrow wavebands integrated daily in Toronto over 5 years. (a) The UV-B flux close to the border with the harmless UV-A band; (b) the flux at the centre of the UV-B range. There is a perceptible rise in the latter for the mean winter value (at the time of maximum ozone thinning) but note that the scale at (b) is 1% of that shown in (a). Reprinted with permission from Kerr & McElroy (1993), copyright 1993 American Association for the Advancement of Science.

has been a study published by the US Academy of Sciences suggesting that for every per cent rise in UV-B irradiation, the incidence of skin cancer would increase by 2%. It was further calculated

that a 3% increase in UV-B would give 18 000 more cases of skin cancer a year in the United States alone (UNEP 1987b). It is important to note that these seemingly alarming figures are for all skin cancers, and not for the small proportion of those that would be the more serious form, melanoma.

To get such figures into proportion, it is worth seeing them in the context of latitudinal changes in skin cancer incidence shown in Fig. 8.5. The effect of a 3% reduction in the ozone layer would be the equivalent of everyone in the United States moving 2° of latitude further south; that is, for example, moving from New York to Washington D.C. Few would regard this as a serious health threat, at least in terms of risk of skin cancer!

A further cause for concern for human health in the face of increased UV-B radiation is the report from work on animals that an increase can suppress the efficiency of the body's immune system (UNEP 1987b). This seems to suggest that stratospheric ozone depletion would cause an increase in other forms of skin infection, beyond the risk of skin cancer, attributable directly to the mutagenic effect of the UV-B, and would include skin infections such as those caused by herpes virus. At present this remains an unsubstantiated hazard. R.M. MacKie (1993), reviewing the effect of UV radiation on the skin, reports that 'in animals excessive exposure to UVR affects the immune system causing immunosuppression, but in man the evidence of biologically significant UVR induced immunosuppression is as yet lacking'.

Adverse effects of tropospheric ozone

While most attention given to changes in atmospheric ozone has centred on its role as a UV screen in the stratosphere, the evidence for rising ozone in the troposphere has other important implications. It may seem rather perverse that the ozone level is falling at higher altitude, where its effect is perceived to be helpful, while at ground level, where its major impact on living things is as a toxic gas, it is increasing! As indicated above, the changes are due to quite different phenomena at the two levels. The stratospheric decline is attributed to the effect of CFCs, while the tropospheric rise is attributed largely to oxides of nitrogen from motor vehicles and a range of industrial processes (see Chapter 9). After the recognition around the middle of the last century that ozone was a naturally-occurring constituent of the atmosphere, it came to be widely believed that it was not only vaguely beneficial, but that it was especially associated with the seaside environment (Ashmore & Bell 1991). This resulted in some confusion, lasting through several generations of seaside visitors, between the smell of rotting seaweed and the supposed smell of ozone! Only since the 1950s, when ozone came to be implicated as an injurious component of the notorious Los Angeles smogs, has its role as a health hazard to human beings and a threat to crop production been recognized. The widely propagated environmental message of the early 1990s that we ought to use 'ozone friendly' aerosols, so as not to damage stratospheric ozone may have blunted appreciation of the fact that ozone itself, as a component of that part of the atmosphere that we breathe, is far from friendly!

Unlike the rather controversial evidence for a significant rise in ground-level UV-B radiation, it is very clear that 'ground level' atmospheric ozone has risen appreciably in certain parts of the world, especially in areas of significant industrial development. Measurements of atmospheric ozone levels in the last century, coupled with a far greater number over recent years, suggest that in Europe at least, over a wide range of low-elevation rural sites, the mean ground-level ozone has roughly doubled over the last century. The cause of this increase is not securely established, but the most likely cause is the rising levels of oxides of nitrogen and of non-methane hydrocarbons, both associated with the extraction and combustion of fossil fuels. Both of those groups of anthropogenic atmospheric pollutants are known to act as catalysts in the photochemical production of tropospheric ozone from oxygen.

The damage caused to crop plants by even low levels of ozone has been amply demonstrated by numerous experimental observations, especially in North America. Table 8.2 shows the estimated losses on various crops resulting from exposure to ozone levels raised from an average ambient

Table 8.2 Estimated percentage of crop yield reductions in the United States resulting from projected levels of ozone increased from a base of 25 to 40 and 50 parts per billion. From Ashmore & Bell (1991).

Species	7-h mean ozone concentration (ppb)	
	40	50
Bean	11.0	18.1
Wheat	9.4	16.4
Peanut	6.4	12.3
Cotton	5.9	10.0
Soybean	5.6	10.4
Corn	1.2	2.6
Sorghum	0.8	1.5
Barley	0.1	0.2

ground-level figure of 25 ppb to 40 and 50 ppb of ozone over a 7-hour daylight period. (Ashmore & Bell 1991). Such direct observations on the effect of ozone alone may not adequately reflect the problems that may ensue from a combination of this with other processes of global change. Limited experimental evidence suggests that a combination of rising ground-level ozone combined with higher CO_2 levels may compound the effect of the ozone alone. Further experimental work on the effect of combinations of elevated temperature, CO_2 and ozone levels are clearly needed. In the meantime it seems that the threat to both crops and natural ecosystems from rising tropospheric ozone may be quite as great as those from increased UVR resulting from thinning of the ozone in the stratosphere.

Further reading

Ashmore, M.R. & Bell, J.N.B. (1991) The role of ozone in global change. *Annals of Botany* **67** (Suppl. 1), 39–48.

NERC (1989) *Our Future World: Global Environmental Research*, 2nd edn. Natural Environment Research Council, Swindon, UK.

UNEP (1987) *The Ozone Layer*. UNEP/GEMS Environmental Library 2, 1–36.

Wellburn, A. (1994) *Air Pollution and Climate Change: the Biological Impact*, 2nd edn. Longman, New York.

Chapter 9

Materials in Motion

In Chapter 5 we looked at the carbon cycle and considered in some detail the ways in which atmospheric carbon dioxide (CO_2) levels are changing as a result of human modifications of this cycle. We saw that CO_2 levels are rising and that this leads to an enhancement of heat-trapping by the atmosphere and will probably lead to a rise in global temperature. But the carbon cycle is not the only biogeochemical cycle affected by human activity and we need to consider other cycles and how they may interact with the developments in the carbon cycle.

As in the case of the carbon cycle, the information that we really need concerns the size of the various reservoirs of elements, the rate at which the element moves from one reservoir to another (flux), and the degree to which human activity is altering either the direction or the speed of this movement. Often, as with the carbon cycle, our species is accelerating a part of the cycle, such as the movement of carbon from the geological reserves (fossil fuels) back into the atmosphere. Nothing is being added to or subtracted from the cycle when viewed at a planetary scale, but the balance of the reservoirs is being shifted by encouraging and accelerating one particular part of the cycle (in the case of carbon by the burning of fossil fuels).

Some elements, like carbon, nitrogen, sulphur and oxygen, have a gaseous form that is present (and sometimes abundant, as with nitrogen and oxygen) in the atmosphere. Other elements, like calcium, phosphorus and lead, do not have a common gaseous form in the atmosphere and their movement in the air is dependent upon solid dust particles, or movement in solution in water droplets. All elements interact with the hydrological cycle (Chapter 2) and are dependent on water transport for part of their global cycles.

Figure 9.1 shows the composition of some of the major reservoirs of elements on earth (the lithosphere, hydrosphere and atmosphere) and provides some indication of the locations where the most common elements reside (Butcher *et al.* 1992). Figure 9.2 shows the composition of two examples of living organisms, a maize plant and a human being. A comparison of Fig. 9.1 with 9.2 gives some indication of what elements are needed in quantity by living things in relation to the sources of supply.

Plants take up carbon from the atmosphere and most other elements from the soil in solution, so the lithosphere is the most important reservoir for them (except for aquatic plants, of course, which rely on the hydrosphere). It can be seen that the important elements required by plants are found in reasonable abundance in the non-living environment (although they may not be easily available to an organism, as in the case of gaseous nitrogen, which is relatively inert). The one element that is in quite short supply, yet is needed for membrane structure, energy transfer mechanisms and DNA in the cell, is phosphorus. This is often a limiting element in both natural and agricultural ecosystems.

Most animals derive their required elements from the food they eat and the water they drink, so a herbivore will be mainly dependent on the elemental constitution of plants as its source of chemical materials. Happily, the chemical constitution of animals is closely similar to that of plants, except that bony vertebrates may need more calcium. One element needed by animals and probably not needed at all by plants is sodium, for this is involved in nerve function in animals—a process lacking in plants. Plants therefore generally contain little sodium and may not supply a sufficient quantity for a herbivore's requirements and the animal may need to resort to licking rocks or soil to supplement its diet. In sodium-poor areas, pastoralists may need to provide a salt-lick for grazing animals. Carnivorous

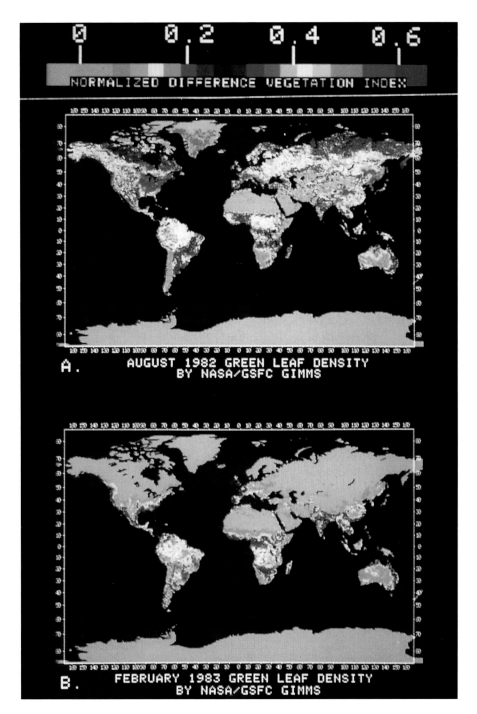

Plate 2.1 Earth satellite image showing the photosynthetically active radiation absorbed by terrestrial vegetation for August 1982 and February 1983. Brown–orange indicates zero absorbance, through to red–violet, high absorbance. Closed forest and intense agriculture generally give high absorbance; the lowest absorbance is either arid desert (as in North Africa) or snow cover and ice–cap (as in Greenland and Antarctica), both lacking any significant plant presence. (From *Nature*, 16 Jannary 1986, cover picture by National Aeronautics and Space Administration (NASA) Goddard Space Flight Centre.)

[*facing page* 136]

Plate 2.2 'Earthrise' over the lunar surface. The earth, with its oceans and cloud cover, contrasts with the lifeless cratered surface of the moon, associated with its almost total lack of any atmosphere. From a colour slide purchased through the BGS, ex–NASA.

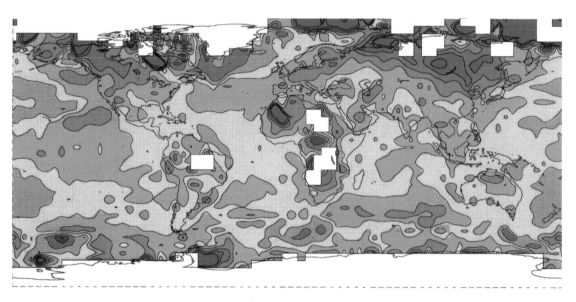

-1.0   -1.0 to -0.75 -0.75 to -0.5   -0.5 to -0.25   -0.25 to 0   0 to 0.25   0.25 to 0.5   0.5 to 0.75   0.75 to 1.0   > 1.0

Temperature anomaly in degrees C (wrt 1961–90)

Plate 5.1 Global map of annual surface temperature anomalies (deviations from global mean trends between 1961 and 1980) for the period 1986 to 1995. Note the tendency for increasing temperatures to be most strongly recorded in the high latitudes, together with some continental tropical and sub–tropical regions (data from Africa is incomplete and not fully reliable). Kindly provided by D. Callum, Meteorological Office, Bracknell, England.

Plate 13.1 Harvard Forest model of the pre–European forest in Central New England. (Reproduced with permission: The Fisher Museum, Harvard Forest, Harvard University.)

Plate 13.4 Harvard Forest historical model: farm abandonment, *c*. AD 1850. (Reproduced with permission: The Fisher Museum, Harvard forest, Harvard University.)

Plate 13.2 Harvard Forest historical model: an early settler clears a homestead, *c*. AD 1740. (Reproduced with permission: The Fisher Museum, Harvard Forest, Harvard University.)

Plate 13.5 Harvard Forest historical model: an 'old field' crop of white pine, *c*. AD 1910. (Reproduced with permission: The Fisher Museum, Harvard Forest, Harvard University.)

Plate 13.3 Harvard Forest historical model: the height of cultivation of farm crops, *c*. AD 1830 (Reproduced with permission: The Fisher Museum, Harvard Forest, Harvard University.)

Plate 13.6 Harvard Forest historical model: hardwoods following on from 'old field' white pine, *c*. AD 1915. (Reproduced with permission: The Fisher Museum, Harvard Forest, Harvard University.)

Plate 13.7 Harvard Forest historical model: the hardwood stand reaches cordwood size, *c.* AD 1930. (Reproduced with permission: The Fisher Museum, Harvard Forest, Harvard University.)

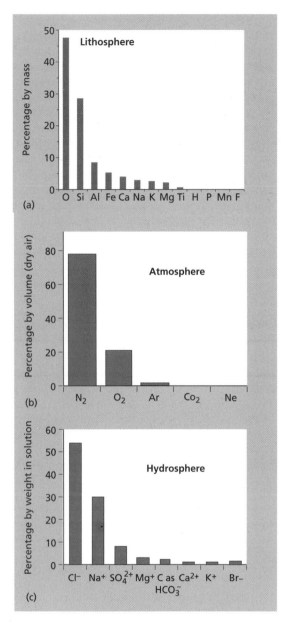

(a)

(b)

(c)

Fig. 9.1 The relative abundance (by mass) of different elements in (a) the lithosphere, (b) the atmosphere and (c) the hydrosphere. Values less that 1% are not shown, but their positions on the axis represents their order of decreasing abundance.

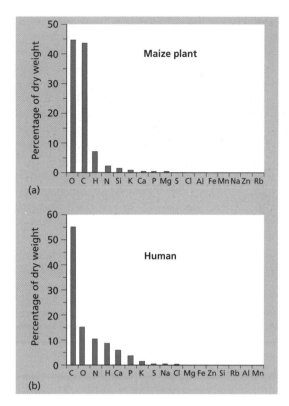

(a)

(b)

Fig. 9.2 The relative abundance (by mass) of different elements in a green plant (maize) and a vertebrate animal (human being). Note the differences in the relative importance of nitrogen, phosphorus and sodium.

animals, of course, feed upon a resource that is very like their own chemical constitution; their only problem may arise in the lack of certain molecules, such as some vitamins, that cannot be manufactured by the animal's own metabolism. Many of these are available from plants and even the strict carnivore takes in a certain amount of plant material, even if only from the gut contents of its herbivorous prey.

Human beings have nutritional demands for elements, like all other animals, and some of these must be supplied to our food organisms, both plants and animals, as fertilizers and feed supplements, but we also demand chemicals for other purposes, particularly our industries. This means that we place a high demand for certain elements upon our environment and we may damage the environment in extracting them. We are often wasteful in our application of fertilizers to our crops and may thus accelerate the flow of some nutrients into the soil and subsequently to streams and lakes. We may release some elements into the environment accidentally during industrial processes, in the disposal of waste, or during transport of raw materials, so creating artificial local concentrations of these materials that may prove toxic to some organisms.

In order to maintain a healthy balance in the earth's ecosystems, and in our own species, it is important to establish whether the global cycles of elements are being maintained and how severely these cycles are being influenced by human activity. To understand the potential impact of humankind on global biogeochemical cycles we shall take some of the more important elements and look in detail at the points where human influences are most strongly felt.

The nitrogen cycle

This is an extremely complex and important cycle for life on earth. Nitrogen is a vital element for all living organisms since it is present in all amino acids, which are the building blocks of proteins. It is also an extremely abundant element in its gaseous form (Fig. 9.1) but is relatively scarce in the lithosphere. It has a low solubility in water, but nitrogen does enter the oceans through microbial fixation (see Fig. 9.3 below). The construction of living things, therefore, is dependent on the tapping of the gaseous resource. Of the non-crustal nitrogen reservoirs, the atmosphere accounts for 99.96% of all nitrogen (Jaffe 1992).

The remainder is largely associated with organisms, either in their living or in their decomposing bodies. In fact, dead organic matter is a much bigger reservoir than living, for almost 90% of the nitrogen in a terrestrial ecosystem may be in the organic matter of the soil and only about 4% in living organisms (the remainder being in an inorganic form in the soil). Within the biomass, about 94% of the nitrogen is in plant material, 4% in microbes and 2% in animals, despite the fact that animals and microbes tend to be proportionately richer in protein (and therefore nitrogen) than plants. But the precise allocation of nitrogen to different reservoirs varies considerably from one ecosystem to another. The main global reservoirs and paths of movement of nitrogen are shown in Fig. 9.3.

Flux values for nitrogen are small in comparison with the size of the reservoirs, so it is difficult to represent these on the same scale as that used in Fig. 9.3. The fluxes are shown in Fig. 9.4, where the units are in Tg yr^{-1} of nitrogen (Tg = g $\times 10^{12}$), while the values in Fig. 9.3 are in g $\times 10^{18}$, one million times bigger). What should be immediately apparent, therefore, is that the fluxes are minute in comparison with the reservoirs. Unlike the carbon cycle, we are not facing problems of certain reservoirs expanding or contracting rapidly; there is no danger of the atmospheric reserves of nitrogen gas becoming depleted or swollen as a result of natural biological or of human activity.

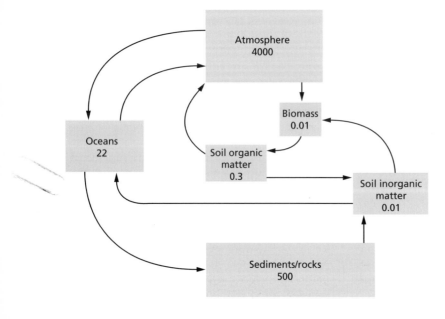

Fig. 9.3 The nitrogen cycle showing the main routes of nitrogen movement and the quantities of nitrogen as Gt (g $\times$ 10^{18}, tonnes $\times$ 10^{12}) of N in the major global reservoirs. Note the dominance of the atmospheric reservoir. The data have been collated from many sources.

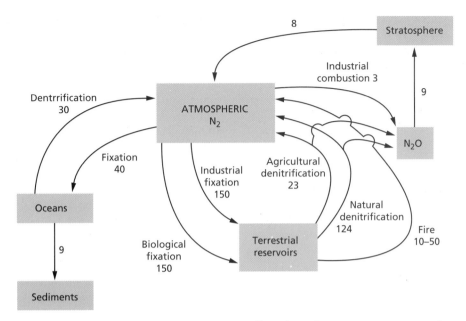

Fig. 9.4 The nitrogen cycle showing the flux rates between reservoirs expressed as Tg ($g \times 10^{12}$, tonnes × 10^6). Note the importance of biological fixation in the flux between atmosphere and terrestrial biosphere. The data are mainly derived from Jaffe (1992).

Nitrogen, like carbon, may be diminishing in soils on a global scale because of changes in land use, but this is currently difficult to quantify.

If we compare the nitrogen fluxes resulting from human activities with those associated with natural biological processes, the role of our species in this cycle becomes more prominent. Many organisms are able to 'fix' atmospheric nitrogen, that is to say convert it from its relatively inert gaseous form, N_2, into a reduced form, NH_4^+, in which it can be incorporated into organic molecules, such as amino acids. Microbes, such as certain bacteria and the blue-green cyanobacteria, are the most important fixer organisms, some working independently in soils, others living within the tissues of host plants, such as beans, alder trees and cycads, or in association with fungi in some lichens. The biological fixation of nitrogen involves the use of enzymes, protein molecules that form complexes with the N_2 molecule and disrupt its stable bonds. The same reaction can be achieved by industrial chemical means (the Haber process), but involves the use of very high temperatures and pressures and is therefore very expensive in terms of the energy needed to

effect the reduction to ammonia. The use of this method for the production of artificial fertilizers is therefore highly energy-demanding. From Fig. 9.4 we can see that industrial fixation of nitrogen is very significant in comparison with global biological fixation, accounting for about 20% of the total terrestrial fixation.

Thus, although the nitrogen flux from atmosphere to biosphere is small, a significant addition to this movement has therefore resulted from human industrial activity. But the reverse process, denitrification, has also been affected. Denitrification generally occurs when oxygen levels in the soil are low and certain bacteria use nitrate ions (charged groups of atoms having the formula NO_3^-) as a means of respiration instead of oxygen. Effectively they use the nitrate ion as a raw material in respiration and release water and nitrogen gas. Natural denitrification is estimated to be a little smaller than biological fixation, but broadly of the same order.

Organic matter in the soil contains proteins and other nitrogen-containing materials and these are decomposed by bacteria and fungi. They initially split the large protein molecules into smaller sections and then into amino acids. Plant physiologists are unsure whether some species of plant may be able to take up organic nitrogen directly, possibly complexed with polyphenols like tannins (Northup *et al.* 1995). The fungal associations with plant roots (mycorrhiza) may assist in this process. But most plants seem to need their nitrogen supply from the soil to be in an inorganic form.

The amino acids themselves are broken down and the nitrogen is released into the soil as ammonium ions, NH_4^+. These are positively charged groups of atoms and they are held in the soil quite efficiently by the negatively charged particles of clay and humus, so they are not immediately washed away. But many plants cannot take up nitrogen in this form, they need the oxidized form of nitrogen, nitrate ions, NO_3^-. But wherever there is a source of energy in nature one can usually find an organism of some sort exploiting it and this is the case with the ammonium ions. A number of bacteria are able to use NH_4^+ ions as a source of energy and they achieve this by oxidizing them to nitrate ions. Obviously, they can only effect this transformation if the soil is well aerated, however, so this process (termed 'nitrification') is impeded if the soil becomes compacted or waterlogged.

But nitrate ions, NO_3^-, are negatively charged and they do not become attached to the clay and humus fraction of the soil nearly as well as the positively-charged ammonium ions, and this means that they are soon washed out of the soil by a process termed 'leaching'. As a consequence, the form in which nitrogen is most accessible to growing plants has a relatively short life within the soil; if it is not rapidly absorbed by the plant roots it is soon lost by leaching. This presents two problems to the agriculturalist. First, the farmer needs to supply nitrogen to the crop either at the precise time when the demand is greatest (i.e. when the crop is growing fastest), or in a form which will release nitrate slowly (for example, as ammonium ions or even as proteins in organic fertilizers). The second problem is that nitrates may not only be lost to the fields by leaching (which is economically unfortunate) but may also pollute streams, rivers and lakes by creating high nutrient conditions leading to excessive and rapid growth of algae. Problems then arise when the algae die, decompose and take up the oxygen out of the water in the process, often leading to the death of fish and other oxygen-demanding aquatic organisms. This excessive enrichment of an ecosystem is termed 'eutrophication'.

Nitrates in water can lead to problems of environmental health in human populations, for the nitrate ion may be reduced to the toxic nitrite ion (NO_2^-) in the gut. This can be particularly serious for young babies and in areas badly affected by high nitrate levels (the European Community has set a safety limit of 50 ppm nitrate in drinking water, but the validity of this particularly limit is disputed) it may become unwise for small children to drink tap water. Some researchers link nitrate levels to the formation of carcinogenic nitrosamines in the gut.

The nitrates also find their way eventually to the oceans and local eutrophication problems may be recorded here. The North Sea, for example, which lies between Britain and continental Europe, receives water discharge from several major river systems that pass through areas of high population density, including the Thames and the Rhine. Such areas are subject to eutrophication and other aquatic pollution problems. The same applies to the Baltic Sea, which is almost completely enclosed by the Scandinavian and other north European countries and receives effluent from all of them via about 250 rivers that discharge into it. An estimated inflow of 1.1 million tonnes of nitrogen arrives in this Sea each year. Figure 9.5 shows the very rapid rate at which eutrophication is taking place at various locations through the Baltic Sea (Rosenberg et al. 1990).

Neither the farmer nor the environmentalist desires eutrophication—it is both wasteful and harmful. But questions have arisen about exactly how the excess nitrate is derived. Is it really just surplus fertilizer applied over-generously by the farmer, or is its origin more complex? Some experimental studies at the University of Agricultural Sciences in Uppsala, Sweden in 1987 (see discussion by Long & Hall 1987) suggested that fertilization was not the most serious source of nitrates in streams. Changes in land use, particularly the conversion of pasture grassland to arable fields, could be a more serious source.

In Fig. 9.6 some of the Uppsala data have been simplified into flow diagrams representing the flux of nitrogen in four different agricultural ecosystems. Lucerne, or alfalfa (*Medicago sativa*), belongs to the Fabaceae (=Leguminosae), or pea family of plants and it has symbiotic nitrogen-fixing microbes associated with its roots. As a consequence its growth leads to a build-up of nitrogen in the soil, for the quantity of nitrogen the crop contains (25 g N m^{-2} yr^{-1}) is greater than the

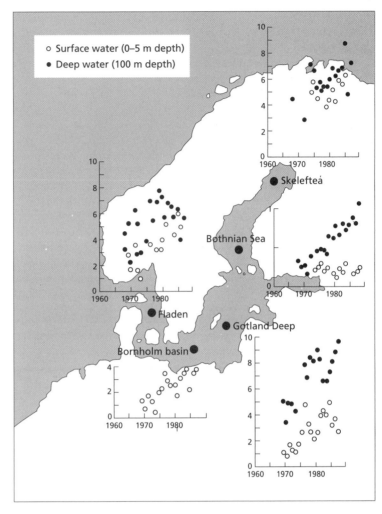

Fig. 9.5 A series of graphs showing the rate of rise in nitrate concentration in the waters of the Baltic Sea. The steep increase in the fertility of these waters (eutrophication) is found in all parts of the Sea, especially in the surface waters. From Rosenberg *et al.* (1990).

amount harvested when the fodder crop is taken in. The additional nitrogen is stored mainly in the soil, in both living and non-living forms. Relatively little nitrogen is lost either by leaching or denitrification.

A non-nitrogen-fixing grass, the meadow fecue (*Festuca pratensis*), yields a similar harvest of nitrogen to that of lucerne, but requires almost as much application of nitrogen in fertilizers (20 g) as the ultimate yield (24 g). Leaching and denitrification, however, remain low.

A crop of barley that has been fertilized yields only half the nitrogen of the pasture, and all of this has been supplied in the form of chemical fertilizers. Leaching losses in this case are considerably greater (1.8 g), as might be predicted. In a similar barley crop which has not received a fertilizer input, however, the yield is dramatically lower, but the leaching losses are still high (1.0 g), ten times the loss associated with the pasture grass. The conclusion is that fertilizing crops may enhance the losses of nitrogen by leaching, but the conversion of a grass field to arable is much more serious and can lead to nitrate losses increasing by an order of magnitude. The flush of nitrates into streams or aquifers during this process results mainly from the sudden breakdown of organic matter in the soil.

Scientists at the Agricultural Research Station at Rothamsted in England have confirmed these conclusions in their studies of nitrate discharges following the ploughing of grassland (Addiscott 1988; Addiscott and Powlson 1989). Agriculturalists had made an untested assumption that about 40% of the nitrogenous fertilizer applied to arable land is lost as nitrate leakage in drainage water.

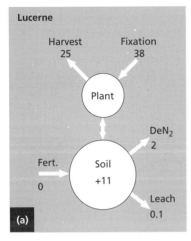

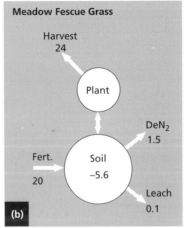

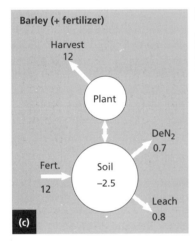

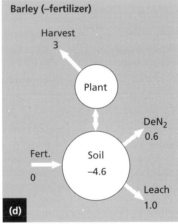

Fig. 9.6 The flux of nitrogen in four different agricultural ecosystems, expressed as g N m^{-2} yr^{-1}. The ecosystems are: (a) lucerne, or alfalfa (*Medicago sativa*), a leguminous herb with symbiotic nitrogen-fixing bacteria in its root nodules; (b) managed (fertilized) grassland dominated by the meadow fescue grass (*Festuca pratensis*); (c) a fertilized crop of barley; (d) an unfertilized crop of barley. Data from Long & Hall (1987). (De N$_2$ = denitrification)

But experimentation and modelling studies at Rothamsted showed that when increasing amounts of fertilizer are added to a soil, the concentration of nitrates in drainage waters builds steadily to a maximum and then levels off (Fig. 9.7). So the flushing of nitrates through soils is a complex issue and land-use changes may be more important than fertilizer applications. In natural ecosystems this has been observed in the dramatic flushes of eutrophication that have resulted from experimental clearance of forests from watersheds in the Appalachians (Likens & Bormann 1995).

A number of implications for agriculture emerge from these studies. Organic manures, rather than chemical fertilizers, have several advantages in that they release nitrogen more slowly and meanwhile the organic matter may improve the structure, the water-holding capacity of the soil and its ability to retain positively charged ions, including ammonium. On the other hand, nitrates will eventually be formed and will be subject to precisely the same leaching process as that operating in the case of inorganic chemical additions. So organic applications do not solve the problem of nitrate losses but may delay them. All nitrate sources should be added preferably at the time of maximum crop growth, when the plant demand for nitrate will be at its greatest. The worst time to add fertilizer is in the fall when no crop is growing and water passing through the soil will take away much of the released nitrate. Most important of all, the conversion of pasture to arable should be avoided wherever possible.

One must ask the question whether fertilizers should be used at all? An inspection of the data in Fig. 9.6c and d quickly shows that yields are great-

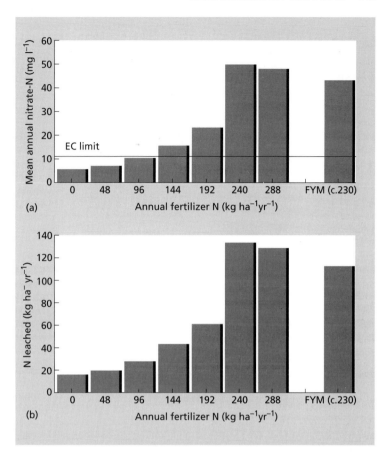

(a)

(b)

Fig. 9.7 Results of an experiment conducted at Rothamsted research station, England, in which increasing quantities of nitrate fertilizer were applied to a wheat crop and the amounts leached were recorded. (a) Concentration of nitrate in the leachate, and (b) total quantity of nitrogen leached per year. The 'EC limit' refers to the European legal limit for N-concentration in agricultural drainage water. FYM = farmyard manure applied during autumn. Nitrate leaching increases regularly with increased applications, levelling off at very high applications. Note that leaching from organic manure is also substantial. Data kindly provided by K.W.T. Goulding. See also Rowlson and Goulding (1995).

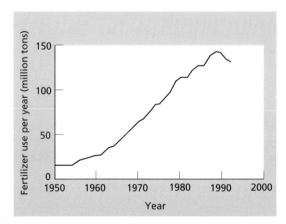

Fig. 9.8 The use of fertilizer over the past 40 years through the world in millions of tons.

ly improved by fertilizer additions, so global food supply would certainly drop if no fertilizers were applied to any crops. On the other hand, a more cautious attitude to fertilizer use is clearly beneficial from both an economic and an environmental point of view, and it is interesting to note that world fertilizer use is actually beginning to drop (Brown 1993), having peaked in the late 1980s (Fig. 9.8). Whether this drop will be sustained may be questioned, especially as the use of fertilizers in China is still expanding rapidly. Fertilizer use in the United States has fallen back since 1980, but is now fairly steady.

The main global growth of fertilizer use was in the decades following 1950, when a number of factors demanded their application. Increased population growth required more food; a lack of additional land meant that productivity had to increase rather than more land being brought into cultivation; new crop strains responded well to increased fertilizer supply; and increasing urbanization led to a breakdown of the traditional use

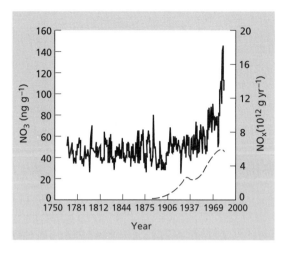

Fig. 9.9 The concentration of nitrates in the Greenland ice core within materials accumulated over the past 230 years (solid line). The dotted line shows the output of gaseous oxides of nitrogen as a result of human activities in the United States over the past century. From Mayewski *et al.* (1990).

of human sewage for agricultural application. Economically, the subsidies governments provided for fertilizers also led to excessive use. It may be that the current drop in use is connected with economic factors, namely the rise in oil prices since 1990, leading to fertilizer production by industrial processes becoming more expensive.

There are other aspects of the nitrogen cycle where human impact has become important on a global scale. Both fertilizers and the dung of domesticated animals release ammonia gas into the atmosphere. This gas is not easily oxidized, but it is very soluble and is soon brought back to earth by rainfall (Ayres *et al.* 1994) and can cause eutrophication in a range of habitats normally low in nitrogen content, such as acid bogs and heathlands. In some areas, such as north Germany and The Netherlands, this is regarded as a serious threat to the conservation and the continued persistence of the heathland habitat. In The Netherlands, for example, the aerial deposition of nitrogen can exceed 60 kg N ha^{-1} yr^{-1}. Nitrates may also find their way directly into the atmosphere and these add to the aerial eutrophication problem. The fact that this has been a fairly recent development in history can be seen from an analysis of nitrates in the successive layers of ice in the Greenland ice-cap (Fig. 9.9). Only since about 1960 has there been a sustained and marked

rise in the fallout of nitrate, corresponding with the increased global use of fertilizers (Fig. 9.8).

The burning of biomass globally, over 80% of which takes place in the tropics in such activities as savanna burning, acts as an artificial denitrification system. Some oxides of nitrogen are produced (see Fig. 9.9) but much of the element enters the atmosphere as gaseous N_2. One estimate of the rate at which this occurs is 10–50 Tg yr^{-1}, which would mean that fire losses of nitrogen to the atmosphere from the biosphere is equivalent to between 7 and 30% of the nitrogen gained by global biological fixation (Lobert *et al.* 1990; Kuhlbusch *et al.* 1991).

Oxides of nitrogen are produced in a variety of human activities ranging from industrial processes, motor car engines, and managed fires (as in the savannas of Africa and Asia). Managed fires of the latter type have recently been estimated to account for about 7% of the global nitrous oxide emission (Cofer *et al.* 1991). These gases are important for a number of reasons. Nitrous oxide (N_2O) has strong infra-red absorbing properties and is therefore a greenhouse gas of some importance (see Chapter 2). It is also capable of combining with ozone in the stratosphere, so contributes to the depletion of the ozone layer (see Chapter 8). Nitrogen dioxide (NO_2) reacts with other compounds in the atmosphere to produce nitric acid which leads to the formation of acid rain and also contributes to photochemical smogs. It is chemically corrosive and acts as an irritant to lung tissue. Virtually all of the NO_2 gas in the atmosphere has been formed as a consequence of human activity (Jaffe 1992).

The conclusion is that, although the global reservoirs of nitrogen are very large and the fluxes relatively small, the impact of humankind on these fluxes is considerable and of sufficient consequence to cause acceleration of natural pathways in some areas (e.g. eutrophication) and the generation of new and potent nitrogen compounds (e.g. NO_2) in other parts of the nitrogen cycle.

We must also, however, face the question of whether rising global temperatures would significantly influence the nitrogen cycle. On the basis of geological records from the ocean sediments, it is apparent that denitrification in the oceans diminished during glacial episodes leading to higher nitrate levels. A consequence of this glacial eutrophication of the oceans would have been increased productivity resulting in the observed drawdown of carbon from the atmosphere (Ganeshram *et al.* 1995). But this remains speculative. If we extrapolate in the other direction, warmer seas could result in faster denitrification, limiting the nitrate supply to phytoplankton and resulting in a diminution of the oceanic carbon sink. This in turn would lead to a further build-up of CO_2 in the atmosphere and enhance the greenhouse effect yet further.

The sulphur cycle

Sulphur ranks about sixteenth in abundance of the elements that comprise the earth's crust—too low to be shown in Fig. 9.1. The crust is, nevertheless, the major reservoir for the element, together with the oceans (in the form of sulphate; SO_4^{2-}

ions, and the ocean sediments; Fig. 9.10). It is an essential element for life, ranking about ninth in abundance in living plants and animals, where it forms a part of the structure of certain amino acids and therefore is found in many proteins.

Because of its presence in proteins, organic fossil materials, including oils and coals, usually contain some sulphur (between 1 and 3% in the case of coal) and there is also a tendency for iron sulphides to accumulate during the process of fossilization. The burning of these materials for energy production releases the sulphur in an oxidized form, the gas sulphur dioxide, SO_2. The reactions of this gas with other compounds in the atmosphere are extremely complex (see Charlson *et al.* 1992) but one of the important products of its solution and further oxidation is sulphuric acid (H_2SO_4), together with suphurous acid (H_2SO_3), both of which (in combination with

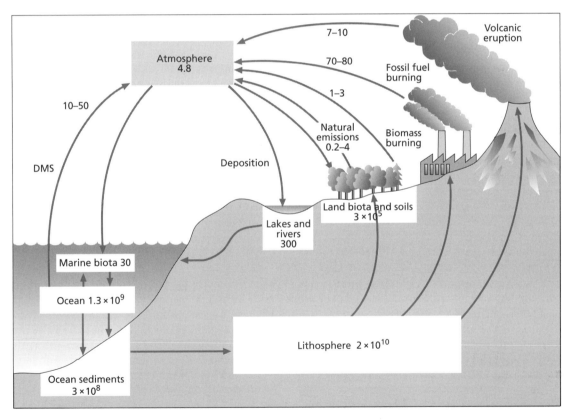

Fig. 9.10 The global sulphur cycle showing the major reservoirs and the flux rates between these. Figures are in Tg (g × 10^{12}, or tonnes × 10^6) S for reservoirs and Tg yr^{-1} for fluxes. Note the importance of human activity on the transference of sulphur from the lithosphere to the atmosphere in comparison with the natural flux from volcanoes. Note also the important emission of sulphur compounds (particularly dimethyl sulphide, DMS) from the marine phytoplankton. Data from Charlson *et al.* (1992).

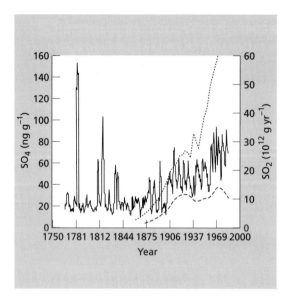

Fig. 9.11 The concentration of sulphates in the Greenland ice core within materials accumulated over the past 230 years (solid line). The dotted line shows the global output of man-made sulphur dioxide into the atmosphere and the dashed line shows SO_2 output from the United States of America. From Mayewski *et al.* (1990).

nitric acid) are major contributors to the problems of acid rain.

The history of sulphate emissions can be traced in the ice layers of the ice caps, as in the case of nitrates (Fig. 9.11) and it is clear that there has been a considerable increase during the last century. The historical record in the case of sulphates, however, is complicated by periodic volcanic eruptions (as in 1783, 1815 and 1912) that have left their sulphurous mark in the Greenland ice (Mayewski *et al.* 1990). Even so, it can be estimated that the output of sulphates globally has increased by a factor of about three times over the past 100 years (Langner *et al.* 1992).

Some of the sulphates produced during fuel burning fall to earth in a dry, particulate form

(Mohnen 1988), while some dissolve to form the acidic droplets that eventually fall to earth as acidified rain. Under extreme conditions, this acidification can be most severe, as in the case of an event in the city of London in December 1952 when the pH (see Fig. 9.12) of water droplets in a sulphurous smog fell to an estimated value of 1.6, rather more acid than lemon juice and, as a consequence, about 4000 people died of respiratory diseases (Pearce 1987). This stimulated the development of legal controls on coal burning within the London area.

The direct impact of acid rain can damage not only human lungs, but also buildings and plants. The leaf of a plant is covered by a protective waxy layer that has several functions including the shedding of excess water, the control of evaporation of water from the delicate superficial cells and the protection of those cells from microbial invasion and from the effects of aerial toxins. Acid droplets have been shown to strip off the wax layer leaving the exposed cells liable to infection and open to pollutants. The effect upon a forest can be most destructive, but it is often difficult to pinpoint the precise nature of the harmful effect of acid rain since the plants are not only physically damaged but also have to cope with acidified soils in which some toxins, like aluminium, can have harmful effects. Forest decline (or *Waldsterben* as the Germans have termed it in Bavaria where effects were first documented) is really a syndrome of various harmful processes instigated by acid deposition (Schulze 1989). Some German work has suggested that acid mist interacts with ozone to produce the foliar damage to trees (Blank 1985).

Freshwater aquatic ecosystems are also influenced strongly by the pH of the precipitation, especially in areas where soils lack compounds like calcium carbonate (lime) and magnesium carbonate which can neutralize the effects of the acids as the water percolates into streams. Canada

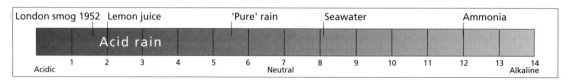

Fig. 9.12 The pH scale of alkalinity. Low pH represents acid conditions and high pH is alkaline, with pH 7.0 being neutral. It is a logarithmic scale, so one point drop in pH means ten times more acid, two pH points drop means 100 times more acid, and so on.

and Scandinavia have suffered particularly badly from lake acidification because of the acid nature of their rocks and soils, and the proximity of industrial areas to the south from which they receive acidified air masses. Figure 9.13 shows the pattern of distribution of sensitive, acidified lakes in the north-east of North America (Schindler 1988). The problem is most serious in these northern areas during spring when snowmelt carries a sudden flush of acid waters into lakes and can cause acute changes in pH, often accompanied by aluminium toxicity and resulting in fish death.

The fact that lakes were becoming more acid has been noticed for many years, but some controversy has existed concerning the relative importance of acid deposition and land-use changes in recent times. An approach to the question of when lakes began to acidify was pioneered by Roger Flower and Rick Battarbee of London University, who examined the remains of diatoms in the sediments of some lakes in Scotland. Diatoms are unicellular algae that have a hard outer coat made of silicav which survives well in lake sediments after the organism has died and is highly sculptured and recognizable even to species level. In addition, different species of diatom are known to have very specific requirements in terms of the acidity of the waters in which they are able to live. So the stratified diatom fossils in lake sediments provide a record of changing pH in the past.

Figure 9.14 shows the record of diatom stratigraphy from a Scottish loch and it can be seen that there have been considerable changes in the abundance of certain diatom species in the upper sediment layers, particularly in the past 130 years (equivalent to about 20 cm of sediment). Species that demand relatively neutral pH have declined and those that tolerate lower pH have increased in abundance. This has allowed the authors (Flower & Battarbee 1983) to reconstruct a proposed pH curve which shows a clear decline within the last century.

Since 1970 SO_2 emissions in the United Kingdom have fallen by 40% as a result of legislative controls, and the diatom profile of very recent sediments in the Scottish lochs have shown a reversal of their former trends (Battarbee et al. 1988). This not only demonstrates that the original acidification was indeed a consequence of acidic emission, but also shows that the acidification is reversible and it is possible to recover the lost flora and fauna of these aquatic sites. A similar study of acid-stressed lakes in the vicinity of the metal smelting works of Sudbury, Ontario, has also shown that reduced sulphur emissions is

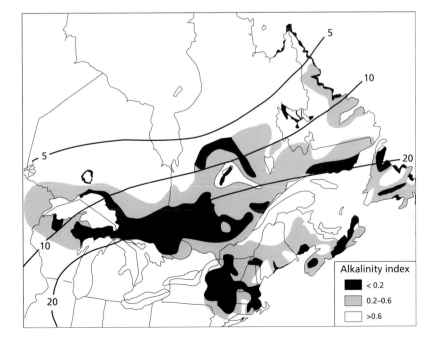

Fig. 9.13 Map of north-eastern North America showing an index of lake acidification (expressed as a ratio of alkalinity to calcium and magnesium ions present). A low index (below 0.2) indicates an area of acidified and sensitive lakes. Shaded areas are somewhat less sensitive and open areas are less sensitive still as a result of their greater alkalinity. The heavy contours represent the levels of sulphate deposition in kg ha^{-1} yr^{-1}. From Schindler (1988).

Alkalinity index

■ < 0.2
▨ 0.2–0.6
□ >0.6

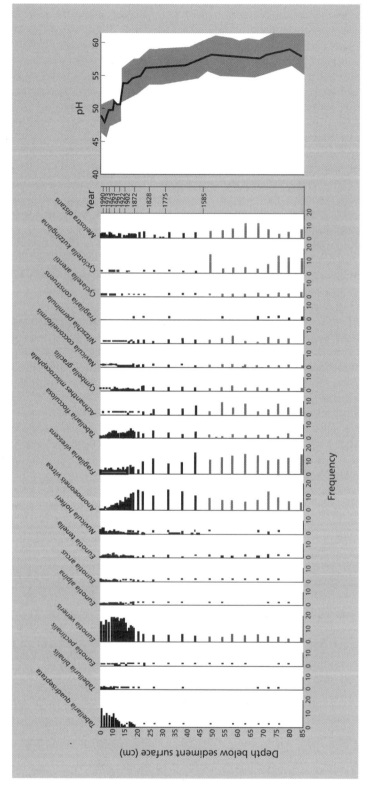

Fig. 9.14 Fossil diatoms preserved in the sediments of a Scottish loch, the Round Loch of Glenhead. Considerable changes in the diatom community are apparent in the upper layers of sediment (since about 1920) and these changes are associated with a recent drop in the pH of the sediments. This is likely to have been caused by acid rain. From Battarbee *et al.* (1988).

accompanied by lake recovery (Gunn & Keller 1990). Acidity has not fallen as fast as might be expected following the lower emissions of SO_2, however. It is likely that cleaner industrial practice has also removed fly ash from stack emission and, since this output is strongly basic in reaction, the neutralizing effect on acidification has also been lost.

Sulphur also finds its way into the atmosphere by routes that are entirely natural and relatively uninfluenced by humankind. Hydrogen sulphide gas is released both by volcanic eruptions and by the less catastrophic output of gases from water-logged swamps and marshes. Over the oceans there is also an output of sulphur into the atmosphere in the form of dimethyl sulphide, probably generated by planktonic organisms (Andreae & Raemdonck 1983). It is possible that some sea birds are able to detect these emissions and home in on them, for where the plankton accumulates the fish will gather (Nevitt *et al.* 1995). Most of the dimethyl sulphide becomes oxidized in the atmosphere to sulphate ions and then dissolves in rain water and is precipitated, adding to the acid rain effect.

Sulphates can even act as nuclei for the formation of rain drops, which means that their presence in the atmosphere generates cloud formation. This could have an effect on global precipitation and also on global temperature since clouds reflect more of the incident energy of sunlight and so keep the surface cooler, and they also radiate heat, cooling the atmosphere. This could be a situation where one pollutant produced by fossil fuel burning (SO_2) is counteracting the effect of another (CO_2). The involvement of plankton in sulphur cycling, however, also complicates the question of responses to climate change. If a change in climate influenced global marine productivity, would this in turn affect dimethyl sulphide production and hence produce a feedback mechanism? And would this be positive or negative? If warmer temperature means less oceanic nitrate and lower phytoplankton productivity, then less dimethyl sulphide would be produced. Current estimates of the flux of dimethyl sulphide from the ocean to the atmosphere (IPCC 1992) vary considerably (from 10 to 50 Tg yr^{-1}), but at the upper end of this range the value approaches the estimate of SO_2 production from fossil fuel burning (70–80 Tg yr^{-1}).

Sulphates have a very short residence time in the atmosphere (a matter of weeks), so they do not have a significant effect upon the sulphur content of the upper layers of gases above the earth in the stratosphere.

The phosphorus cycle

All of the cycles considered so far have an important gaseous form (CO_2, oxides of nitrogen, SO_2 and dimethyl sulphide), but there is no significant gaseous form of phosphorus, so the atmosphere does not represent a major reservoir for this element, nor an important route by which it can enter the global cycle. Its main reservoir is in the sediments and rocks of the earth (Fig. 9.15) and from these it becomes available to living organisms with assistance from the hydrological cycle (see Chapter 2).

The data presented in Figs 9.1 and 9.2, representing the abundance of different elements in various parts of the living and non-living world, show that phosphorus is an important element to both plants and animals, yet is relatively scarce when compared with the other major requisites of life. It lies eighth in the list of the elements needed by plants and sixth in those needed by animals, yet its abundance even in its richest source, the lithosphere, is only about 0.1%. It is not surprising, therefore, that plant and animal productivity in many habitats (including agricultural crops, lakes and marine ecosystems) is often limited by the availability of phosphorus.

Phosphorus in required by living cells for the formation of phospholipids, which are an essential component of cell membranes that control the movement of all chemicals into and out of the cell. It also plays a vital part in the energy transfer reactions of all cells in the molecule adenosine triphosphate (ATP), and it forms a central part of the structure of deoxyribonucleic acid (DNA), the chemical basis of the genetic code. In addition, it is a major component of bone structure in vertebrate animals. The element therefore plays a key role in all living organisms.

The oceans contain an important reservoir of phosphorus (Jahnke 1992) (Fig. 9.15) and the biological productivity of the surface layers is often limited by the rate at which new supplies of phosphorus arrive by upwelling of water masses from the deep. Areas of the ocean where such

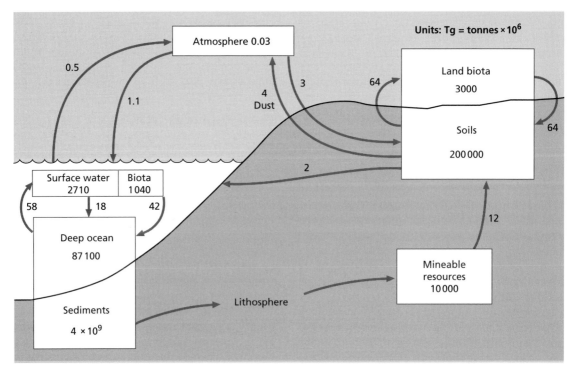

Fig. 9.15 The phosphorus cycle showing the major reservoirs and the rates of flux between them. Units are in Tg (reservoirs) and Tg yr^{-1} (fluxes). Note the general scarcity of this element when compared with the others considered in this chapter and the lack of an important atmospheric reservoir. From Jahnke (1992).

upwellings occur are highly productive and are associated with important fishing industries. Fish-eating birds that exploit these areas act as a route for the transfer of phosphorus from ocean to land, for they deposit their fish meals as faecal output in their roosting and nesting sites. This 'guano', especially from South America, has been an important source of fertilizers in the past. Since climate change is closely connected with global oceanic circulation patterns, any shift in the climate could change the pattern of upwelling and thus modify the oceanic phosphorus cycle (see Chapter 6).

In terrestrial systems the weathering of rocks, particularly those containing apatite minerals with their phosphate components, releases phosphorus into the soils. The behaviour of phosphorus in soils is complex because it can exist in a number of forms depending on the acidity of its environment. In acid soils it may be bound with iron

and aluminium and become unavailable for plant uptake. But high pH can also create problems because the phosphates are then in the wrong form. Managing the supply of phosphorus to a crop is, therefore, a difficult process requiring additional inputs derived from phosphate mining in order to maximize productivity.

A typical cereal harvest involves the removal of between 15 and 35 kg P ha^{-1} (Smil 1990), and this needs to be returned either by natural rock weathering in the soil or in fertilizer additions. The former is generally too slow to cope with this rate of removal, so fertilizer application is usually necessary if productivity is to be maintained. The mining of apatite minerals only began in 1847 (in Suffolk, England), but has since become a major extractive industry, especially in the United States, Russia, Kazakhstan and Morocco. As in the case of nitrates, only a small proportion finds its way into the human body, which is the intended ultimate destination of such fertilizers. Nevertheless, about 10% (2.5 Tg) of the phosphorus found in all the living animals on earth is probably located within our species.

Wastage of phosphorus in leaching from soils and in human and domestic animal sewage is a serious problem, not only because of the limited

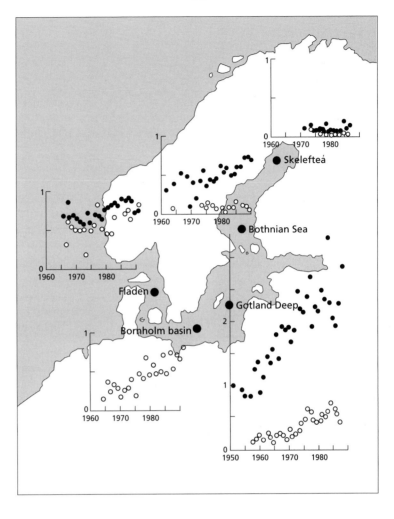

Fig. 9.16 A series of graphs showing the increasing concentration of phosphates in the Baltic Sea over the past 30 years. Highest levels of pollution are currently in the south of the Sea (especially in the surface waters). The northernmost part of the Sea is still relatively low in phosphates.

supplies of available terrestrial phosphorus, but also because of the eutrophication problems created by local excess of the element in aquatic ecosystems. The pollution aspect was made worse by the development of polyphosphate detergents for use in domestic and industrial washing processes. Figure 9.16 shows the build-up of phosphates in the Baltic Sea in northern Europe and, as in the case of nitrates (Fig. 9.5), there is evidently an accumulation of this element in both surface and deep waters. Phosphorus conservation demands close attention, for we cannot afford to be profligate and wasteful with such an important and relatively scarce element.

Trace elements

Many of the global environmental problems caused by our species have resulted from an acceleration of part of a natural cycle of an element, resulting in its rapid build-up in one particular reservoir. So, carbon and sulphur are moving rapidly from lithosphere to atmosphere, and nitrogen and phosphorus are being moved into the hydrosphere, creating eutrophication problems. These are not toxic materials created by man, but unusual concentrations that are resulting from human activities. A similar effect is found among a range of potentially toxic trace (uncommon) elements that have been used by humankind for a number of industrial and other purposes and have been allowed to accumulate in particular sites where they may now cause problems of toxicity.

The mining and smelting of metals has created local concentrations of potentially toxic elements since prehistoric times. Tin, iron and lead have

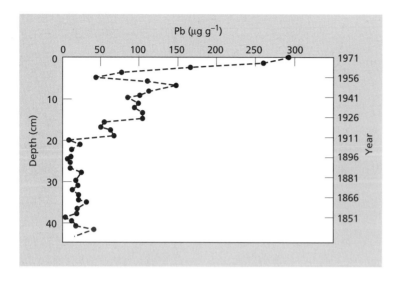

Fig. 9.17 Lead concentrations in the superficial sediments of Lake Washington, Seattle, USA. Note the increase in lead in the lake since the beginning of this century. From Alderton (1985).

been mined in Europe and Asia since Copper and Bronze Age times and the spoil from such operations provides a source of metals that can be leached into water supplies, rivers and lakes. The problem has long been with us, but has become increasingly acute during the industrial revolution and as intensive industrial activity has spread around the world. This intensification of pollution can be seen when lake sediments are analysed and a picture of the history of metal discharge from the catchment is displayed (Alderton 1985). Figure 9.17 shows such an analysis of sediments from Lake Washington, close to the industrial centre of Seattle. The background level of lead, up to about 40 μg g^{-1}, reflects the lead content of the soils of the area from which a constant natural leaching of this element into the lake is occurring. Then, around 1890, a considerable change takes place and the lead content of the sediments rises rapidly, corresponding to the development of industry in Seattle. Smelting currently releases about 8000 kg of lead each year into the atmosphere of this area, much of which is deposited in dust and in rainfall and some of which is then washed into the lake.

Lead levels in urban environments are often very high because of the use of this metal as a petroleum additive and also, in the past, in paints. The contamination of urban soils can reach very high levels, as in Westminster in London, England, where concentrations of 1000 μg g^{-1} have been recorded (Culbard *et al.* 1988). The

absorption of lead by adults in the United States has been estimated at between 4 and 11 μg per day and for children it ranges between 5 and 22 μg per day. The presence of lead-based paints in a child's environment can raise this absorption rate to 100 μg a day, which represents a serious health risk (Thomas & Spiro 1994). Children are particularly at risk of lead contamination because of their greater likelihood of ingesting dust and soil on toys.

Marine sediments around our coasts document the increasing release of metals associated with human activity, as in the Santa Barbara Basin in southern California (Fig. 9.18). Here mercury, released from a variety of industrial activities, begins to increase in concentration after about 1880 as human activity in southern California intensified. What is less easy to explain is why the mercury levels in the early 19th century sediments are higher than those of earlier, prehistoric times. Perhaps this reflects increased soil erosion associated with early agricultural settlement of the area.

Mercury is a toxic element that has made an impact on human health when released into the environment in high concentrations, so has caused particular concern to environmental scientists. It is also unusual among the metals in that it can be transported through the atmosphere in gaseous form and its contamination and dispersal through this medium is the most important route of mercury pollution (Benjamin & Honey-

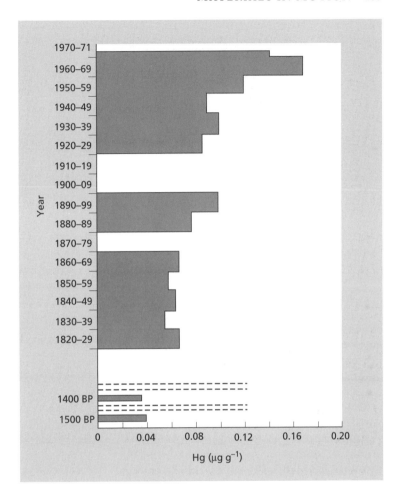

Fig. 9.18 Mercury concentrations in the marine sediments from the Santa Barbara Basin, southern California. Note the increase in mercury in sediments formed since about 1890. It is also apparent that the sediments of the 19th century contained more mercury than those of more ancient times. From Alderton (1985).

man 1992). Its use in paper manufacture as a fungicide has created serious problems in various parts of the world, particularly Japan and Canada. Mercury released into waters may find its way into biological food chains, especially through shell-fish, and this can lead to toxicity problems, blindness, and even death in human consumers. Fungicidal seed dressings of methyl mercury in the 1950s played an important part in the *Silent Spring* of Rachel Carson, when populations of seed-eating song-birds in North America and Europe became severely depleted. The consumption by people in Iraq of seed that had been treated in this way led to at least 500 deaths.

Mercury, lead and other trace metals are such a regular feature of our background environment that the human diet is bound to include some of them. The World Health Organization draws up recommendations regarding the limits of trace metal intake that the body can be expected to cope with and in the case of mercury this is $0.5\ \mu g\ g^{-1}$. The human body is capable of eliminating some of the toxins that it inevitably acquires, but this means that human sewage is relatively rich in these rejected toxins, including heavy metals. Any proposed use of sewage as a source of fertilizer, or even an organic base for growing bacteria as a potential food resource, must bear in mind this problem of metal toxicity. The same applies to the reclamation of land formerly used in sewage treatment, for the soils of such sites are usually contaminated with heavy metals.

Industrial development around the world continues apace, and the demand for metals, including trace metals, can be expected to grow accordingly. Figure 9.19 shows the growing world demand for these metals (Brown *et al.* 1990) and illustrates the increasing problem that will be

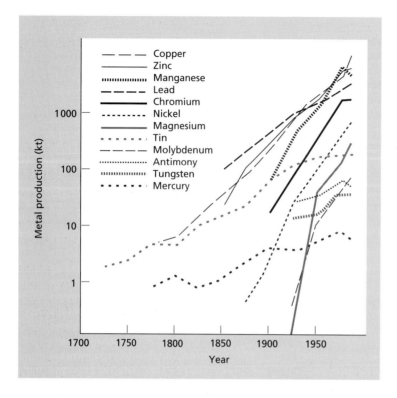

Fig. 9.19 Annual world-wide production of certain metals over the past 200 years. All have been rising sharply this century. From Brown *et al.* (1990).

posed by their leakage into the environment. Perhaps the most serious risk is the possibility of leakage of plutonium into the environment, for this artificial element, produced from uranium for use in nuclear power plants, is extremely toxic and carcinogenic. Many people feel that the risk of plutonium pollution is so serious that its use as an energy resource should be questioned (see Chapter 10). Whatever the element, however, the avoidance of losses and the recycling of waste materials containing these valuable toxins makes both economic and environmental sense. We should aim at closing our part of their biogeochemical cycles instead of leaving it open to losses.

There are occasions, however, when disposal can constitute less of an environmental threat than recycling. This debate became prominent in Europe in 1995 with the decision of Shell Oil Company to dispose of an obsolete oil storage platform from the North Sea oilfield by sinking it in deep water in the Atlantic Ocean. This proposal, quite reasonably, caused alarm among environmentalists who were concerned about the concept of using the deep ocean as a waste-

disposal site simply on the grounds of economic attractiveness.

The reasoning one should apply to the analysis of this kind of situation, however, must take into account the contribution that such an incident of pollution would make to the background level of the polluting materials at the site. Nisbet and Fowler (1995) have pointed out that the oil platform contained many metal wastes (probably amounting to hundreds of kilograms) of cadmium, mercury, zinc, lead and nickel, together with some radioactive material, the steel of the structure, 40 tonnes of oil and 100 tonnes of silt. But the thermal vent environment of the deep ocean trenches is extremely rich in many of these metals. Often the sediment already contains 20–40% iron. This is one of the few ecosystems on earth where there is a productive biological system operating at very high concentrations of metals. It is, therefore, one might argue, one of the few places on earth that is unlikely to be disturbed by the addition of relatively small quantities of these elements. Disposal on land, however, could lead to very much more serious pollution of ecosystems that contain very low levels of these elements, and that are highly sensitive to their toxicity. Disposal at sea, however, does create a precendent and could lead to regular dumping of successive obsolete platforms. We have often in the past been complacent about how much of our waste the natural environment can absorb without damage.

Land disposal was eventually decided upon as a result of pressure of opinion from environmental groups, despite the opposite recommendation made by many leading environmental scientists. The case does, however, mean that oil companies need to consider the expense and the problems of decommissioning such structures before their initial erection.

Pesticides

In many cases, the effect of human activity upon natural cycles is simply to accelerate them, or certain parts of them, and thus to create bottlenecks and concentrations of certain compounds that subsequently have an adverse effect upon our own or other useful species. But we have also created certain compounds with the precise aim of disrupting ecosystems, especially the destruction of those organisms that interfere with our particular plans for managing those ecosystems. Agricultural crops, forestry plantations, herds of domesticated livestock, urban settlements of human beings, are all ecosystems that we manage very heavily in order to achieve high productivity or, in the latter case, healthy and comfortable conditions. But the conditions we create are also ideal for the expansion of populations of those other organisms that can take advantage of them—the pest and disease organisms which infect either our own species or our domesticated animals and plants. We then need to add a further factor to our management practice, namely the control of these pest populations which often takes the form of introducing toxins or pesticides into the ecosystem.

The control of pests is as old as our species, and the need to persecute wider ranges of species became more acute with the domestication of other organisms. Not only did we have our own parasites and predators to contend with, but also those of our associated animals and plants. We inherited their pest problems, and often contributed to them by herding them or growing them in dense monocultures and sometimes taking them into biogeographical regions or to climates where they were not well provided with resistance. Pest control essentially entails the reduction, or even elimination, of the populations of pest organism and this has traditionally been achieved by a wide range of measures including hunting and poisoning. The toxins used in the past have often been naturally occurring materials that we have concentrated in order to achieve our aims of specific destruction. Metals such as copper and mercury have already been referred to as fungicides that can be used as foliar applications to plants or as dressings on seeds that reduce infection by surface fungi. Arsenic and strychnine have often been used to poison vertebrate pests, such as foxes and moles, or birds of prey and carrion feeders that have often been perceived as enemies of domestic flocks. The use of such indiscriminate poisons has led to widespread deaths of unintended victims, from seed-eating song-birds to household pets. And the accumulation of these elements in food chains leads inevitably to increasing levels being found in the human body. Organic compounds, both pesticides and waste products, especially those that are soluble in fats, can also accumulate in food chains, as shown in Fig. 9.20.

An obvious way forward out of this dilemma of self-poisoning is to seek compounds that are toxic specifically to target organisms, the pests, and are harmless to other species, including ourselves. Antibiotics form a good example of this type of approach, since these are compounds which are toxic to microbes at concentrations which are too low to be harmful to humans or other host animals. Naturally, the development of antibiotics has proved of major importance in the control of certain microbial pests, but their widespread use has also led to complications. Microbes have short generation times, and their fast rates of reproduction permit them to evolve quickly to new conditions, including the presence of potentially harmful compounds in the environment. Added to this, once they have developed resistance to a drug they can pass the genetic capacity through the population by a process of transformation, or exchange of genetic material between individual microbes. Once a compound becomes common in the environment, the development of resistance is almost inevitable. The widespread

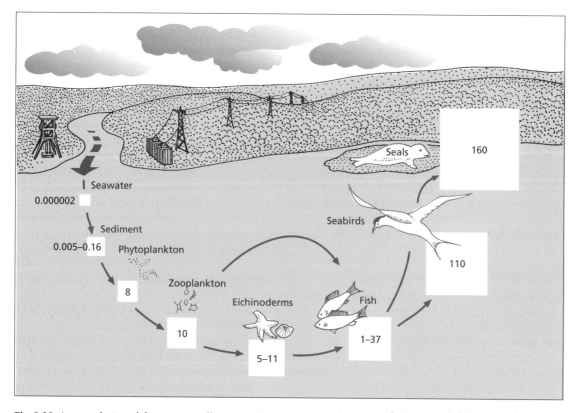

Fig. 9.20 Accumulation of the organic pollutants PCBs (polychlorinated biphenyls) in marine food chains on the German coast. These fat-soluble compounds are stored in animal bodies, hence pass from prey to predator and build up to higher concentrations in the longer-lived animals. PCB content is in mg l⁻¹ for the sea water and sediment and mg kg⁻¹ for the animals (Böhlmann 1996).

use of antibiotics has therefore become counter-productive since we have provided conditions in which we have encouraged the evolution and spread of drug-resistance. This is particularly a problem when antibiotics fed to domestic animals (to prevent disease under crowded conditions) has led to resistant microbes that may subsequently be transferred to human populations.

Often the attack against disease organisms has been focused on the vector, the animal that carries the disease, rather than the disease itself. This has led to the development of a very wide range of chemicals that are insecticidal, including many organic compounds, some of which have had, and continue to have, profound effects on the environ-

ment. An example is provided by the chlorinated hydrocarbons (organochlorine compounds), a group of organic materials that includes DDT. Although DDT was first synthesized in the laboratory in 1874, it was not until 1939 that its insecticidal properties were first appreciated and, since it appeared relatively harmless to humans, it was then used in abundance as a means of controlling insects and lice during the Second World War. Shirts impregnated with DDT provided protection against lice, fleas and ticks, and the application of DDT to swamps at concentrations of only 1 lb per acre was sufficient to control mosquitoes (carriers of malaria) and yet to leave the water drinkable without apparent harmful effects (Mellanby 1967). It is undoubtedly true that many lives (perhaps running into millions) were saved and great military advantages gained by the use of DDT in the control of such diseases as louse-borne typhus. Post-war agricultural expansion then led to its wide use as an insecticide in crop pest control.

DDT is fat soluble and accumulates in the body fats, including milk. Concerns began to emerge in the 1950s when residues were detected in cows' milk and later even in human milk, coupled with the discovery that some carnivorous birds (from fish-eating grebes to bird-feeding predators like the peregrine falcon) were suffering from ill-effects. These included reproductive problems, such as egg-shell thinning and disturbed reproductive behaviour. By the 1970s concern had risen to a point where many developed countries were imposing a ban on the use of DDT in the interests of human environmental health. But this has not solved the problem of its widespread distribution because the compound is still used (and undoubtedly saves lives) in some countries, coupled with the fact that it is extremely persistent in the environment, with a half-life of about 100 years. This means that the bulk of the DDT produced and distributed over the earth in the 1950s and 1960s is still with us and has not yet been degraded.

Not only are such compounds as DDT still with us, they continue to accumulate in certain regions of the world. The concentration of persistent organochlorine compounds in food chains is well known, but there is also a tendency for these materials to build up at certain latitudes of the earth. The process leading to this is known as 'global distillation'. A survey of organochlorine pesticides in tree barks from all over the world (Simonich & Hites 1995) has shown that many of the volatile forms of these compounds, that is those which are easily converted into a vapour, are most concentrated in latitudes between 40° and 60° north and south of the equator, despite the fact that in many of these temperate areas organochlorine pesticide use has long been banned. It is in tropical countries that they are currently used in any abundance. The explanation may well be that such volatile compounds as hexachlorobenzene evaporate in the tropics and are transferred by global atmospheric circulation to the temperate areas where they condense and are absorbed by, among other things, tree bark. Less volatile materials, such as endosulphan, do not show this latitudinal concentration effect and tend to remain in the region where they are being used.

The most important question in such a situation is whether the persistence and accumulation of such pesticides is potentially hazardous to human health or, indeed, to the well-being of other animals. Evidence is emerging that DDT, or the compound DDE that is formed from DDT in the body, may affect the reproductive organs of humans and, in particular, influence the viability of sperm in the male (Sharpe 1995). Deformities in male alligators in DDT-contaminated waters have led to questions being raised about the role of the compound in hormonal interactions in vertebrate males in general, and recent work (Kelche et al. 1995) points to the operation of DDE as an oestrogen (female hormone) imitator in males, possibly leading to reduced fertility and even more serious conditions such as testicular cancer. It is quite possible that the rising rate of testicular cancer observed in the world, together with abnormal penis development, failure of testes to descend, and low sperm counts, all represent the consequences of excessive use of DDT in the mid-20th century.

The lesson to emerge from such studies is that future generations will have to live with the outcome of our current policies with respect to environmental contamination. Erring on the cautious side at the present may prevent the kind of unpredicted outcome that resulted from the use of DDT. It is also unwise to think in a parochial manner about such issues because no country is immune to the effects of the policies of its neighbours, or even of distant regions. The need to regard the earth as a single ecosystem in such studies is becoming increasingly apparent.

Further reading

Butcher, S.S., Charlson, R.J., Orians, G.H. & Wolfe, G.V. (eds) (1992) *Global Biogeochemical Cycles*. Academic Press, London.

Schulze, E.-D. (1989) Air pollution and forest decline in a spruce (*Picea abies*) forest. *Science* **244**, 776–783.

Socolow, R., Andrews, C., Berkout, F. & Thomas, V. (eds) (1994) *Industrial Ecology and Global Change*. Cambridge University Press, Cambridge.

Turner, B.D. (ed.) (1990) *The Earth as Transformed by Human Action*. Cambridge University Press, Cambridge.

Chapter 10

Energy

All living organisms need energy. Without it no plant or animal can grow, or move or reproduce, so right from the very origin of living things the supply of energy proved a crucial key to the evolutionary advancement of life. The very first organisms probably used the chemical energy that surrounded them. Both inorganic and organic compounds can interact to release energy and this can be tapped by living creatures. For example, some bacteria, such as *Thiobacillus*, can make a living from the energy provided by the oxidation of iron (II) to iron (III) in the natural environment (such as in acid mine waters). These are said to be *chemoautotrophic* (autotrophic means self-feeding, that is not needing other organisms to produce the 'food'—see Chapter 2).

The evolution of pigment systems allowed certain living organisms to avail themselves of a much more reliable and abundant source of energy—the sun. Photosynthetic pigments revolutionized the development of life, because they opened up the tapping of the almost infinite energy resources of the sun. This process, along with its by-product of oxygen gas has had, as we saw in Chapter 2, a profound impact on the development of our atmosphere. It has also resulted in a steady stream of energy-rich molecules being produced, upon which the vast majority of the world's living organisms are now dependent. During the course of the earth's history the process of photosynthesis has occasionally led to the formation of very considerable reservoirs of energy-rich materials, at times and in locations where the consumption of herbivores and the degradation of decomposers has not matched the rates of productivity. These geological reservoirs are still with us in the form of peats, lignites, coals and oils, and they still represent an energy resource when they can be unearthed and tapped.

Most animals demand energy simply to supply their immediate requirements to do work. So a butterfly opens its wings while settled on a flower and draws energy-rich nectar through its tubular mouthparts while it absorbs the radiant energy of the sunlight directly. Both sources provide a means of enhancing its ability to fly to the next flower or to set off in search of a mate. Some animals seem to think ahead in terms of their energy supply, storing up energy reserves either in their bodies or in hidden caches ready for anticipated hard times. A migrating bird accumulates fat reserves before setting off on its journey, and a squirrel may build up a store of food ready for winter hibernation. But our own species has developed increasingly complex systems of energy storage and use in order to modify our immediate environment and make it more comfortable.

The use of fire is an early example of the non-food exploitation of an energy resource. Here, the human being is taking materials that are rich in energy (ultimately dependent on the photosynthetic fixation process), such as wood or animal fats produced from the consumption of vegetable materials, and oxidizing the matter by combustion in air, thus releasing heat and light. Both the heat and the light would, of course, be an extremely valuable resource within the cool, dark confines of a primitive shelter or a cave. Not only would cooking and space-heating become possible, but also light for extending the day's social activities, forming a physical focus for communal activities and even for the development of leisure and artistic faculties, as in cave painting. Some of these uses of energy are not the simple, direct ones, therefore, that are associated with energy use in all organisms, but are indirect uses. It is often not so much the energy itself that is needed, but the services that the energy can provide—in this case light, but in more recent times transport, industry, communications, etc.

The use of fire dates back well into our ancestry, probably into the times of *Homo erectus*, perhaps 1.75 million years ago. Although the first uses of fire as an energy resource by prehistoric peoples probably involved the use of biomass such as straw and wood, the use of fossil fuels may well have been adopted at an early stage in human cultural development. Peats are the most obvious and perhaps the most available fossil fuels and their combustibility would soon have become evident to intelligent beings in need of an energy resource, especially in high-latitude cold climates where biomass may have been in short supply. But even coal may have been used quite early in human prehistory. Charcoal from palaeolithic settlements dating from the last glacial in southern France has been shown by electron microscope studies to possess the deformed character of wood cell structure that is usually associated with compressed coals (Thery *et al.* 1995). Outcrops of Jurassic coal are known in the area and the palaeolithic peoples used the soft black rock for carving, so it is not surprising that they chanced upon its energetic value and subsequently exploited it in what was then a treeless landscape.

The more direct use of solar energy resources is probably even older than that of fire. Drying animal skins, fish, fruit and clothing in the sun is not a process that demands very high levels of ingenuity. Following the development of domestication and agriculture (which is in itself a form of energy diversion into a human food-chain), the use of solar energy for grain-drying prior to storage may have required more sophisticated techniques for dealing with larger quantities. Similarly, the drying of hay for winter feeding of animals would have involved the invention of racks and stook systems to improve air penetration and to ensure rapid drying before fungi could attack the crop.

Wind energy has long been used as a means of transport and also as a means of performing mechanical work, such as grinding corn and pumping water. The harnessing of water and wind power for such mechanical purposes could well be regarded as the first stage in the industrial revolution. But it was the use of steam as an intermediary between combustion and the performance of mechanical work that really heralded the industrial revolution proper. The principle is simple. Heated water boils to produce steam that occupies a much larger volume than the original

water. Further heat expands the steam and the outcome is a pressure that can be harnessed in a piston that is forced into a thrusting motion, easily transferred to a rotating motion by driving a wheel. This is the essence of the engine invented by James Watt in 1769, but the idea had been developing through a series of prototype stages during the 18th century. The potentiality of this system of energy harnessing received further impetus in 1884 with the invention of the steam turbine by Charles Parsons. Here the steam creates a direct output of rotating motion, thus greatly improving efficiency.

The success of such inventions in fuelling the industrial revolution was in part due to a process of positive feedback. The James Watt engine permitted a much more efficient system of water pumping that allowed the development of mining where water accumulation had previously created insuperable problems. So the machine effectively fed itself, generating more coal by its activities. It is no exaggeration to say that these 18th century concepts and experiments provided the basis for modern environmental changes. Although more sophisticated devices, such as the solar cell, are now available, it is still true that the steam turbine is a major force in the modern energy industry and is likely to remain so for some time, whether fuelled by coal, oil, gas or nuclear energy.

Sources of energy

Ultimately, the energy available to us comes either from the sun, from the earth's interior, or from the gravitational forces of sun and moon that generate the tidal flow of the oceans. The sun provides a direct source of radiant energy and also a range of indirect sources, including the biomass that is currently accumulating as a result of contemporary photosynthesis, fossil biomass (coal, oil, gas) that is derived from photosynthesis in the past, and the physical turbulence of the atmosphere and hydrosphere created by the differential heating of the earth's surface. The latter effect results in the generation of winds and waves and the maintenance of the hydrological cycle of evaporation, condensation and the flow of water under the influence of gravity. Gravity, this time of the sun and moon in concert or opposed to one another, is also responsible for the surges of the oceans backwards and forwards over their beds as the free-moving waters are pushed and pulled by its force.

The earth's interior is a source of geothermal energy as the molten core of the planet gradually cools and releases its energy to the atmosphere. There is also the energy of molecular decay processes in which unstable elements break down to release energy in the form of ionizing radiation. Such radioactive elements are found in the rocks of the earth's interior. Of the current daily balance of energy input and output over the earth, the sun is responsible for over 99% and the geothermal and tidal energy for the remaining 1% (Myers 1994).

All of these sources of energy are being tapped for human exploitation but by no means evenly. Figure 10.1 shows the current state of global energy use. Around 80% of the energy used by our species is derived from fossil fuels, coal, oil and gas. These are non-renewable energy resources; once they have been used up they cannot be replaced. When they were first used to fuel the steam engine it could not have been anticipated that the reserves were sufficiently limited to create a problem. But the growth in energy demand is such that the finite nature of fossil fuel supply must now be taken into consideration (quite apart from the effect their burning has upon the injection of geological carbon back into the atmosphere—see Chapter 5). A major problem is the accelerating growth in demand for energy around the world. It has been estimated that the demand for energy will be 50% higher by the year 2000 than it was in 1980 (UNEP 1993) and this is likely to place an increasing strain on the availability of fossil fuels.

Before 1950, the commercial consumption of energy in the world was increasing by about 2.2% per annum. In the 1960s this rose to 5.6%, but the sudden increase in oil prices in the early 1970s, together with the accompanying realization that resources were limited, led to a slowing down in the steepness of the curve. In the 1970s increased consumption was still running at 3.5% per annum and in the 1980s, 2% (Tolba & El-Kholy 1992). But such figures still mean that more energy is being demanded each year.

Fossil fuels

Global use of fossil fuels has been increasing over the past 150 years and especially in the last 50 years. We have already seen in Chapter 5 that this creates particular problems in that the element carbon is rapidly transferred from the lithosphere (the earth's crust) to the atmosphere, and there the oxidation product of fossil fuel combustion, CO_2, is a major contributor to the greenhouse effect. We also saw in that Chapter that the CO_2 can be extracted from the atmosphere by solution in the sea and also by plants in the photosynthetic process, but these sinks cannot absorb the gas fast enough to cope with current rates of release mainly from the burning of fossil fuels. The residence time of CO_2 in the atmosphere could run into hundreds of years, so the current build-up will take centuries to clear whatever happens from now on.

Perhaps the most obvious question we should ask concerns how long the reserves of fossil fuel will last. This is difficult to answer precisely because of all the variables involved. We need to know how much fuel is available (that is how much is both present and recoverable) and also how fast we are likely to use it up (which depends in part on how fast we turn to other sources of energy). Even the estimates of recoverable reserves are constantly changing, partly because of improved survey information, but also because

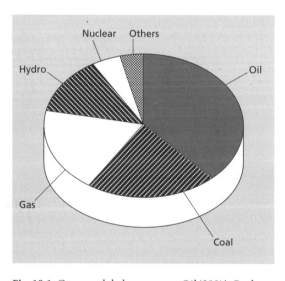

Fig. 10.1 Current global energy use. Oil (38%); Coal (21%); Gas (19%); Hydro (13%); Nuclear (5%); Others (4%). Compiled from various sources.

'recoverable' is a term that involves both techno-logical and economic considerations. What may not be recoverable or not worth recovering this year may be attainable and economically worth while in 10 years' time. An estimate by British Petroleum in 1991 put the recoverable reserves as follows:

1 Coal: 572 000 mtoe
2 Oil: 136 000 mtoe
3 Natural gas: 109 mtoe

The units used here, 'mtoe', stand for 'millions of tonnes of oil equivalent'. The use of this term enables one to make meaningful comparisons between different types of energy reserves. One million tonnes of oil is approximately equivalent to 1.5 million tonnes of coal. The reserves of coal are thus large, perhaps as great as 220 times our annual consumption (Myers 1994). But this does not mean that they would last us for over two centuries, because our consumption is increasing and, given the current high level of usage of oil and gas (see Fig. 10.1) and given the possibility that these may well be used for other purposes in future (as feedstuffs, or in the chemical industry), coal exhaustion could occur sooner than is immediately apparent.

Exactly how we use fossil fuels as an energy resource in the future is also an important consideration. A cleaner use of coal (that is, one that generates less CO_2) would be to strip out the hydrogen from the organic components of coal—consisting of carbon, hydrogen and oxygen— and combust only the hydrogen. This would lead to the generation of water as a waste product, but not CO_2. The total energy output would be lower, but the reduced efficiency of energy recovery (only about 25% that of full coal combustion) might be worth while in environmental terms (Beyea 1991). This would be a more attractive alternative if hydrogen were to become the major fuel for transport, replacing petroleum in the engines of vehicles. The number of energy transfer stages could then be cut and the economics might look more feasible. But our coal reserves would not then last as long.

A question mark still hangs over shale and tar oils which are not yet exploited as energy sources because of the economics of extraction and use. If the technology becomes available and the economic climate is appropriate, then this resource (roughly equal to the coal reserves) could come on

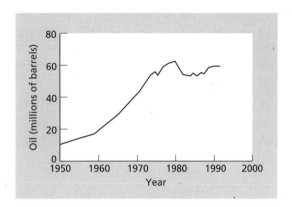

Fig. 10.2 Changes in world oil production on a per day basis (Flavin 1992a).

tap and greatly extend the fossil fuel reserves of the earth (Beyea 1991). The outcome of their use, however, could prove disastrous for any attempt at global control of CO_2 output (Sundquist & Miller 1980).

World oil production actually seems to have stabilized over the past few years (Fig. 10.2), never having fully recovered from the disruptions to supplies in the 1970s (especially 1979). The Gulf War of 1990 may also have helped to stabilize demand and the long-term prospects are not likely to rise substantially above these levels because of the inherent unpredictability of supplies resulting from political factors (Flavin 1992a). Natural gas production, on the other hand (Fig. 10.3), continues to rise steadily, especially as a result of Russian production (Flavin 1992b) and because of its environmental advantages. The combustion of

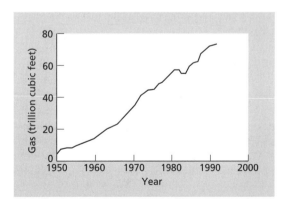

Fig. 10.3 Changes in world natural gas production on a per annum basis (Flavin 1992b).

gas results in the release of less carbon to the atmosphere than in the case of coal and it is also clean in terms of its sulphur content, so its climatic and acid-rain impact is minimal for a fossil fuel. It is difficult to estimate how long this source of energy could last, but even with a doubling of use over the next 20–30 years, Flavin (1992b) suggests that output could be maintained for several decades. Nisbet (1991) claims that world reserves should last for 60 years in total.

One of the problems with natural gas is the possibility of leakage, because methane (its major constituent) is an extremely powerful greenhouse gas (approximately 25 times as effective as CO_2; see Chapter 5). Gas pipelines inevitably leak, especially older ones, and this is a serious disadvantage when considering its use. Gas, on the other hand, emits the least amount of CO_2 per unit of energy released of all fossil fuels. Coal, for example, emits about twice as much CO_2 for every joule of energy supplied in its combustion. Hence a shift from the use of coal to gas can reduce CO_2 emissions very considerably.

The continued use of fossil fuels is inevitable in the short term, perhaps even until the bulk of economically recoverable reserves have been harvested. It may prove possible, however, to reduce the global impact of their use by cleaning exhaust gases, especially by removing CO_2 from them. This is technically possible, but the economics of a 90% CO_2 removal would result in an approximate doubling of the cost of the electricity generated (Beyea 1991). One approach could be to pump the exhaust CO_2 to the bottom of the ocean where it would dissolve and be lost to the atmosphere, but this could prove very expensive.

Nuclear energy

At first sight, there are many apparent advantages associated with the use of nuclear reactions as a source of energy for human consumption, but the early enthusiasm has gone into rapid decline. Nuclear power stations produce virtually no CO_2, so they do not contribute to the greenhouse effect, apart from that which may be generated during mining and transportation of fuel. The clean energy image led to considerable expansion of nuclear energy production in the early 1980s, but since that time its contribution to world energy supply has stabilized (Fig. 10.4). The main problems that

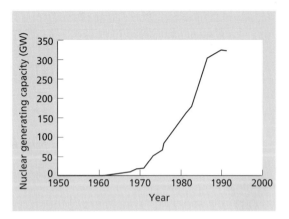

Fig. 10.4 The electrical generating capacity of the world's nuclear installations (Flavin & Lenssen 1992).

have led to this change are considerations of safety and the disposal of radioactive nuclear waste. Added to this, the active life of a reactor is limited and decommissioning becomes necessary after perhaps 20 years, which adds considerably to the expense involved in this method of energy production. The far-reaching consequences of accidents in older reactors was brought to public attention by the incident at Chernobyl in Russia in 1986, which resulted in the widespread dispersal of radioactive materials. Since that time public fears over nuclear installations and waste disposal have led to a steep decline in the popularity of nuclear energy. Quite apart from the problems of human error and mechanical failure in the normal operation of nuclear establishments, there is always the possibility of natural catastrophe, warfare or terrorism leading to unforseen disasters against which any operation is difficult to render secure.

The risk factor in the development of nuclear power results from the problems of heat transfer within the system. Heat generated within the nuclear reactor itself is first absorbed by a primary circuit of water passing around the system under pressure (see Fig. 10.5). This is then transferred to a steam-generating secondary circuit which extracts the energy, some of which is converted into electricity via steam turbines and the remainder discharged to the environment via condensers. The pressurized system of water-cooling, coupled with the inherent dangers of radioactive material, are the main source of risk, since any

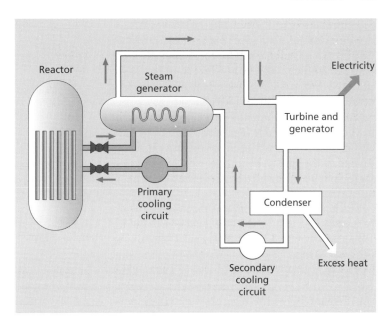

Fig. 10.5 Diagram of the structure of nuclear electricity-generating installation.

rupture of cooling pipes could lead to overheating in the reactor. Current models duplicate cooling systems so that a single failure need not create problems, but all the systems run close to one another, so it could be argued that rupture in one pipeline could damage adjacent ones (MacKenzie 1994). Continued improvements in safety design and efficiency will undoubtedly be the focus of much attention in coming decades.

Besides the risks of accident or deliberate sabotage damaging the cooling systems of nuclear installations, there is the inevitable production of regular waste during the energy-generating process. Low-level waste, which includes liquid and solid materials lightly contaminated by short-lived radionuclides, account for about 90% of the waste material generated, and there are opportunities for reducing the output of this material very considerably if appropriate techniques are applied. Over the past 10 years the discharge from the Sellafield reactor in the United Kingdom, for example, has been reduced by a factor of six for beta-radiation and by 20 times for alpha-radiation (Brown 1992), and these reductions can be detected in the recent sediments of the North Sea (UNEP 1993). There are fears, however, that even low-level release could have very serious health implications, possibly resulting in leukaemia in children. There have also been suggestions that the concentrations of leukaemic children that have been observed in the vicinity of nuclear plants could be due to the exposure of fathers to ionizing radiation if they work at the stations. But attempts at controlled analysis of the data currently fails to produce any substantial evidence that this is the case (Doll *et al.* 1994).

Intermediate-level wastes include materials contaminated by longer-lived radionuclides, such as plutonium and isotopes of uranium. Their disposal currently involves packaging in specially designed containers and burial at designated sites. In the past (pre-1983) they were also dumped at sea, both in the North Atlantic and in the Pacific, but this practice has been suspended. There are very reasonable fears concerning the long-term survival of such containers, given the possibility of unforeseen corrosion or the occurrence of catastrophes such as earthquakes that could liberate the contaminated material. The ethical questions raised by the export of such unwanted and politically-embarrassing packages from the richer, developed countries to areas (like Native American reservation lands) or countries (such as West Africa) that are in need of hard currency for development purposes, are now being raised and debated.

The disposal of high-level wastes, i.e. materials containing significant quantities of long-lived radionuclides, sufficient to result in the raising of the temperature of their surroundings, is even more problematical. Deep disposal (several hundred metres) in geological formations on land, possibly in a liquid, vitrified form are currently being examined, but are not likely to be viable until the early part of the next century. The reprocessing of waste in order to regenerate more fuel by concentration is an attractive option and can be economically viable (Berkhout 1994), but only certain centres are equipped to conduct this operation so waste has to be transported, often across large distances, leading to further risks of accident, theft or sabotage.

The majority of nuclear power stations currently in operation around the world were built before 1986, which means that many will shortly be decommissioned (estimated at 256 installations by the year 2000; Tolba & El-Kholy 1992). This process will result in high risks of exposure during the dismantling of the reactors and will be very expensive. The cost of decommissioning a single plant could amount to $480 million. A new generation of reactors is therefore due and, if nuclear energy is to replace the use of fossil fuel, very many more facilities will have to be built. Given the current public concern about the risks involved, it is very doubtful whether sufficient appropriate sites, acceptable to the public, could be located. Nuclear energy, despite its early promise, seems to be a blind alley as far as a future global energy supply is concerned.

Solar energy

Figure 10.6 shows the distribution of solar energy over the surface of the world. Using the sun's energy directly can be achieved both by technologically simple and by more complex methods. Laying out fruits, or grain, or fish or meat to dry in the sun is a primitive but effective means of directly tapping solar energy. It is not difficult to enhance the efficiency of such a relatively 'low-tech' approach by building equipment in which parabolic reflectors can be used for concentrating the solar radiation and thus generating higher temperatures and speeding the process of drying. Or solar ovens can be constructed for cooking. Such methods may help to reduce the demand on biomass for building fires in areas where the stripping of woodland and scrub can degrade land, as around the edges of the Rajasthan Desert in India.

The use of parabolic reflectors on a larger scale

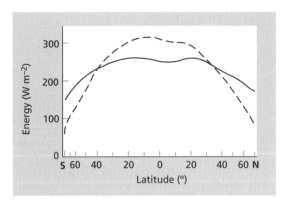

Fig. 10.6 Absorption of solar energy over the surface of the earth (– – –) and radiation from the earth (——).

as a means of concentrating solar heat onto tubes containing fluid, such as oil, is also being investigated, particularly in Australia and in the United States (Anderson 1994). The heated oil (up to 390°C) can then be used to heat water that evaporates to drive turbines. The Australian National University has constructed a 400 m diameter reflector as a prototype and it is planned to erect a bank of 28 of these to supply 2 megawatts of electricity to a small town in the near future. In California, a total of nine power plants along these lines (but using linear troughs rather than circular reflectors) has been constructed in the Mojave Desert, and these have a combined generative capacity of 354 megawatts of electricity, about 2% of the local grid. This currently accounts for about half of the total solar electricity generated on earth. Perhaps the most critical aspect of improving the efficiency of such a system is to ensure that the fluid-carrying tubes absorb as much as possible of the heat focused on them and do not reradiate it as heat. An increase in the efficiency of absorption here means that focusing need not be so precise and the size of the parabolic reflectors can be reduced, thus allowing higher density of units per land surface area. Smaller systems will also be easier to tilt to the angle of the sun.

Water heating and space heating by direct solar power can be achieved using very simple equipment, such as absorptive panels through which water drains and warms in the sun, or by using the greenhouse effect of glass through which short-wave radiation passes more easily than long-wave. If a 'greenhouse' structure is combined with heat retentive materials, such as brick, within its confines, the heat can be stored and only gradually released during the night, providing a well-balanced thermal environment. Many buildings, even in the cooler temperate regions of the earth, have been successfully built on such principles.

Solar ponds work on a similar principle except that water is used as a short-wavelength transparent material and a long-wavelength insulator in the place of glass. The idea seems first to have emerged from observations on the Sinai Peninsula where pools of salt water close to the Gulf of Eilat but separated by shingle banks became highly saline as water evaporated (about 16% salinity). When storms brought the relatively less saline water from the Gulf (about 6%) into these pools, its lower density led to its forming a separate layer over the dense salty waters. In this very stable density gradient, sunlight penetrates to the bottom of the pond (only 1–2 m) and heats the dark basal mud. Heat is conducted into the dense lower water, but the stable gradient of density does not permit the full development of convection currents, so the basal waters may become heated to over 60°C while the surface remains relatively cool (15–20°C). Exploitation of this system in specially constructed 'solar pools' can be used for driving turbines and generating electricity.

A possible exploitation of this system on a massive scale could be achieved by using the Dead Sea between Israel and Jordan and supplying less saline water by pipe from the Mediterranean Sea (Sifrin 1983). The Dead Sea is about 400 m below the level of the Mediterranean in a deep rift valley, so the transfer of water over 115 km by canal and tunnel would not be energetically expensive—indeed, it could be used to generate hydro-electric power in the final descent into the valley. The fresher water would lie over the saline layers and create a solar lake in which lower layers could be raised to high temperatures by solar penetration. The level of the Dead Sea would be raised by about 10 m in 20 years of operation and this would result in the flooding of about 100 km² of salt plain. But this area has dried out in recent times (since about 1930) because of the reduced water input from the River Jordan as water has been extracted from this source upstream for crop irrigation. So the solar pond scheme would only bring

the Dead Sea back to its former level. It would, however, create problems for the current chemical plants that extract materials from the salty sediments around its shores, and the entire development would depend upon close co-operation between Israel and Jordan, which has been politically unlikely until relatively recently.

The use of photovoltaic cells is an alternative method of direct solar power use, and the economics of this process has been steadily improving in recent years (Hubbard 1989; Anderson 1994). A major problem, however, is the area required for solar collectors. It is difficult to envisage a situation where large areas of land surface are given over to solar collectors, even in agriculturally unproductive regions, because in deserts and 'wilderness' areas the competition from wildlife conservation interests in likely to be strong. The use of floating platforms at sea, or even in space, has been examined.

Another problem with solar electricity is that it can only be generated during the daytime, and energy demand by people is often greatest at night. Storage is therefore essential, even if only short-term. Batteries do not prove very effective and the alternative, pumping water to a raised reservoir from which hydro-electricity can be generated when required, is expensive and energetically inefficient. Compressed gas or hydrogen production could provide alternative storage systems and the operation could prove effective at a local (e.g. office-building) level (Beyea 1991). On a larger scale, especially in a well-integrated system such as in the United States, the supply of electricity could follow the sun westward and distribute the electricity over a national grid.

Although solar systems are generally clean in operation and do not generate CO_2, the production of photovoltaic cells could give rise to new pollution problems in their construction and disposal.

Other energy sources

A range of other sources of energy is available and many of these have been tapped effectively in different parts of the world. Each of these has certain advantages, but also some disadvantages, so there is no simple solution to the energy supply problem that is devoid of environmental risks or damage.

Hydropower is effectively a means of tapping the combined effects of sun and gravity. It is solar power that causes water to evaporate and to travel through the mobile atmosphere until it is eventually precipitated as a result of cooling. Often this cooling takes place over higher ground and the resultant water moves under the force of gravity back to lower levels, generating kinetic energy from the potential energy it has gained from its new elevation. This kinetic energy is converted into mechanical energy and ultimately into electrical energy as the moving waters turn turbines and run generators. The process is therefore devoid of the problems of waste generation, but it does require the construction of dams, the flooding of valleys and the loss of habitats for wildlife, or for farming or settlement by people. Landscape changes and wildlife conservation implications are increasingly regarded as important issues in the developed world, while in some dryland areas the supposed advantages gained by flood control and water conservation for irrigation have not always materialized. In Egypt, for example, the Aswan High Dam has controlled the flooding of the Nile, but this has led to reduced fertility in the Delta as well as problems of silting and water-weed growth in the slower-flowing distributaries. Dams may also suffer from siltation and in-filling and can also isolate sections of a river corridor, preventing the migration of fish and other aquatic animals. Hydropower, however, despite the associated problems, is likely to increase in importance, especially in those developing nations that are short of local fossil fuel supplies. At present, hydropower accounts for about 13% of the world's generated electricity.

Wind power is also generated ultimately by the sun, for the movement of air masses is dependent on the unequal heating effects of the sun on the earth's surface. Wind and water power have long provided energy resources for human activities. The first windmills date from the mid-11th century in north-western Europe and they represented a revolutionary development in energy technology, but their design remained relatively static until the last few decades when they have re-emerged as a potentially important and clean energy-trapping system. The problems relate largely to the noise generated by the gear-boxes of the wind turbines and the visual impact caused by the large and mobile structures erected in serried ranks in exposed sites. One wonders whether the 11th-century environmentalists of north-west Europe had similar objections to the development of the quaint structures that once ground the corn and drained the marshes of that region. A further objection is the hazard that the rotating blades represent for birds in such regions. There is evidence that large birds, such as birds of prey, may be at risk, as has been demonstrated at Altamour Pass, California, and Tarifa, on the southern coast of Spain. But both of these sites are renowned for their populations of birds of prey and such problems should have been anticipated. It is clearly necessary to ensure that the siting of wind farms should avoid bird migration routes and regions where birds congregate or regularly hunt (Webb 1994).

Tidal power involves the construction of barrages across estuaries and tidal channels so that turbines can be driven as the tidal ebb and flow sweeps through the region. Like freshwater hydro schemes, the technology can be harmful to fish and other aquatic life that migrates through those waters and can become caught in the turbines. Special provision is required for such fish as salmon that may migrate regularly through the area and, of course, for the shipping that will also need access. As in the case of inland dams, estuarine barrages may alter the dynamics of water movement, causing changes to patterns of silting and this can lead to habitat changes for the invertebrates and the birds of the mudflats that are affected.

It is possible to derive energy from the oceans in the form of wave power, and also by exploiting the temperature differences in the ocean between the surface layers and the cold ocean depths.

Temperature gradients with depth are also apparent in the solid crust of the earth, the temperature increasing by approximately 1°C with each 30 m in depth. Water pumped underground can be heated in this way and either used directly or converted into electricity. Currently, such geothermal energy is used by 27 different countries.

A final source of energy is the vegetable matter that surrounds us, particularly the wood of shrubs and trees. This, of course, owes its energy richness to the solar power from which the plants have constructed their organic materials and the free availability of solar power makes this source of energy attractive as a renewable energy resource.

Biomass is perhaps the most ancient energy resource exploited by our species and the use of fire for its energy release has played an important part in our social and technical evolution. In many parts of the developing world, such as southern Asia, it remains the most important source of energy for food preparation and space heating. Besides the option of direct wood combustion, there is also the possibility of microbial conversion of vegetable produce into a liquid transportation fuel, such as alcohol, and this renewable resource could supplement the available supplies of gasoline and oil. The emphasis here is on the use of biomass as an energy supplement, for the total replacement of fossil fuels by biomass energy would require far more land surface than could reasonably be devoted to that purpose. Beyea (1991) has calculated that the replacement of gasoline and diesel by biomass-generated alcohol in the United States transportation system would require the conversion of 400 million acres (162 million hectares) of land to biomass production. This is roughly equivalent to the entire area currently occupied by US agriculture and is clearly unacceptable. Proponents of biomass energy, on the other hand, point out that marginal land and unproductive areas can be converted to biomass production, but this is often the land that is currently used for wildlife conservation, so biomass generation would have repercussions in that direction. Much research remains to be carried out on the extent to which biomass energy production can be integrated with wildlife and landscape conservation. The lessons of medieval Britain would suggest that sustainable yields of fuel wood can be maintained within forests that remain rich in wildlife (Rackham 1986).

The combustion of biomass does, of course, generate CO_2, so its immediate effect is to add to the greenhouse effect. But since each hectare of forest harvested for this use would revert to further forest development there will be a downdraw of CO_2 from the atmosphere equivalent to that generated by the combustion. In other words, the long-term effects of biomass energy use are effectively carbon-neutral. Solar energy is being tapped via the photosynthetic process and is ultimately dissipated as heat, while carbon is being recycled through the atmosphere and plant material.

Efficiency of energy use

Whatever system, or combination of systems, is used for energy generation in the future, the importance of energy efficiency will continue to increase. Any energy saved is energy that does not need to be generated, and this must save money, resources and the environment. Transport is clearly important here, and much opportunity remains not only to develop transport systems that cause less pollution, but also ones which are efficient in their energy use. Public transport systems can be more effective in this respect, but only where they are fully utilized, for running a system that is virtually empty, but highly convenient to the few passengers that travel, is not energetically efficient. Where demand for a particular transport service is low, individual transport becomes more energetically efficient.

The efficiency of energy use and conservation in buildings is currently being improved, but brings new types of problem with it. The use of double-glazing and wall insulation can greatly improve energy conservation, but also reduces air-flow and ventilation, sometimes leading to hazards from internal pollution and safety problems in an emergency such as fire.

Recycling of materials can save energy. Some materials, such as glass and plastics, are energetically expensive in their initial construction and energy may be saved if they are re-used. But the energy saving can be reduced if energy-demanding processes, such as sorting and cleaning of recycled material, is required. When carrying a used bottle to a bottle bank we must ask whether the energy we are using in its transport exceeds the energy saved by its being recycled. This is not the only factor in the equation, however, for if the bottle is not recycled it may well end up in a landfill waste site, in which case it will contribute to other environmental problems.

Recycling has a solid logical base that is sadly not always realized in terms of its economics. In the case of the recycling of paper from newsprint, for example, the economic costs of separation, collection, cleansing of ink, reconstituting and resale, may not prove an economic alternative to the felling of new softwood plantations for pulp. The full equation should, of course, also include the cost of combustion or landfill as an alternative system of disposal. But combustion could be coupled with a space-heating facility or power generation (in urban situations), while waste paper in landfill sites will create new problems of methane generation. Since the costs and the benefits may come out of and go into different pockets, it is often difficult to determine an overall optimum strategy for the recycling of a particular material.

Future energy supply patterns

The development of human civilization and its consequent demand for energy supplies over and above those provided by ingested food was largely supplied by biomass sources. Wind and water provided novel and valuable additional supplies of energy during historic times and often permitted degrees of industrialization in society. The industrial revolution of the past two or three centuries, however, has been largely fuelled by fossil energy reserves and this remains the case to the present day. As our per capita energy demands continue to rise, however, the finite supplies of fossil fuels begin to decline, and this fact alone means that we must look elsewhere for our long-term energy requirements. The additional concerns raised by the environmental consequences of energy consumption, particularly the elevation of atmospheric CO_2 through fossil fuel burning, provide even greater impetus to the search for alternative energy sources.

Projections suggest that fossil fuels will continue to supply about 70% of the world's energy needs well into the next century (Fells 1992), but increased energy-use efficiency and cleaner technologies for energy extraction will become increasingly important. Nuclear power is almost certain to become more widespread in the short term, but public anxieties may well limit the value of this energy resource as a permanent feature of our civilization. Renewable energy resources, such as biomass, hydro, wind, tidal and solar power will all play increasingly important roles in local energy supply, but it is unlikely that they will contribute more than about 20% of global energy needs within the next few decades.

The need for global energy policies is very apparent, but very difficult to achieve. It is important that the damage caused to the environment as a result of different types of energy generation should be evaluated in financial terms so that appropriate fiscal measures can be taken via taxation to discourage the more harmful processes. 'Carbon-tax' is one such measure. It is very apparent that leaving energy production to develop under the influence of the selective pressures of the market place can only lead to short-term inefficiency and environmental degradation in the interests of rapid profits.

Further reading

Myers, N. (1994) *The Gaia Atlas of Planet Management*, 2nd edn. Gaia Books, London.

Nisbet, E.G. (1991) *Leaving Eden: To Protect and Manage the Earth.* Cambridge University Press, Cambridge.

UNEP (1993) *Environmental Data Report 1993–94.* Blackwell, Oxford.

Wyman, R.L. (1991) *Global Climate Change and Life on Earth.* Chapman & Hall, London.

Chapter 11

Biological Resources in a Changing World

It is a very evident fact that the human species is both very common and most influential upon our planet and its environment, but it is difficult to assess our impact in precise terms. We look to the earth for the provision of our needs (food, energy, clothing, building materials, etc.) and we rely heavily upon the biological productivity of the planet to supply these, but just how much of the earth's productivity do we use? And can this be sustained?

One of the factors that can easily be overlooked when undertaking such calculations is the extent to which our agricultural and forestry activities modify the natural vegetation of an area and change its productivity. For example, when we convert an area of tropical forest to grassland and use it as pasture, we harvest a crop of meat or dairy produce that we can calculate relatively easily, but we have reduced the overall productivity of the site and this must be taken into consideration when the total human impact is calculated. Using this approach, one analysis by Vitousek *et al.* (1986) is summarized in Fig. 11.1.

In this analysis the annual global primary productivity is estimated at 224.5 pg (picogram = 10^{15} g), of which 59% is derived from land-based productivity, 41% from the oceans and an insignificant 0.004% from freshwater. Human beings and their domestic animals each year consume about 5 pg of this productivity directly and also use a further 2.2 pg as timber and firewood (total 7.2 pg). The creation of biologically barren areas within which we live and move (cities and highways) removes 2.6 pg of potentially productive terrestrial land from the equation, and the creation of biological wasteland (desert) as a consequence of land mismanagement in dry regions accounts for a further 4.5 pg of lost productivity each year. There remain the large areas of the earth's surface that we have modified (usually reducing overall productivity even if diverting greater proportions towards ourselves). These include arable agricultural land and pastoral land that may otherwise have supported more productive (but from our point of view less immediately useful) ecosystems. The potential productivity lost in this way amounts to 4.58 pg. Thus the total productivity diverted or used by our species adds up to 60.1 pg each year, or 27% of the world's annual productivity. The bulk of this impact falls upon the productivity of the land rather than the oceans, and one can calculate that 39% of terrestrial productivity is diverted or used by humankind (Diamond 1987). This gives us some idea of how very influential we are upon the biological energy balance of the planet.

Agriculture

Although our species evolved into a rather broad ecological niche, hunting a wide range of animals and gathering a great diversity of plants for our sustenance, this way of life is now confined to only a very few cultures in remote locations, such as the Amazon forest. Even there such a way of life is usually accompanied by some form of environmental management and the exploitation of particular species of animal and plant as reliable and sustainable resources. There is a fine line of division between increasing reliance upon and management of a particular wild species and the process of domestication. The Palaeolithic hunters of north-western Europe 10 000 years ago followed reindeer herds on their migrations, much as wolves do today. The relationship became a close and interlinked one, although certainly not exclusive. The native Americans of the mid-West formed close links with the American bison, and the European Mesolithic cultures with the red deer and even with the hazel tree, a reliable source of nuts.

All of these peoples managed their environ-

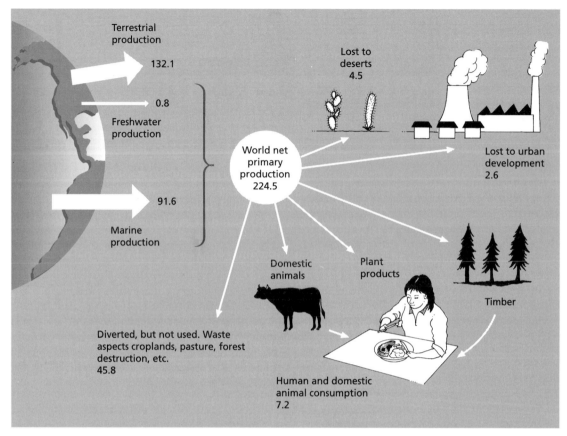

Fig. 11.1 The role of *Homo sapiens* in appropriating or diverting the world's primary production. The units used are picograms (pg = 10^{15} g) per annum. Based on data from Vitousek *et al.* (1986).

ment to encourage the healthy growth of populations of their prey, often by burning vegetation to stimulate new vegetation growth rich in minerals and proteins, thus providing good forage for their animal supporters. Human cultures, in this way, radiated into a range of more specialized niches, each associated with particular sources of food and certain geographical areas. Any such narrowing of a niche does, however, create new limitations for an organism, confining it to the range of the food species. So, although our species has very broad physiological tolerance to a wide range of climatic conditions (especially in view of our ability to control our own local environment by wearing clothes, building shelters and heating them), a specialist culture can become restricted by the more sensitive requirements of its prey.

The domestication of plants and the development of arable agriculture provided a means of more efficient harvesting of natural productivity. The plants involved, of course, provided a source of food (or fibre) but they were also selected for the ease with which they could be propagated, cultivated and harvested. The timing and the location of the first agricultural communities has occupied the research activities of archaeologists for centuries, but it is clear that the idea originated in many different parts of the world and involved the use of various types of plants. It is also clear that, wherever the location, the time of agricultural development is often in the early part of our current interglacial, around 10 000 to 8000 years ago. We know that this was a time of rapid climate change (see Chapter 4), but it is still a remarkable coincidence that the same idea caught on in so many parts of the world more or less at once. In the Middle East, wheat, barley, lentils and chickpeas were domesticated around 10 000 years ago; in eastern Asia, rice, millet and perhaps cabbages were brought into cultivation about 9000 years ago; and in Central America, beans and chili peppers were domesticated by 8000 years ago.

It seems reasonable to suppose that global changes were in some way connected with this shift in human behaviour patterns, and one possible suggestion is that the rising carbon dioxide (CO_2) levels of the current interglacial (see Chapter 5) may have provided sufficient stimulus to the growth of many of these plants to make their cultivation worth while (Sage 1995). The main rise in CO_2 concentration in the atmosphere, taking atmospheric concentrations from 200 to 270 ppm, was complete by 12 000 years ago and this would have had a substantial stimulatory effect on those plants with a C_3 photosynthetic system (which includes the crop plants listed above). It would also have provided them with more effective water balance properties (by permitting them greater flexibility in stomatal closure), which may have been important in the dry areas where agriculture often had its origins. If this theory, put forward by Rowan F. Sage (1995), proves correct, it shows how early human cultural development was dependent on global change, whereas now our cultural development provides major impacts upon global change.

One consequence of domestication is that agricultural communities become increasingly dependent on their domesticated species and, to a certain extent, become limited by them. Once fully domesticated, the species of animal or plant could be taken out of its normal geographical range and transported to new areas, wherever it was capable of sustaining a population, given the protection and aid of the human agency. Selection of particular genetic variants, or even careful breeding programmes, could be used both to increase yields in general and also to provide better performance under the new conditions experienced, such as cooler climates, lower water availability, higher salinity, new pests and parasites, new food sources (for animal domesticates), and so on. The capacity of a species to be manipulated in this way is dependent on the genetic variability of the wild stock and this importance of the genetic resources of our wild plants and animals is still not fully appreciated. In many respects, this forms one of the main utilitarian arguments in favour of conservation.

Darwin (1868) noted how rapidly animals such as pigeons were able to change under the influence of domesticated selection, and the remarkable variation in form that has been achieved by breeding in organisms that range from dogs to cabbages is an expression of the genetic variability within any species. Figure 11.2 shows the range of form in cabbages, all of which have been derived from selection of certain required traits from just two main parental species, *Brassica oleracea* and *B. cretica*. The genetic range is present in the wild ancestors, but most of the characters displayed here would be a positive disadvantage if they were to be expressed under natural conditions, rendering the plant too obvious to grazers, or too palatable to be adequately protected. These genetically controlled characters are therefore recessive in the wild and can become evident only when the selective pressures that would normally suppress them have been removed under the protective influence of domestication (Lester 1989). Animals and plants are rich in such unexpressed genes and breeders are often at pains to bring them into the open.

There are, however, limits to the extent of possible modification of any species and we can often see the marks of an organism's evolutionary history and geographic origins when we study its growth, development and performance. In a study of the growth of a number of important temperate crop plants, for example, Watson (1971) showed the differing patterns of leaf emergence through the season (Fig. 11.3). Leaf emergence is expressed in terms of leaf area index (LAI), which is a measure of the area of leaf tissue of a species covering any given area of ground. An LAI of three, for example, would mean that every square centimetre of ground is, on average, covered by three square centimetres of leaf tissue. Or any given point is likely to have three layers of leaves above it. From Fig. 11.3 it can be seen that, at its peak biomass, a crop like barley can have an LAI as high as 10, meaning that every point on the ground can be expected to have 10 layers of barley leaves above it. Obviously, such a high LAI should be efficient from a light-trapping point of view, though an LAI of 4–5 is effective in trapping about 80% of the incident light.

Wheat was one of the first plants to be domesticated in the Middle East and its ancestors grew (and many still grow) in these regions of Mediterranean climate with cool, moist winters and hot dry summers. The wheats are annual

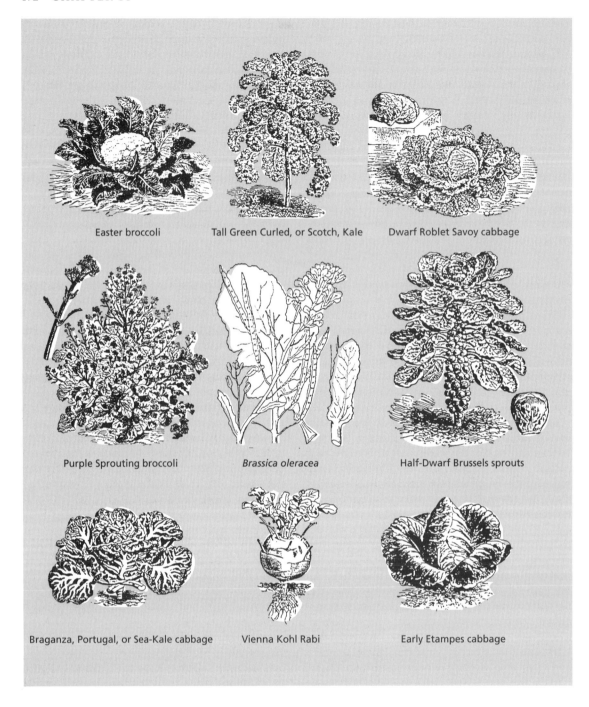

Easter broccoli

Tall Green Curled, or Scotch, Kale

Dwarf Roblet Savoy cabbage

Purple Sprouting broccoli

Brassica oleracea

Half-Dwarf Brussels sprouts

Braganza, Portugal, or Sea-Kale cabbage

Vienna Kohl Rabi

Early Etampes cabbage

Fig. 11.2 The range of form found in the domesticated varieties of a wild plant, the cabbage. The ancestral wild species, *Brassica oleracea* (or possibly its close relative *B. cretica*) has produced a remarkable range of forms under the selective forces of horticulture. From Lester (1989).

grasses, completing their life cycle within one season and relying upon their seeds for survival through the unfavourable summer drought, and it is this dependence on seed that ensures a heavy allocation of reserve food into the seed and the

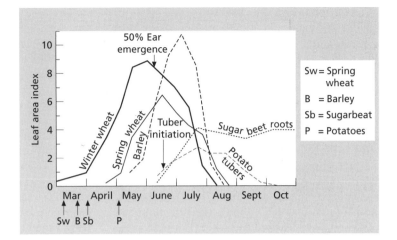

Fig. 11.3 The development of leaf canopies in a series of crop plants grown in a northern hemisphere temperate region. Canopy extension is expressed in terms of leaf area index, that is area of leaf per unit area of ground. Arrows represent sowing dates. Data from Watson (1971).

production of high numbers to ensure that at least some make it through the summer and have enough resources to establish themselves in the following wet period. Germination and establishment may occur in the autumn, in which case the young seedling remains quiescent through the cool winter, or in the spring. These variants form the basis of 'winter wheat' and 'spring wheat' respectively (see Figs 11.3 and 11.4). The winter-quiescent form is in a strong position to expand its canopy rapidly in spring, which is a great advantage over 'spring wheat'. It reaches its maximum LAI of about 9 in late May and then declines rapidly. This pattern of leaf expansion is ideal for the Mediterranean climate, where winters are

relatively mild (mean minimum January above −12°C) and summers are hot and dry, and it is also very appropriate for many parts of the temperate world, such as North America, Europe and central Asia (Rosenzweig 1985). Where winters are particularly cold, the spring wheat may be more appropriate since winter survival is not required, but spring establishment is slower.

Barley expands its canopy a little later than wheat and is rather more sensitive to heat and drought in the early summer than is wheat. In the Moscow region of Russia, for example, wheat is practically at its northern limit as far as ripening seed is concerned while barley is generally more reliable (Parry 1990). If conditions were to become

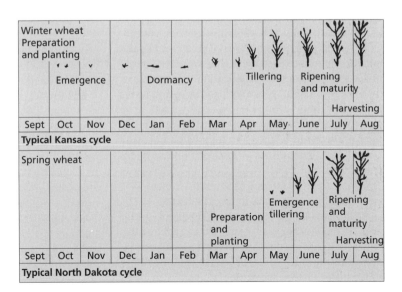

Fig. 11.4 The developmental sequence of winter and spring wheat in Kansas and North Dakota respectively. Data from MacDonald & Hall (1980).

warmer, however, the balance could shift in favour of wheat and the pattern of agriculture could change.

The other species considered in Watson's experiments also display something of their evolutionary history. Potatoes originated at high altitudes in the South American Andes and have been in cultivation for at least 2000 years. The crop was introduced into Europe in the 16th century, probably initially into Spain, and has since proved valuable in most temperate areas, extending as far north as Alaska and southern Greenland, but has not proved appropriate for the lowland tropics, perhaps in part because high night-time temperatures cause excessive respiration, though pests also cause problems. Canopy extension does not occur until late in the season, around May, which reflects its montane origins. Sugar beet is a temperate, biennial species that is also late in forming a canopy but, like the potato, is of value because during the autumn it stores up food reserves for the winter in its underground stems (Tivy 1990).

These examples of domesticated plants show how useful species have been transported to new areas of the world by human agencies both in ancient and more recent times. Rather remarkably, the bulk of the food supplied to humans from agriculture is based upon only about 16 species of plant (see Table 11.3, p. 194), so we have a limited resource base in this respect. This fact, when combined with the climatic limitations of these species (both in terms of geographical range limits and productive potential limits), must lead to concern about the effects of climate change on food production. Given the pattern of leaf canopy expansion and fruit production or food storage within our crops, those areas that are most appropriate for their growth at present may not prove to be so in the future. We must bear in mind, however, that there is a great residual variation of genetic material within each crop species and this may allow us to adapt our crops to new conditions. Even failing this, we can probably overcome most problems by substituting one crop for another. We will not necessarily be looking for new crop species in order to face the future.

At present, the most promising approach to the prediction of future agriculture comes from the construction of computer models in which we can combine estimates of likely climate change with information about the response of crop species to alterations in their seasonal regime. In this way, one can generate maps displaying the areas in which crop species will probably perform most efficiently in response to a proposed change (Gates 1993). Such modelling is still in a very crude state. The detailed prediction of climate change for an area given, say, a doubling of CO_2 is still very tentative, especially when detailed information about cloud cover, season length, precipitation and soil moisture are required. It becomes even less reliable when we introduce the further variable of how a particular crop will respond to such conditions, especially if pests and pathogens need to be taken into consideration, which they most surely do. Some initial models display what one might broadly expect, however, that the belt of wheat-growing in North America, for example, will shift northwards into Canada, perhaps extending right up to the Hudson Bay (Rosenzweig 1985).

Probably the simplest approach to the prediction of future agricultural potential is the use of analogues. If we can predict by modelling the future climate of a region, we can then seek the closest current analogue to that climate and from this estimate the future potential for crop growth. Using this technique, a doubling of CO_2 should convert the climate of Iceland into one resembling that of present-day Scotland; Finland will resemble Northern Germany and Denmark; the St Petersburg region of Russia will have the climate of the north-western Black Sea (Parry 1990).

It is also possible to predict the effects of warming on the altitude at which crops can be grown. In the European Alps, for example, a rise of 1°C would elevate the cultivation limits by about 150 m, while a rise of 4°C would raise the limits by about 450–600 m, effectively resulting in the Alps having the climate of the present-day Pyrenees. This will also raise the tree-line in the mountains, leading to changes in forestry. Because of the geometry of mountains this will effectively mean that less ground area would be available for forestry, but only if the lower limit of the forest, which is determined by human clearance for pastoral activities and settlement, keeps pace with the upward movement of the forest belt.

In lower latitudes, the inter-tropical convergence zone around the equator will extend, leading to greater levels of precipitation spreading out into regions that are currently dry, such as the Sahel zone in the southern Sahara (see p. 117). But whether this results in major agricultural changes will depend upon the extent to which the increased temperature leads to higher evaporation rates compensating for rainfall gains. The overall picture for Australia suggests that the total productive potential will increase by 20% in over half of its total area, the north-west region benefiting most (Parry 1990).

Trying to develop overall projections concerning future agricultural potential is made more difficult by the complexity of additional factors that will undoubtedly come into play in this process. Soil degradation by misuse may well reduce the potential, while urban sprawl may result in productive land being wasted. There is also the problem of some potentially productive lands being swamped by rising sea levels and hence lost for human use. Some possible trends have been postulated, however, and these will now be briefly outlined, but it must be recognized that they comprise only initial projections based upon a number of somewhat insecure assumptions. A fuller critique of these assumptions and the methods used in such studies is to be found in Parry (1990).

Canada. Large temperature rises are expected, especially in the north, though changes in precipitation are still debated. The potential for grain production should increase, perhaps by as much as 6% by the year 2000. Wheat and maize should both benefit from climate change, with consequent declines in barley, oats and soybean.

United States and Mexico. The temperature increase will be greater in the north than in the south. The States will become drier overall, but Mexico may benefit from additional precipitation. Both areas may see a decline in grain production, perhaps by about 3% by the year 2000, involving both maize and wheat. This assumes, of course, that the current crops cannot be adapted, by the selection of new varieties, to cope with the new conditions.

South America. Temperature increases will generally be smaller in the low latitudes, and projections suggest that most parts of South America will also be drier. Wheat, maize and soybean production will probably fall slightly, but less so further south where grain production may increase overall.

Europe. Large increases in temperature will be experienced in the north with smaller increases in the south. The north will also become wetter, but it may be drier in the south. There is likely to be a decline in overall grain production including wheat, maize, barley and oats.

Africa. Smaller increase in temperature, and probably wetter in most areas except the interior. Wheat production may fall overall.

Russia and western Asia. Large increases in temperature and high increases in grain production (perhaps 6% by 2000), involving wheat and maize. Barley and oats production will fall.

Eastern Asia. Moderate temperature increase, but may be drier. Wheat production will fall but rice production will increase.

India. Small increase in temperature and wetter with stronger monsoons. Lower grain production but an increase in rice.

If we examine current trends in world grain production (Fig. 11.5) it can be seen that the total land area under grain cultivation has effectively stabilized. Historically, up to 1950 increasing grain production was a consequence of greater areas of land being taken into cultivation. Between 1950 and 1980, the bulk of the increasing grain production (80%) resulted from higher productivity on the land already cultivated. Since then virtually all increase in production has come from more efficient or intensive farming of existing arable land (Brown 1992). Since global population has risen through this period, the area of grain production per head of population has actually fallen (Fig. 11.6). This will continue as the world's population grows by around 90 million each year.

It would appear that the world has virtually reached a limit as far as the areal expansion of grain farming is concerned, with the possible exception of Brazil, and even where potential exists, the conservation consequences of further expansion of cereal growing are serious. It is important, therefore, that the current areas of fertile land should not be lost, yet several factors are

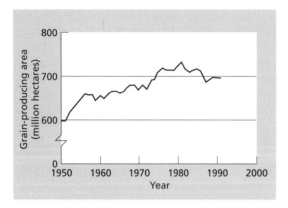

Fig. 11.5 Changes in the area of the world's land surface given over to grain production since 1950. From Brown (1992).

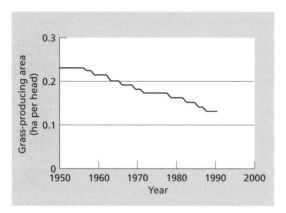

Fig. 11.6 Change in the area of the world's land surface used for grain production per head of global population since 1950. From Brown (1992).

leading in this direction. In China there has been a loss of cropland of about 10% since 1976 as urbanization and industry has developed over productive areas. Similar processes are at work in Thailand and in Egypt, where productive land is particularly limited (to the Nile Valley and Delta).

Detailed models of future world food production have been constructed by Cynthia Rosenzweig and Martin Parry (1994) and their projections for a greenhouse world with double the current levels of CO_2 (predicted by some models to be achieved by about the year 2060) are, at least in total terms, relatively reassuring for the developed nations but rather less so for the poorer nations of the world. Making reasonable assumptions concerning future population growth (rising to 10.3 billion), economic growth, technological

progress and trade liberalization, they calculate that global cereal production (wheat, rice, maize, millet, sorghum, etc.) will decrease by only a small degree, less than 5%. But this does assume improvements in farm technology (including irrigation) which inevitably will be more easily secured in the richer countries rather than the poorer ones. The outcome is likely to be that the overall number of people at risk of hunger will increase by up to 20%.

This may be regarded as a comparatively benign outcome (see Reilly 1994), especially for those lucky enough to inhabit the developed world, but many of the assumptions made could still prove incorrect, such as the response of plants to long-term elevated CO_2 levels (see Chapter 7). Such factors as canopy competition for light and root competition for water and mineral nutrients could reduce the expected stimulation of photosynthesis under high CO_2 conditions and thus reduce the expected stimulation of crop growth built into the Rosenzweig and Parry model. In addition, their projections do not take into account the possible effects of the El Niño Southern Oscillation (ENSO) (Pittock et al. 1994) (see Chapter 6). In a study of maize yields in Zimbabwe, it was found that the best predictor of yields was a study of sea-surface temperature in the eastern equatorial Pacific, on the other side of the world. The El Niño event is so closely tied into global climate variations, such as drought in southern Africa, that it impacts strongly on Zimbabwean maize production and consequently upon famine in that country (Cane et al. 1994). The same value of ENSO studies for the long-term forecasting of other crops, such as rice and cotton production in Peru (Rosenzwieg 1994), offer opportunities to predict failure and prepare for them. But such events also need to be built into the long-term modelling of global food production.

Much will also depend on the control of soil erosion to ensure that productive areas of the world are not lost by misuse. Pimentel et al. (1995) have estimated that about 10% of the energy consumption of the United States is currently spent in offsetting the loss of nutrients, water and crop productivity caused by soil erosion. Periodic stripping of vegetation from the surface of the land, which is often involved in agricultural practices, is bound to increase the risk of wind and water erosion, and up to 80% of the world's agricultural land suffers moderate to extreme erosion in this way. During the last 40 years, nearly one-third of the world's agricultural land has been lost by soil erosion, more than 10 million hectares per day, and this cannot be afforded if the world continues to grow in population and sustain the effects of the greenhouse conditions. The risks are particularly high in the continental areas of the United States, where the experience and the lessons of the 'Dust Bowl' of the 1930s do not yet seem to have been learnt.

Climate change could also affect the quality of the food materials that are being produced, but little attention has so far been paid to this aspect of the problem. In the case of wheat, for example, high growth temperatures (in excess of 35°C) can result in the grain crop producing a low-strength dough when it is subsequently used in bread manufacture (Blumenthal *et al.* 1990). This is probably because the heat-shocked embryos in the growing grain produce special protective proteins and this affects the quality of the food materials that eventually result. It is important that more attention should be given to such possible changes in quality as this may also affect the outcome of food-prediction models.

Desertification

Climate change, accompanied by increasing pressures for additional human food supplies, can lead to land degradation and loss of productive potential. This has already been observed in many of the world's drier regions where the process has been given the term 'desertification'. The dry lands clearly demand special attention when confronting the various issues raised by global change.

Approximately 40–45% of the earth's land surface is occupied by what could be termed 'desert' (the term can be used both for the hot and the cold unproductive regions of the world, but is not here used of the ice-covered regions; see Table 11.1). Deserts are characterized by dry conditions and very low productivity, often less than 0.3 kg m^{-2} yr^{1}, yet about 13% of the world's human population reside there. The demands placed upon this ecosystem are, therefore, considerable, and they often exceed the capacity of the habitat to supply such requirements in terms of

Table 11.1 The extent of the world's dry lands (United Nations Environment Programme 1992). The aridity index is obtained by dividing the mean annual precipitation of a site by its mean annual potential evaporation.

	Aridity index	Percentage of world land area
Dry–subhumid	0.50–0.65	10
Semi-arid	0.20–0.50	18
Arid	0.05–0.20	12
Hyper-arid	<0.05	8

food production. Images of drought, famine, mass migration, bare landscapes and blowing sand have created a concept of desert spread that seems inexorable. But, although the deserts are, in certain respects, fragile, they are also resilient natural ecosystems—the product of a long evolutionary process in which their inhabitants, both plant and animal, have been subjected to fluctuating degrees of disturbance and aridity.

It is important to understand initially that deserts form a perfectly natural part of the earth's cover. Even if human beings had never evolved there would still be extensive deserts over much of the globe, particularly in a belt around the world at about 30°N and S of the equator (Fig. 11.7). These are the zones where the air masses that rise, hot and moist, over the equatorial regions descend once more, having shed their moisture in the low-latitude tropics. These descending air masses produce high-pressure, dry conditions, especially when they are accompanied by prevailing winds that move from the continental land masses out over the oceans. Deserts may actually fringe the oceans, especially where the oceanic currents bring cold water to the region, thus reducing evaporation and humidity. The continents themselves create high-pressure systems in summer that help to generate these dry winds. In Asia, the deserts extend further north because of the effect of the extremely large continental land mass and the rain-shielding effect of the Himalayas, preventing the moist tropical monsoon winds penetrating into this region. The outcome is a cold desert.

The geographical conditions that create deserts can thus be clearly defined, and there are situations where these conditions may abutt quite sharply onto regions with very different climates.

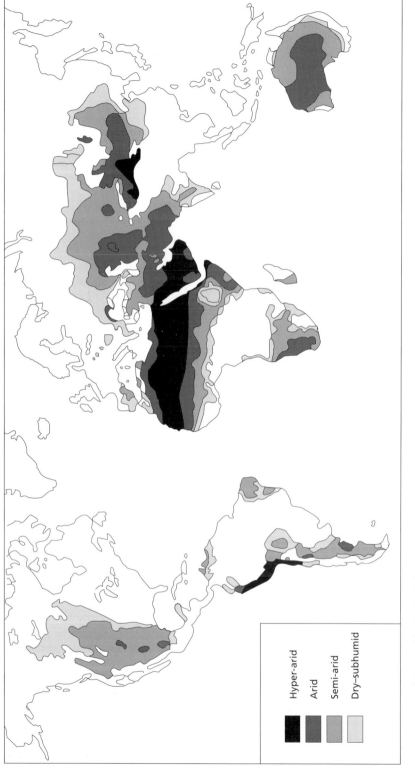

Fig. 11.7 Map of the world's dry lands based on the data of UNEP. Definitions of aridity terms are given in Table 11.1.

Hyper-arid

Arid

Semi-arid

Dry-subhumid

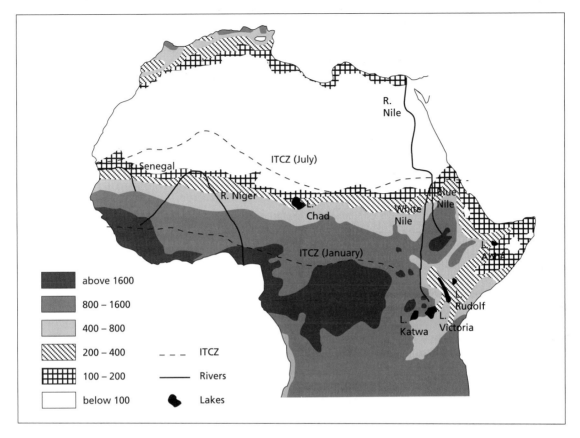

Fig. 11.8 Rainfall map of the northern part of Africa showing the steep gradient of rainfall south of the Sahara Desert. ITCZ = Inter-tropical Convergence Zone. Mean annual rainfall is given in mm.

This can be seen very clearly in the Sahel region that forms a zone immediately south of the Sahara Desert in Africa (Fig. 11.8). In this region, the annual rainfall decreases by approximately 1 mm for every 1.4 km one travels north, and this very steep climatic boundary renders the area very sensitive to shifts in climatic patterns.

Classifying climates is difficult and often fairly arbitrary boundaries have to be drawn, and this is particularly true of dryland climates where the interaction of temperature with precipitation (to determine how much water is available in the soil) makes it especially difficult to define such terms as 'arid' and 'semi-arid'. About 43% of deserts are 'hot deserts' with average summer temperatures above 30°C, while 24% are 'cold deserts' with average winter temperatures below 0°C (Allan & Warren 1993). The United Nations Environment Program (UNEP) (1992) devised an aridity index to assist in the classification of drylands in which they take the mean annual precipitation and divide it by the mean annual potential evaporation (see Table 11.1, p.177).

Climate change and deserts

We saw in Chapter 4 that climate varies quite markedly with time and in Chapter 5 it was suggested that changes in the atmosphere (the greenhouse effect) could cause a new order of climatic changes in the near future. The climatic sensitivity of deserts might well be expected to have rendered them subject to considerable changes in the past and might also expose them to alteration in the future.

There is, indeed, a wealth of evidence that indicates very different climates in the past within those regions of the world that now bear deserts. The geomorphology of deserts, for example, suggest that water has flowed with much force over areas that are now quite dry. The deep, steep-sided gullies of the 'badlands' of western North America and of the Negev Desert in Israel speak of erosion by active water transport, and geologists regard these as the outcome of shifts to wetter climates some 70 000–40 000 years ago—during the earlier part of the last 'glacial' of the high latitudes. At the height of the 'glacial', around 20 000 years ago, some of our current desert regions became dry and experienced moving sands. This was the case in the Rajasthan Desert (Singh *et al.* 1974), where sand deposits of this age are found beneath sites that later became lakes. Lake levels were low in the Arabian peninsula (McClure 1976) and dune relics are found in Nigeria (Allan & Warren 1993) that date from the period of greatest ice extent in the high latitudes.

But at the time when glaciers were finally retreating in the high latitudes, some 10 000 years ago, the strength of the monsoonal winds increased in the tropics, perhaps because the warming of climate led to low-pressure systems and instability developed (Kutzback 1981). The effects of the climate change have left their mark in the geological formations of that time. In Africa, for example, a major tributary to the Nile was in existence at this time, forming a valley that ran west to east through what is now Sudan (Pachur & Kropelin 1987). The extra discharge of freshwater down the Nile and into the Mediterranean resulted in major changes in the layering of waters in that sea, with lighter freshwater forming a cover over the top of the denser salt waters. This had the effect of creating a stable system in which there was little mixing of water layers and the deeper waters became short of oxygen, leading to the production of black, anoxic sediments, or sapropels, that can still be detected in cores from the eastern Mediterranean (Rossignol-Strick *et al.* 1982). At this time the excess water replenished the deep aquifers beneath the Sahara, so that the wells of that area today are actually drawing upon water that arrived at these depths about 10 000 years ago and cannot now be replaced by current rainfall.

Changes in vegetation, responding to the increased wetness of 10 000 years ago, are recorded by the pollen grains in the stratified sediments of the lakes forming at that time. Trees immigrated into the Rajasthan Desert of India. In Africa, Lezine (1989) has collated the information available from studies in the Sahel, showing that the belts of rainforest, woodland and savanna all moved north by perhaps as much as 500 km into what is now the Sahara Desert. But conditions then became drier once more as the present pattern of deserts began to establish itself, and from 4500 years ago onwards aridity is evident in the pollen record and in the falling lake levels.

Human beings were also present in the Saharan 'drylands' through these post-glacial changes in climate and vegetation and they sometimes left their own records of the contemporaneous environment in the form of rock paintings (Roset 1984). These are to be found in many scattered locations through northern Africa and they originated at many different times. Those older than about 10 000 years often show animals of an 'antelope' type consistent with the geological records of a dry climate. Between 10 000 and 7500 years ago there are pictures of giraffe and many 'big game' animals of savanna environments. From 7500 to 4000 years ago there is a pastoral phase in which cattle are the commonest object of the art. The presence of cattle herding in the central Sahara is surprising and would not be appropriate today. A cow demands about 40 litres of water a day (compared with 3 litres for a goat and 0.4 litres for an oryx) and this could not now be supplied in many of the areas where this pastoralism once took place. After about 4500 years ago the cattle pictures are replaced by those of horses and then camels, reflecting the cultural response to increasingly arid conditions.

So, geological, biological and archaeological evidence all point to marked shifts in climate and environment within the last 10 000 years. The archaeological record has continued to provide indications of continuing dryness and desert spread; the desolate cities of Greek and Roman North Africa that once provided grain for the Roman Empire, the sand-buried cities of the Persian and Medean Empires in eastern Iran, all supply the images that we currently associate with the word 'desertification'. The precise causes of these changes within the last two or three millenia, however, are a little more difficult to unravel, for the very presence of agricultural activities, both pastoral and arable, on the scale that was evident in, for example, North Africa could have contributed to the observed environmental degradation.

There is no doubt that 'ecologists' of the period

regarded the environmental degradation of the Mediterranean area in the 4th century BC as the consequence of human mismanagement of resources. Consider the words of Plato in reference to Attica, Greece:

In the primitive state of the country, its mountains were high hills covered with soil, and the plains of Phelleus were full of rich earth and there was abundance of wood in the mountains. Of this last the traces still remain, for although some of the mountains now only afford sustenance to bees, not so very long ago there were still to be seen roofs of timber cut from trees. ... Moreover, the land reaped the benefit of the annual rainfall, not as now losing the water which flows off the bare earth into the sea, but ... receiving it into herself and treasuring it up in the close clay soil.

Jowett 1892

Woody vegetation grows only slowly under dry conditions, so the removal of a harvest, whether by natural herbivores, human hand, or domestic grazers, is difficult to sustain at any but a low level. Total removal may also, as Plato points out, lead to erosion of soils, especially on slopes, as a result of the destruction of the binding properties of roots and the disturbance of the surface crust of lichens and algae that often stabilize bare soils in these regions. The goat is often blamed for the process of vegetation destruction, but this animal is usually the last stage in a long sequence of events in which fewer and fewer grazing animals can find enough to consume as herbaceous plants are removed. The goat has a distinctive foot musculature that enables it to climb trees; it has a wide dietary tolerance; and it has a very high fecundity (Dunbar 1984). So it can exploit the final resource, the scrub vegetation that is left when all else has been removed, and its vigorous depradations at this stage have left it with a reputation for destruction that is only partly deserved.

We have examples of undoubted dryland mismanagement and consequent disasters in the present century, as in the American experience of the Dust Bowl of the Great Plains (see the account of Thomas & Middleton 1994). Pioneer farmers ploughed up ancient grasslands to grow wheat, beginning in the 1870s, but these early days were times of good rainfall and fairly low population density. Further settlement increased pace between 1924 and 1930 and larger areas were brought under the plough. Then long-term unreliability of the climate made itself felt and drought arrived in 1931 leading to widespread wind erosion and dust storms so graphically described by John Steinbeck in his *Grapes of Wrath*. By 1937, 43% of the 6.5 million hectare region at the centre of the Dust Bowl had been damaged by erosion. The emigration of the human population as a consequence of this experience involved the lowering of human densities in the great Plains between 25 and 50% over much of the area during the 1930s, and this trend continued into the 1950s (see Wigley *et al.* 1981). But the lessons of soil conservation were not well learned, for a similar period of erosion was experienced once more during the early 1970s as a result of further ploughing of grassland coinciding with dry conditions (Glantz 1994).

Links between overcultivation and the spread of drylands have been extensively investigated and some authorities, such as the World Bank, have concluded that decreasing yields of some crop species can be associated with mismanagement (Fig. 11.9). Often the problem has arisen as a result of the replacement of traditional dryland, rainfed methods by alternative systems using irrigation and high-intensity techniques. It is difficult, however, to distinguish between inappropriate management and the coincident impact of recent drought on such agriculture. Irrigation (once highly favoured by development agencies) has certainly created problems in many dry areas, leading to an initial increase in crop yields but eventually causing salinization, in which the evaporation of excess water brings salts and lime to the soil surface often reaching concentrations that few plants can tolerate, raising soil pH and reducing the availability of surface soil water to plant roots.

The idea of human-induced desertification thus dates back to Plato and was given immediacy by the Dust Bowl experience, but the word seems to have been coined in 1949 by Henri Aubreville, a French forester writing a report on the changing vegetation of tropical Africa (Glantz 1994). It is a term normally applied to the degradation of drylands as a result of overgrazing, overcultivation, vegetation removal or even ill-managed irrigation. The United Nations study of the subject (United Nations Conference on Desertification (UNCOD)—see p. 183) came up with the follow-

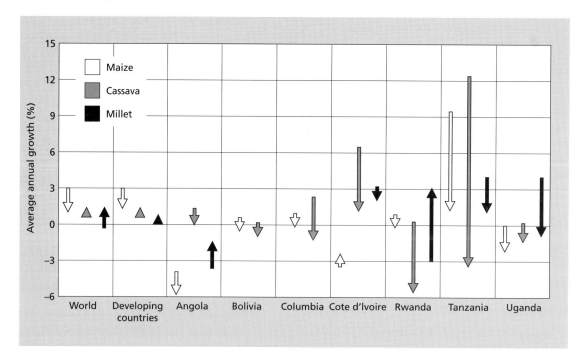

Fig. 11.9 Yield changes in certain selected staple crops between 1970 and 1990. Data from World Bank sources (Thomas & Middleton 1994).

ing definition. 'The intensification or extension of desert conditions ... leading to reduced biological productivity, with consequent reduction in plant biomass, in the land's carrying capacity for live-stock, in crop yields and human wellbeing'. Other definitions have specifically involved the activity of humankind in the development of these condi-tions. Although the problem is far from being a modern one, we have become more familiar with its consequences in recent times, at least in part due to the improved global communications that allow us to hear of, and mentally participate in, disasters in remote areas.

The role of drought in the process of desertifica-tion has been much debated. It can certainly be claimed that drought serves to reveal desertifica-tion most graphically, as was shown in the history of the Sahel during the 1970s. The Sahel, as we have seen, is a long strip of dryland running east–west along the southern edge of the Sahara desert. It crosses many political boundaries and in the 1970s several of the States involved had rela-tively recently become independent following

colonial rule. For several thousand years it has been occupied by human populations who have burned the vegetation to encourage grass growth for stock grazing, have felled acacia trees for fuel, charcoal and gum arabic, and in recent times have used firearms in the destruction of much of the local wildlife, including the birds that disperse tree seeds in their guts. Then, in the 1950s and 60s, a series of wet years (see Fig. 11.10) led to increased settlement and farming, gradually extending northwards into areas that had previ-ously been too dry. Populations of humans and their domesticated animals, including herds of cattle, boomed.

In the summer of 1968 the rains failed and so did the crops and the natural grass vegetation. The actual cause of the drought was the failure of the monsoon to penetrate as far north as usual, per-haps as a consequence of the weak jet stream of air mass circulation that held back the tropical air masses (Fig. 11.11) (Winstanley 1973). When rain was still in short supply for the following two years the result was the death of stock and a short-age of grain for the population. About three mil-lion people in the western Sahel needed urgent food aid by 1970. In the next three dry years per-haps as many as 100 000 people died and 3.5 mil-lion animals (Myers 1994), especially the cattle

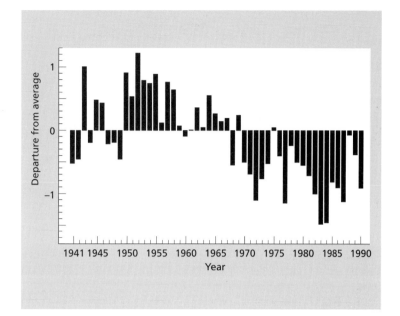

Fig. 11.10 The pattern of rainfall in the Sahel region of Africa over the past 50 years, showing the variation above and below average values. The 1950s and early 1960s were relatively wet years, but subsequently the climate has been drier.

that are so poorly adapted to drought conditions. Since that time there have been years when the drought has lessened, but the general pattern, as shown in Fig. 11.10 is one of persistently low rainfall that has been maintained now for well over 20 years.

The future of the drylands

The problem of desertification is by no means limited to the Sahel, but this region has proved important in bringing the subject to the world's attention. The United Nations set up a UN Conference on Desertification (UNCOD) in 1977 and from this emerged the aims of global awareness of the problem, the collation of information on the subject, and the commencement of a programme to reverse the desertification process. One of the results of the Conference was the production of a global map of those areas most at risk from continuing desertification (Fig. 11.12). This was based upon observations of climate conditions and land-use factors operative at that time. Since then we have become increasingly concerned with the possible effects of global climate change and we need to consider the future of the drylands in the light of a further two decades of research and experience.

From 1950 to 1979 the world saw a small but consistent rise in global temperature (see Fig.

5.16) and detailed records from around the world have shown that this rise has not been uniform. Some regions have become much warmer whereas others have become cooler. A map showing the deviations ('anomalies') of temperature from the global mean values is given in Plate 11.1 (facing p.136) and this provides us with a basis for more detailed projections about the future of the arid lands. Plate 11.1 shows that the areas which are warming most rapidly are generally the high latitudes, particularly the Arctic and Antarctic, together with Siberia, Canada and Alaska. On the basis of tree-ring studies, Siberia seems to have had hotter summers during this century than at any time since 914 AD (Briffa *et al.* 1995). But some of the lower latitude 'hot spots' include those areas that are presently deserts. The midwest of North America, the west coast of South America, both the east and the west of the Sahara in northern Africa, the western part of southern Africa, and central Australia, are all regions that have increased their mean temperatures in excess of the general global rise.

If the predicted greenhouse world continues to develop and we see a continuation of the trend towards higher global temperatures, and if the present pattern of global anomalies continues, then the drylands may well be heading towards further problems. Increasing temperature, if it is accompanied by continuing aridity, may further

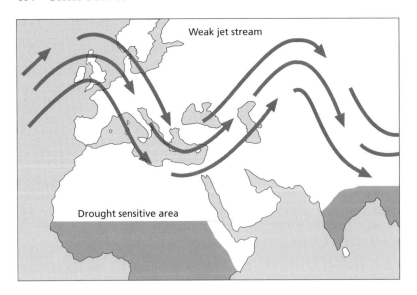

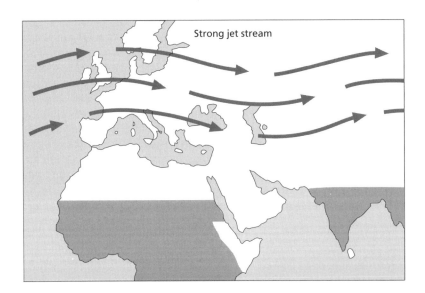

Fig. 11.11 Changing patterns of air mass movement in the 'jet stream' and its consequences for drought in the Sahel and India. When the jet stream is strong the monsoons (shaded area) penetrate further north in Africa and India, bringing rain to drought-prone regions. A weak, meandering jet stream, on the other hand, blocks this northward movement of the monsoons and results in drought in the Sahel and north-west India (Rajasthan). After Winstanley (1973).

reduce the primary productivity of these regions and place pastoralists under even greater stress. Migrations of people and stock out of the arid cores and into the desert fringes will place added strain upon the resources of these, already pressurized, habitats and may well lead to a continuing story of famine and social upheaval.

Viewing the desert fringes from an ecological standpoint, however, it would be wrong to give the impression that we are dealing with a sensitive, fragile ecosystem that is easily and irreparably degraded by human activity. Desert vegetation has evolved in a harsh and unpredictable physical environment and it is adept at recovery following catastrophe, whether induced by drought or overgrazing or both. Areas currently degraded and those that will undoubtedly be damaged in the future can recover, but the carrying capacity of this low-productivity ecosystem is very limited and this applies to the human species just as much as to any other animal that seeks to exploit the habitat as a source of food or an energy

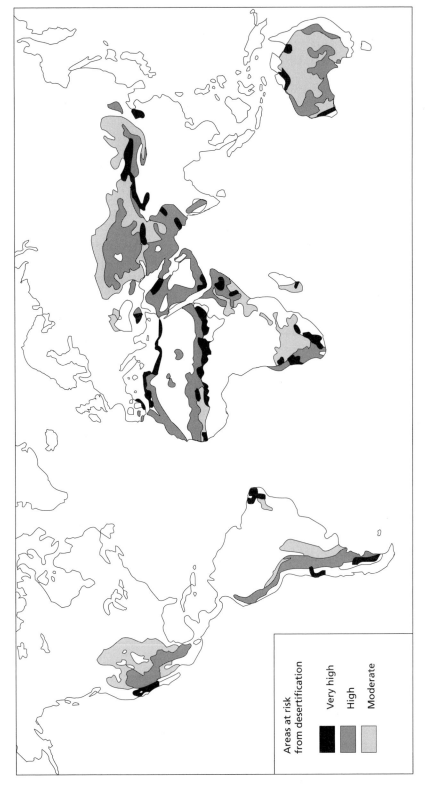

Fig. 11.12 The areas of the earth most at risk from desertification, as determined by the 1977 United Nations Conference on Desertification (UNCOD).

Areas at risk
from desertification

Very high

High

Moderate

resource. The misuse of the desert fringes is and will remain (whether or not the climate changes) a social and a political rather than an ecological or an agricultural problem.

Forest resources

Our exploitation of biological resources is not confined to our need for food. Around the fringes of deserts, the exploitation of vegetation as an energy resource for cooking and heating is apparent. The use of plant material for other purposes, such as drugs, perfumes, fibre for clothing, paper for communications, and timber for building, is a basic component of almost all human cultures. Trees have a particular value for humankind for a whole range of purposes and this has led to very extensive exploitation of the forests of the world. There is an additional reason, however, why forests have often been a focus of destructive management, and that is the difficulty of maintaining a sustainable production of food from them under conditions of high population density. Most domesticated plants are light-demanding, so clearance of shade-producing trees is required. Most domesticated animals can be herded at a greater density and with higher yields of meat, milk and hides if they are maintained in grasslands rather than forests. So the forest ecosystem has long been declining under the impact of agricultural practices.

The history of forests within the United States provides a clear example of this process (Fig. 11.13). Native American populations certainly had some impact upon the forests, clearing areas for agriculture and burning other regions in their hunting activities, but there is still considerable argument about their overall influence on the extent and composition of the forests. It is possible that their use of fire, in particular, may have had some effect upon the developmental processes of the forests, favouring those species that were fire resistant, but natural fires may well have been selecting for such species even in the absence of human populations. The arrival of European colonists led to population expansion, more widespread agriculture and hence forest clearance for settlement. This began in the east, especially along the coastal regions, but soon spread inland, especially clearing areas adjacent to the prairies.

Between 1850 and 1920 the western forests began to be clearfelled for logging.

One consequence of forest clearance is that the United States has been a net importer of wood since 1940. The loss of old-growth, virgin forest has resulted in irreversible damage, since second-growth forest has a different composition and structure from the former ancient woodland and many species may be severely reduced or rendered extinct in the transition. Forest conservation as well as forest management is thus becoming a major issue in American politics (see Chapter 13).

In Europe a similar situation exists, with virtually no primary forest surviving, while in temperate Asia considerable pressures are now being exerted upon the extensive coniferous forests of Siberia. Overall, however, forest loss appears at present to have stabilized in the temperate zone, mainly because of the implementation of reforestation programmes (Barbier *et al.* 1994). Perhaps most attention, however, is currently focused on the tropical forests where clearance for timber and agricultural development is proceeding apace, both in Africa, South-East Asia and in South America.

About 6% of the earth's land surface and about 36% of the land area within the tropics is occupied by tropical forest, but the clearance rate is exceedingly high. It is difficult to estimate this rate accurately, but satellite imagery has assisted in the process and a recent estimate by the UN Food and Agriculture Organization produced a figure of $154\,000$ km^2 yr^{-1}. This means approximately a square kilometre cleared every three minutes of the day and night. About half of the world's tropical forest is situated in the Amazon Basin of South America, and here it is feared that the rate of forest clearance is proceeding in an exponential fashion. There are several reasons why such changes are regarded as detrimental both to the global environment and to the future of our species. Forest loss, especially in a region of the world that probably contains about half of the earth's complement of plant and animal species, is going to involve a loss of biodiversity. The removal of forest will reduce soil stability and lead to soil erosion, thus leading to the silting of the rivers and the destruction of agricultural potential in the land. Nutrients will also be lost as the great reservoir of elements necessary for life become miner-

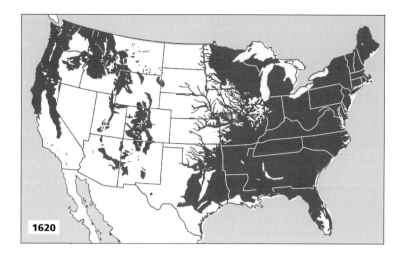

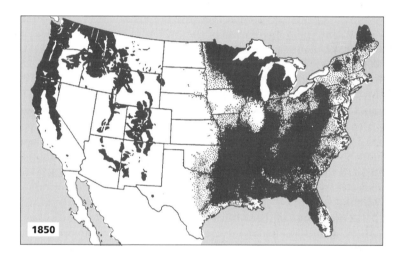

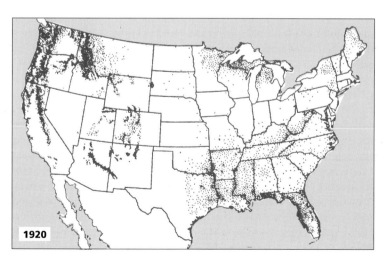

Fig. 11.13 The extent of virgin forest in what is now the United States of America in 1620, 1850, and 1920, showing the progressive destruction of American old-growth forests.

alized and released into the physical environment where most will be lost by leaching of the soils and erosion. This is particularly critical in tropical forests where the major reservoir for most elements is the biomass rather than the soil. That important element, carbon, will be returned to the atmosphere as a result of the decay of vegetation and soil organic matter, adding to the greenhouse effect. Since the vegetation replacing the forest is often grassland (with a lower biomass) there is no equivalent reabsorption of carbon as the ecosystem regrows. And forest removal will also affect the hydrological cycle, possibly leading to local climatic changes.

In a simulation study to model the effects of clearance on an area of Amazonian forest, it has been shown that evaporation and precipitation diminish following clearance, because the interception and the transpiration of the trees is lost and the resulting grasslands behave differently with respect to incident light, reflecting more back into the atmosphere than is the case with a forest canopy. The overall outcome was a higher temperature because, although more heat is reflected, the lower evaporation reduces this method of heat loss. The soils have lower infiltration of water, rendering erosion more likely. Rainfall shows greater variability, but is, overall, lower than when the forest is present (Lean & Warrolow 1989).

The protection of the world's forest reserves is clearly an important global issue, but it is somewhat hypocritical for the developed countries of the earth, most of whom have already exploited and destroyed their own forest resources, to demand that the developing nations should be deprived of the opportunity to raise their own standard of living by harvesting and exploiting their own natural resources. Hopefully, these nations will learn from the mistakes of the past in temperate areas and manage their remaining forests in a sustainable manner to ensure long-term rather than short-term profit. In Indonesia, for example, commercial logging at its present rate could only survive for another 30 years at a maximum, and already many of the most valuable timber trees are scarce. If forests are to provide a reliable livelihood for the native peoples, then exploitation has to be tempered with conservation, leading to a sustainable yield. During the last decade such countries as Ethiopia, Liberia,

Bolivia, Guatemala, Bangladesh and Vietnam have seen their sawn timber production decline by 40% because of former over-exploitation. Replantation schemes account for only very small areas of tropical forest (less than 2%) and have often failed either because of lack of financial inducements on the part of government, or the use of the wrong types of tree (often exotic and unsuitable species), or the harvesting of the next generation of trees far too early. Some form of community forestry in which the local population profits from and therefore takes an active interest in the maintenance of its own forest areas, seems the only way ahead (Ascher 1995).

Fisheries

The exploitation of the natural environment by our species is not confined to terrestrial ecosystems; we also look to the marine world as a potential source of food. As we saw at the beginning of this Chapter and in Fig. 11.1, the oceans are responsible for a large proportion of the total net primary productivity of the planet (about 41%). But, although our exploitation of terrestrial food resources depends largely upon plants and domesticated animals, the food we obtain from the seas consists of neither of these. Oceanic primary production is largely the outcome of planktonic photosynthesis and the very small organisms responsible cannot be harvested economically. But we can harvest some of the larger animals higher up the food chain, particularly the fish.

We must regard fish as populations of wild animals, and fisheries are essentially a modern extension of the ancient hunter–gatherer system of resource exploitation. Fish farming, employing such species as salmon and trout, represents a major step towards domestication, but these activities still form a relatively small proportion of our fish catch. Fish represent an important form of protein for the world's human population, accounting for about 16% of the animal protein supply overall. In many developing countries, however, fish may provide as much as 40% of the animal protein resources (Kane 1992). In some industrial nations, such as Norway and Japan, fish provide the main source of animal protein. It is important, therefore, to protect the world's fish resources from the harmful effects of pollution and of overfishing, especially when other global changes, such as increased ultraviolet radiation (as stratospheric ozone is lost), increasing ocean temperatures, and changing ocean movements as the ice-caps melt, may influence fish stocks in unexpected ways.

The exploitation of the world's fish resources resembles that of many of the world's forests in certain respects, especially in that wild, relatively unmanaged ecosystems are being exploited for their produce. Just as there are many species of tree, there are many species of fish naturally inhabiting the aquatic bodies of the earth, both fresh and salt water, and some of these are preferable to others as far as human exploitation is concerned, both in terms of palatability and in ease of extraction. In the case of enclosed, freshwater bodies, the control of extraction and even the opportunities for restocking, allows for efficient management. As far as the open oceans are concerned, however, the international nature of the resource leads to complications for management and there is a danger of short-term competitive gain becoming a stronger force than long-term rational management of fish stocks as well as those of other exploited marine animals, such as crustaceans, whales and seals.

Since the 1930s it has been recognized that international management has become necessary and a range of advisory bodies has been set up to try to co-ordinate marine animal exploitation, including specialized bodies like the International Whaling Commission, the North Pacific Fur Seal Commission, the International Pacific Salmon Fisheries Commission and the International Commission for the Conservation of Atlantic Tuna. In addition, there are Commissions dealing with specific geographical regions of the world's oceans, the NorthWest Atlantic Fishery Organization, the International North Pacific Fisheries Commission, the Japan–South Korea Fisheries Commission, and so on. Some of these have their own research and monitoring teams and set quotas based on concepts of sustainability in fish crops. More recently, Exclusive Economic Zones (EEZ) have been set up that are organized by coastal states and these have become responsible for determining quotas.

Changes in fish populations, however, are very evident and suggest that past policies are failing. In the North Sea between Britain and the European mainland, for example, the once-abundant herring and mackerel have declined and smaller species are now being exploited. This can have wide repercussions since it will affect other components of the marine ecosystem, such as the sea-bird populations that feed on smaller species of fish. The question of sustainable management and marine conservation is clearly an urgent one and the more rigorous enforcement of quotas is required. But the monitored data and the theoretical models on which such quotas are determined also need to be overhauled and assessed. The structure of the ecological food webs needs to be more thoroughly understood and specific fish need to be targeted for their potential yield.

The ecological principles operating here are precisely the same as those upon which land-based agriculture is founded. Photosynthetic organisms—green plants—are responsible for the primary fixation of energy and these are fed upon by herbivorous animals, which are in turn preyed upon by a series of carnivores. Detritivores and decomposers mop up the residual energy found in the dead remains of any organisms that are not thus consumed and also in the faecal and excretory products of the animals. Because energy is lost at each transfer along a food web, it is energetically more efficient to harvest at an early stage in this web. In a terrestrial system, for example, it is more efficient to harvest and consume lettuces than the rabbits that eat them; and it is certainly more efficient to hunt and eat rabbits than to rely on, say, foxes or buzzards as a source of food.

In aquatic ecosystems, however, there are some important differences in food-web structure and organization that complicate questions of exploitation. The primary producers are mainly planktonic algae and are individually too small in their body size to provide an appropriate source of direct food. (The future may cause us to alter attitudes in this respect, but direct harvesting of oceanic phytoplankton would result in serious problems both for the conservation of fish stocks higher in the food chain and possibly in the global carbon cycle—see Chapter 5). Many of the herbivorous animals that feed upon the phytoplankton (the zooplankton) are also small and unattractive as food. So many of the fish upon which we rely are ecologically situated quite high up the food chains, making them energetically relatively inefficient as a source of food. Salmon are a good example of this (see Fig. 11.14). On the other

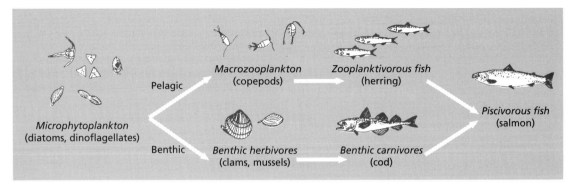

Fig. 11.14 Food web for the salmon, a marine predatory fish.

hand, the biomass of these higher trophic levels is often greater than the lower ones because the latter have such a small body size, a rapid turnover and short life expectation. This contrasts with the lettuce–rabbit–fox system in which the biomass of foxes is much smaller than that of lettuces or rabbits.

As a general rule, the energy transfer efficiency between trophic levels is assumed to be about 10% (May 1976). This means that 10% of the production of one trophic level (say, that of the primary producers) passes on to the production of the next level (the herbivores). In a survey of data concerning such transfers, Pauly and Christensen (1995) have analysed 140 transfers in 48 different aquatic ecosystems and have found a mean value of 10.13%, so the well-worn assumption seems to hold good for most aquatic ecosystems. Using this figure, together with the catch figures for particular fish species in different systems, it is possible to calculate how much of the primary production is actually being harvested by human predators when they fish. Pauly and Christensen have carried out these calculations for many different oceanic and freshwater ecosystems and have been able to obtain a global fisheries figure. The area of the water bodies of the earth totals 364 million km². The mean primary productivity of aquatic systems is 126 g C m⁻² yr⁻¹. The mean level of fish catch is 0.26 plus a further 0.07 g C m⁻² yr⁻¹ that is discarded as unusable (21% of the total catch). After considering the trophic level at which the caught fish feed, it is possible to work back to the amount of the primary production effectively harvested via these fish, and it comes to 8%. So, the current level of global fisheries is actually harvesting a total of about 8% of the available oceanic primary productivity.

By way of comparison, we could consider the harvesting of sheep from grassland. The sheep in temperate upland areas are quite efficient and consume about 20% of the net primary productivity of the grassland, but most of this is lost in the respiration of the sheep together with the faeces. The actual harvest of sheep will be only about 1% of the energy ingested, so the sheep farmer is recovering only 0.2% of the primary production of the ecosystem.

So the figure of 8% for fishery exploitation of oceanic primary production can be seen to be extremely high and it is even more a matter for concern when the uneven nature of fisheries around the globe is taken into consideration. Most fisheries are concentrated upon the near-shore continental shelves and in the oceanic upwelling areas of the oceans. Table 11.2 shows the breakdown for different aquatic ecosystems and it emerges that less that 2% of the available primary production is harvested in the open ocean, while as much as 35% may currently be harvested in the non-tropical shelf areas and 25% in the upwellings. It is to be seriously doubted whether such exploitation is sustainable. Since so many of the declining fish stocks are also those from high trophic levels (like the salmon), the outcome is bound to be an increasing emphasis on the lower levels (shrimps, krill, etc.) and this will inevitably have its effect on the composition of aquatic ecosystems, biodiversity and the conservation of marine mammals and birds (Beddington 1995).

The data also demonstrate that the open ocean is, relatively speaking, under-exploited compared with other waters, and this may be a future source of increased fishing intensity. The trouble is that

Table 11.2 Global estimates of world aquatic primary production (PP) (g C m^{-2} yr^{-1}), the current fish catch, including discards (g m^{-2} yr^{-1}), and the proportion of this required to sustain current fisheries (based upon Pauly & Christensen 1995). In calculating proportions, a conversion factor of 1 : 9 for C : wet weight of fish has been used.

Ecosystem	Area (10^6 km^2)	PP	Catch (g m^{-2} yr^{-1})	Mean percentage PP
Open ocean	332.0	103	0.01	1.8
Upwellings	0.8	973	25.56	25.1
Tropical shelves	8.6	310	2.87	24.2
Non-tropical shelves	18.4	310	2.31	35.3
Coastal/reefs	2.0	890	10.51	8.3
Freshwater	2.0	290	?4.3	23.6
Means (or totals)	(363.8)	126	0.33	8.0

the open oceans have low planktonic productivity and the fish stocks are therefore less dense and less economically attractive (UNEP 1993). An alternative is to 'domesticate' fish and to bring them into an aquaculture system of farming. This has already begun for both freshwater and marine species, especially in Japan, China and India (see Fig. 11.15) and approximately 10–12% of the world's fish harvest comes from aquaculture. This is likely to increase in the future, some estimates proposing a doubling of aquaculture harvests by the year 2010 (Tolba & El-Kholy 1992). There are numerous technical problems associat-

ed with aquaculture, including fish disease and problems of excess food and antibiotics affecting local ecosystems in the vicinity of sea cages (Goldberg 1993). A salmon fish farm of 0.8 ha containing between 50 000 and 100 000 fish produces about 100 tonnes of faeces and uneaten food in a year (Barinaga 1990) and this can influence the nature of food webs in the vicinity since it represents a considerable energetic input to local ecosystems. The stimulation of microbial activity can also lead to anoxic conditions developing on the ocean floor with a consequent death of the local aquatic communities.

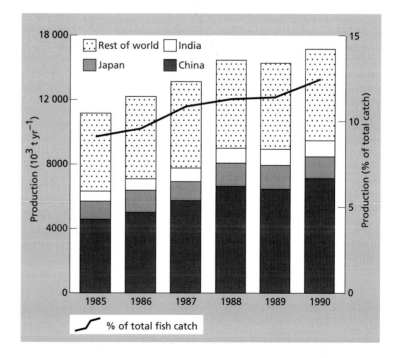

Fig. 11.15 Global production of fish by aquaculture in millions of tonnes per annum. Total global fish catch is approximately 100 × 10^6 tonnes (see Fig. 11.16), so aquaculture accounts for about 10–12% of the total (UNEP 1993).

The fact that fishing is approaching (and may have exceeded) its sustainable biological and economic limits is apparent from overall data on global catch. Figure 11.16 shows the rising global catch over a 40-year period up to 1990 and it can be seen that there has been a steady rise over that period. The steep rise in the 1980s is largely due to increased intensity of fishing for Peruvian anchovy, South American sardine, Japanese sardine and the Alaskan pollock (Kane 1992). It is also apparent, however, that marine fishing has recently begun to fall. When this is expressed in terms of fish catch per person (thus allowing for population growth during this period) it is apparent (Fig. 11.17) that fish catch per head of global population has been fairly steady for some while and is now beginning to fall back. The future may well be determined by economics. If sea-food prices continue to rise the exploitation will also remain high, which will be most harmful to the world's poor (Kane 1992).

Monitoring the catch of fish on a world-wide scale is difficult, but it is even more difficult to monitor remaining stocks in the oceans Obviously, such information is vital if we are to understand the impact of current fisheries upon wild stocks and their natural replenishment. Even in the case of large and apparent marine animals, such as the whales, monitoring stocks has proved a very difficult, and often contentious task. Take the well-watched grey whales that migrate along the coast of California, for instance. The counts achieved there are thought to represent just 15%

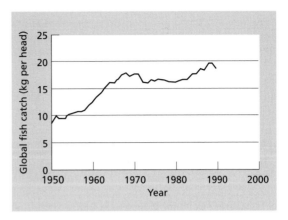

Fig. 11.17 World fish catch expressed in terms of kilograms per head of global population over a 40-year period. From Kane (1992).

of the total population, given the problems of visibility, distance, observer fatigue, etc. But this can only be a crude guess regarding what proportion of animals are actually observed. It is reckoned that 4000 sperm whales visit the Galapagos Islands regularly, and from this a world population of 400 000 animals can be extrapolated, based on the fact that the American whalers of the 19th century caught just 1% of their sperm whales in that area. But conservationists argue that such figures are so unreliable that it is unwise to base a harvesting programme upon it, especially in the case of such slow-breeding creatures who are consequently prone to population decline and extinction (Schmidt 1994).

One reassuring aspect of the fishing problem (in contrast to the whales) is that fish have a high fecundity and should prove resilient following exploitation if they are given adequate chances to recover. A single female may produce as many as a million eggs and if the mortality falls just fractionally, from 99.998% to 99.996%, for example, it should result in a doubling of recruitment to the population (Gulland 1993). But other factors, such as pollution, could make this much more difficult than it appears and there is a danger of complacency in the light of such data, as was the case with the over-exploitation and extinction of the passenger pigeon and very nearly with the fur seal. It is also the case that sharks and rays have much lower fecundities and may prove highly susceptible to extinction from over-fishing. There are also numerous cases of unexplained and sudden

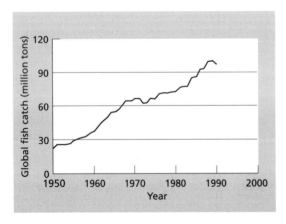

Fig. 11.16 Global catch of marine and freshwater fish over a 40-year period up to 1990. From Kane (1992).

declines in certain species, such as the Peruvian anchovy and the Californian sardine, whose population collapses could be associated with El Niño events (see Chapter 6).

One feature that emerges strongly when considering the world's fisheries is that the fish and other food-source stocks are limited and can very easily be exploited beyond their capacity to keep pace. Much more information is needed on the biology of the individual species (in conjunction with reliable prediction concerning oceanic changes with climate change) if we are to continue to draw upon the biological resources of the oceans for the support of our species.

Biodiversity and conservation

Many of the processes discussed in this chapter are ones which lead to habitat changes or to alterations in the population levels of certain species, which could eventually bring them to extinction. It is a fairly generally accepted principle in modern thinking that extinction is a 'bad thing' and should be avoided at all costs (and sometimes the costs can be pretty high, as in the case of the Californian condor). But there are two points that should be borne in mind before jumping to conclusions about extinction.

In the first place, extinction is a perfectly natural process that has been taking place ever since life evolved on this planet. The replacement of one species by another is part of the evolutionary process itself and ensures that the assemblage of forms currently in existence includes only those best fitted to survive in the present set of conditions. When conditions change, the losers will leave the stage. It is also a fact that the history of extinction has not been a smooth affair during the course of the earth's history. The pattern of extinction has been varied, with certain points in time, such as the end of the Permian and the end of the Cretaceous showing abrupt episodes of mass extinction, possibly brought about by catastrophic changes in the earth's environment (see Chapter 2). But extinction is a process that we should expect even on a healthy planet.

Second, there may be species whose demise would cause us very little distress. The smallpox virus, for example, is now effectively extinct as a 'wild' organism. It survives only in a small number of laboratory cultures and the debate over whether even these should be destroyed is a complex one. There is the medical argument that some unknown source (buried in a crypt or frozen in the permafrost) may survive and we may need a captive sample to develop a future vaccine. There is also an argument based on the pure research value of retaining the species. Again, as in the case of nuclear weapons, one could argue that it should be retained as a deterrent in case someone of evil intent might withold an illicit colony and misuse it. So smallpox for the present survives in a bound state.

But there are other species, particularly viruses and bacteria, that we spend much time and money in trying to eradicate. Few would regret the extinction of the HIV virus, for example. Many other pests and parasites are not really welcome in our environment, from ragweed to tapeworms. It is also salutary to recall that some of the animals whose conservation many of us support would not be particularly welcome if one had to live in their close proximity. The Indian tiger is a magnificent animal at a distance and most conservationists would unhesitatingly support its protection in its now restricted wild habitats. But few of us would like to conduct pastoral farming or bring up young children within its range. Most temperate-dwelling conservationists would think carefully before encouraging the reintroduction of wolves in their local neighbourhood, but we demand such tolerance of others.

So, our opposition to extinction needs to be considered carefully, rather than adopted without thought. On the other hand, the rate of known extinctions over the last century, quite apart from those that are unknown, does give cause for concern. There are several reasons why various people support, in general, the maintenance of species diversity and the avoidance of extinction.

1 There are ethical and religious arguments relating to the rights of animals and plants to an existence. Such arguments are beyond the sphere of science to comment upon, but most people would regard it as reasonable to argue that other species besides ourselves have a right to exist and should not be deprived of that right. Most of the world's great religions (especially Buddhism, but including Judaism and Christianity) reserve an important place and role for species other than humanity and teach responsible attitudes to nature in their doctrines. Philosophies and reli-

gions differ, of course, in their evaluation of different species, especially in relation to human life.

2 There is the aesthetic argument. Plants and animals provide us with many scientifically-unmeasurable pleasures. They enrich our lives both through direct contact and via literature, poetry and art. Their distinctly measurable psychological value, whether as cut flowers, as cultivated gardens, as pets, or simply as a concept of a wildness apart from ourselves, are medically well attested.

3 There is the simple, but powerful argument that 'extinction is forever'. In other words, the process is irreversible so once effected we cannot recreate that species. Most of us are probably sufficiently conservative to hesitate before throwing something away, and losing a species does strike us as something we should recoil from, if only to avoid the subsequent discovery that we have made a grave mistake.

4 Since the ecosystem involves the interrelationship of many different species, linked by energy flow and nutrient cycling processes, and since the population levels of different species are often determined by the presence and abundance of other species (food species, predators, parasites, competitors, etc.), the loss of one species may set off a cascade of other extinctions. If ecosystems are modified or lost in this way, there could be repercussions of a wider nature, affecting local, regional or even global conditions and very possibly influencing human populations. Species conservation is thus part of the whole process of protecting the natural environment.

5 There is the utilitarian argument that a species may prove to be of direct use to us—in agriculture, medicine, pharmacology, industry, theoretical research, and so on. This is the argument that is most frequently quoted by scientists because they can defend it most easily in their own language and terms. It is also an argument that is most easily appreciated by the general public, by governments and by authorities responsible for the allocation of research funds. These are certainly many instances of new uses being found for living organisms, such as the use of yams in the manufacture of contraceptive drugs, the value of the Madagascar periwinkle in the treatment of leukaemia, the possible use of the armadillo in leprosy studies, the extraction of dopamine drugs from cycads, useful in the treatment of Parkin-

son's disease. The natural products of plants and animals have been used in folk-pharmacology throughout the history of our species, sometimes with remarkable success, and there is undoubtedly still a great deal more to be learned, both from ancient and traditional sources and from modern survey work and screening.

The number of plant species currently used for agricultural production is very low, 16 species accounting for most of the world's primary agricultural output (Table 11.3). Undoubtedly this could be improved and new species are continually being sought. The same applies to animals, since we have long depended on a small number of traditional species and breeds. There may well be opportunities to domesticate and harvest wild game species in East Africa, for example, where native species are often more drought- and pest-tolerant than introduced animals. Or we could make greater use of less familiar animals, like ostriches and llamas, as a means of managed meat production, even in temperate areas.

Besides the conservation of individual species with a view to their future use in agriculture and medicine, one should also bear in mind the value of conserving genes. Even within a single species one may find a wide range of variants, both morphological and physiological, that represent modifications of the basic genetic pattern in response to particular environmental pressures. In the case

Table 11.3 The 16 most important of the world's food plants and their global production in millions of tonnes (from Campbell-Platt 1988).

Sugar cane	800
Wheat	500
Maize	400
Rice	400
Sugar beet	400
Potato	300
Barley	220
Cassava	150
Sweet potato	120
Soyabean	100
Sorghum	80
Grapes	70
Plantain ⎫ Banana ⎭	60
Oats	55
Tomato	50
Total	3700×10^6 tonnes

of cereal crops like wheat and barley, for example, there still persist many wild ancestral forms in the Middle East which contain diverse genetic adaptations to drought, salinity, pest-resistance, low nutrient intake, and so on, that could be incorporated into commercial varieties so that these traits can be combined with high productivity of grain. Advances in biotechnology have made such genetic shuffling far easier than in the old days of prolonged breeding programmes, so if the genes are still available the necessary exchanges can usually be conducted. The maintenance of the genetic pool of a species may involve widespread collection of individuals and their maintenance in botanical gardens and zoos, the storage of tissue or of seed or sperm and eggs in specially designed banks, and the conservation of wild populations in their natural habitats.

Mexico is the probable source area of one of the world's most important crop plants, maize (*Zea mays*). This plant, domesticated at least 5600 years ago, is now the staple food of hundreds of millions of people. It is also a food plant for domesticated animals and has wide use in industry in the manufacture of adhesives, alcohol,

paper and construction boards. In 1979 a new (fourth) species of wild maize was discovered in Mexico, *Zea diploperennis*, which has opened up a wide range of possibilities for putting back some of the disease resistance into the modern, highly modified varieties of maize. This new species is endemic to one small area of Mexico, just 360 ha in extent (see Fig. 11.18) (Guzman & Iltis 1991), and its survival and discovery is a remarkable piece of good fortune for plant breeders and hence for a large proportion of humanity. The area in which it grows has been designated an International Biosphere Reserve, but there is now, of course, the problem of how to manage the reserve in order to ensure the survival of the species. *Zea diploperennis*, like so many of the grass species that are the ancestors of our cereals, are essentially weeds, growing as opportunistic invaders of disturbed habitats when the more robust species have suffered a set-back, often as a result of destructive human activity. Simple protection of this reserve and the exclusion of humans would undoubtedly lead to the extinction of this valuable plant. A management regime similar to that which has prevailed up to the pre-

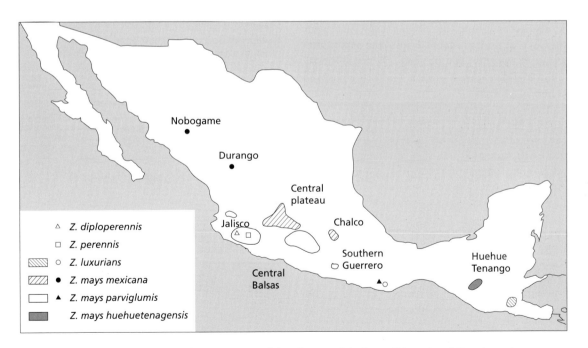

Fig. 11.18 Map of Mexico showing the very restricted distribution of the four wild species of *Zea*, the maize genus. From Guzman & Iltis (1991).

sent is obviously required, including human disturbance. Conservation, in the modern world, can rarely mean leaving a species or a habitat to its own devices.

It is, perhaps, ironic that so much of the world's genetic wealth is stored within its poorest countries. Companies made rich by the development of agriculture and pharmacology are now seeking new genes in many of the developing nations. The United Nations Food and Agriculture Organization (FAO) has become increasingly conscious of our agricultural and medical debt to the resources of the developing world for the provision of new genes for globally important crops and drugs. Legislation is clearly necessary to protect seed source, plant material, and even the folklore of these countries from economic exploitation by richer nations (MacKenzie 1991), because too often in the past the resources have been removed without fee, yet the native farmers may have to pay to stock their fields with new breeds and varieties.

One final utilitarian reason for the conservation of biodiversity is the valuable role many species of animal, plant and especially microbe, will be found to play in the process of cleaning up this planet now that we are becoming aware of the mess we have made of it. Predatory animals may be needed as biological control agents for pest species that we have carried from one part of the world to another. Plants that are resistant to heavy metals and other toxic materials can be used to rehabilitate waste tips, to stabilize soils and recreate a productive ecosystem. But most useful of all are the microbes that can be used to degrade organic compounds, including toxic and carcinogenic ones, together with the hydrocarbons released during accidental oil spills, and render them harmless in the natural environment. Micro-organisms are somewhat neglected in discussions of biodiversity and conservation (Hawksworth 1994), simply because they are so often overlooked in the study of ecosystems. Their general role in ecosystems is, of course, appreciated by ecologists, but their potential for exploitation both in industry and in the field makes the study of their biodiversity and their conservation a fast-developing field for the future.

Further reading

Allan, T. & Warren, A. (eds) (1993) *Deserts: The Encroaching Wilderness*. Mitchell Beazley, London.

Barbier, E.B., Burgess, J.C. & Folke, C. (1994) *Paradise Lost: The Ecological Economics of Biodiversity*. Earthscan, London.

Parry, M. (1990) *Climate Change and World Agriculture*. Earthscan, London.

Solbrig, O.T., van Emden, H.M. & van Oordt, P.G.W.J. (eds) (1994) *Biodiversity and Global Change*. CAB International, Wallingford.

Thomas, D.S.G. & Middleton, N.J. (1994) *Desertification, Exploding the Myth*. Wiley, Chichester.

Chapter 12

Paradigms and Politics

In this book so far, the main themes surrounding the issues of global environmental change have been largely addressed from a scientific viewpoint, despite the earlier admission in Chapter 1 that 'it would be an oversimplification ... if we were to assume that science provides the only approach needed to the global problems we shall discuss'. Although, in as far as possible, this approach has been a 'balanced' one, it is now vital to enter some important caveats about the essential validity of the science itself. In fact, we shall go further and consider the viewpoint that the science as such has really very little to do with the questions under debate. There are two main reasons for this.

First, science, like all approaches to understanding and knowledge, is deeply affected by what are generally called 'paradigm shifts', or movements in the dominant frameworks of thought. The American historian of science, Thomas S. Kuhn (1962, 1970), defined paradigms as 'universally recognised scientific achievements that for a time provide model problems and solutions to a community of practitioners'. A paradigm thus shapes a scientific community through certain shared criteria for the choice of problems which, if the paradigm remains unquestioned and unchallenged, may be assumed to have set solutions.

'Global environmental change' represents one of the dominant modern 'green' paradigms, creating both the current scientific agenda and demanding certain solutions. Some noteworthy headline words and phrases, which, not surprisingly, also crop up regularly throughout this book, tend to be associated with this paradigm, including 'equilibrium', 'sustainability', 'climate change', 'global warming', 'pollution', the 'tragedy of the commons', 'environmental degradation', 'population pressure' and 'biodiversity'. We may already, however, be entering what Kuhn

has termed a 'crisis phase', in which new data, new theoretical thinking, and new speculation all necessitate a re-assessment of the dominant paradigm. This is precisely why we must now re-evaluate what has so far been presented in the first eleven chapters of this book, which is, of course, being written and marketed under the 'global environmental change paradigm'. Obviously, students and scholars will want to buy books that serve the core agenda of the dominant paradigm, because it is this paradigm which inevitably controls the content of their examination papers, education courses and current research programmes.

Secondly, all science functions within a political context, and we must therefore try to deconstruct this context to see the extent to which the current paradigm, or dominant viewpoint, is merely a political construct. Indeed, the slogan 'putting politics first' has recently become a central theme in the developing discourse on political ecology (Atkinson 1991; Bryant 1991, 1992). Important questions arise. Who actually placed global environmental change on the world agenda? Why did they do so? Is the agenda agreed and shared by all, rich and poor, North and South, scientist and lay person, biologist and art historian? And who is using the paradigm, and for what purposes? In this respect, it is important to remember that intellectual 'colonialism' may be far more insidious than mere political and economic colonialism, and that economic and political concerns will often cynically pay 'lip-service' to a dominant paradigm for purely commercial and electoral purposes (as with the increasing range of 'green' alternative merchandize found in supermarkets and stores).

Nandy (1983) has argued that colonialism is 'a state of mind'. As Jewitt (1995) writes: 'It follows that colonialism can survive the physical act of decolonization, and that colonial-style policies can be enacted and be forcefully policed by post-colonial states'. Jewitt goes on to show that this was true of forest policies in India between 1947 and 1990, the greater extent of the post-colonial Forest Department's control over local people enabling large-scale timber extractions in the 'national interest', in spite of resistance to scientific forestry by those dependent on the forests. In this particular instance, the colonial 'mind-set' was perpetuated through what Jewitt calls the 'inner Other', that is, the continuation of the colonial viewpoint by those indigenous forest and government officers who were educated and trained under the colonial paradigm.

More powerful and significant, however, is the continuation of the external process of colonialism and colonization through the use of newer global scientific paradigms. This is, again following Jewitt (1995), 'a project of the imagination—a classification, celebration, or normalization of "Others"', representing the desire by the colonial powers 'to shape others according to their own image and likeness, and to impose on them their particular way of living' (Nehru, in Elwin, 1964). Such 'projects' need not, however, be confined to people and countries; under the 'green' paradigm, in particular, recent important 'Others' have included 'tropical rain forest' and 'tropical biodiversity', two of the late-20th century 'Orientalisms', to use Edward Said's (1993) famous term.

We must therefore ask two further crucial questions. How far is global environmental change really about the advanced countries of the North keeping their control (hegemony) over the poorer countries of the South? And, how far is global environmental change about scientists persuading concerned governments to continue to fund their expensive research, despite the fact that few people really want it, or actually benefit from it? High technology remote sensing, for example, is a case in point. Who benefits from it? The military certainly do; the scientific community also; but who else? In reality, it is extremely difficult to find a single instance where remote sensing has truly helped poor people on the ground, and one seriously wonders what might have been achieved if the money had been spent on a less 'scientific' approach.

The 'science' of global environmental change may therefore represent one of the more insidious forms of modern 'Otherings', or 'Orientalisms'.

'Whatever prank betides!'

The relatively swift scientific and political rise of the global environmental change paradigm, or 'Othering', itself merits closer examination. The very idea that 'environmental change' should attract special attention is a strange one, because all the sciences, and particularly biology, ecology and geography, take change to be the norm. Chapter 1 of this book actually begins with the assertion: 'The earth is not a static place'. A far more revolutionary concept would be one that claims that we experience global stasis, without change. Just imagine the impact of such newspaper headlines as: 'World never to change again!', 'Boston climate totally predictable!', 'China's population static!', and 'AIDS can't spread'. We must therefore ask another important question: why has the concept of global environmental change taken on such a great scientific and political significance in the latter part of the 20th century?

It was, after all, only 18 000 to 14 000 years ago that an ice sheet covered most of what is now Canada, and Alaska was joined to Siberia, the water removed from the oceans to form the Pleistocene ice masses lowering sea levels eustatically on a world-wide basis. Such changes had profound effects on the human populations of North America, helping to pattern, for example, the 13 language isolates (single languages or small groups of closely related languages) which existed there (Rogers *et al.* 1990). Moreover, the harbour at Cenchreai, Corinth, from which the Apostle Paul sailed for Ephesus in 53 AD now lies beneath the 'wine dark' sea; by 1300 AD the rabbit (*Oryctolagus cuniculus*) was beginning its inexorable assault on the character of the British landscape; by the 17th century, Europe was largely deforested; and, the newest sexually transmitted disease, the acquired immune deficiency syndrome (AIDS), has been attacking human populations from the mid-part of the 20th century onwards (Learmonth, 1988). There simply is no stasis; only complex change at all levels of the biosphere, throughout time.

This essential concept of 'change' has been recognized in all forms of knowledge, especially since the middle part of the 19th century, which witnessed the profound paradigm shift from 'creationism' to 'evolution'. The much revered

New England poet, Emily Dickinson (1830–1886), lived through this paradigm shift in her small world at Amherst, Massachusetts (Wolff 1988), and yet, in a remarkable poem, probably written in the very year, 1859, that Charles Darwin encapsulated the debate in his *On the Origin of Species*, she wrote:

What if the poles should frisk about
*   And stand upon their heads!*
I hope I'm ready for 'the worst'—
*   Whatever prank betides!*
                    The Complete Poems,
                        1970, Poem 70

Of course, we now know that Dickinson was right. The Yugoslav mathematician and astronomer, Milutin Milankovitch, performed detailed calculations between 1912 and 1941 to show that the earth's position in space, its tilt and its orbit around the sun, not to mention its 'frisky' poles, have all changed. But the crucial line in the poem is perhaps: 'Whatever prank betides!', a sentiment so at odds with current 'green' attitudes to change, whereby we seem to wish to halt change itself and to seek global stasis and security, a hopeless and illusory goal. Change may actually be essential for us.

This desire for stability in a changing world is nevertheless very deep, and it is surely one of the main reasons for the enormous and burgeoning interest in global environmental change. Its associated 'buzz' words, such as 'sustainability', 'balance', and 'equilibrium', are used, as adjectives, to make the idea of growth, and of change itself, more palatable. However, given the intrinsic reality of change, it is more likely that change should be warmly embraced and welcomed, because change may give far greater resistance and persistence for human beings than so-called stasis and stability. These latter concepts may in contrast spell doom.

For example, in the debate on 'enhanced greenhouse warming' and general global warming, we tend to have forgotten that the traditional human responses to climate change have never been to try to control the direction and degree of that change—surely an impossible task even for modern scientists—but to respond to the change through human adaptation, through migration, and by developing new crops, animals and systems of habitation. The transfer of the scientif-

ic emphasis, and of the funds provided by concerned politicians, from such 'small' sciences as livestock and crop breeding to those 'big' sciences which claim that they can predict and control the direction and degree of climate change might eventually prove to be the biggest chimera of them all.

In this respect, it is perhaps politically significant that the term 'economic migrant' has become a negative concept during the last two decades, because humans have traditionally survived change by becoming 'economic migrants'; it is a normal response to change, albeit a traumatic one. Migration only becomes impossible in a world with fixed boundaries, in which even the political entities are fossilized on the map. Here we see another reason why we have become obsessed with the concept of global change—many world polities, particularly the older advanced countries, like Germany, the United Kingdom and the United States of America, simply do not want it, and cannot cope with it, even though all of them were actually created by such 'economic migrations'.

We are thus living in a changing world which often prizes stability over change, and where, when the idea of change is eventually accepted, special adjectives and nouns are used to modify and to soften the essentials of change.

'Sustainability' and 'equilibrium' as dangerous concepts

Two such words littering the debate on global environmental change are 'sustainability' and 'equilibrium' (Adams 1990). In these concepts, the idea of change is accepted, but it is then immediately corrected and ameliorated by the qualifying notion that the direction and degree of change are 'sustainable', and not self-destructive, and that systems are likely to settle down to an original, or to a new equilibrium level, at a certain point in the future.

These palliatives come in many versions. For some, the systems are seen as self-correcting systems, with many negative feedbacks, which will naturally, through time, return the systems to a given equilibrium (*the static equilibrium state*). Others envisage systems alternating between two or more equilibria (*stable limit cycles*), while yet others see systems as staggering from one equilib-

Table 12.1 Some differing time frames of 'sustainability' for contrasting human groups.

Hunter–gatherers	1 day to 1 week
Peasant farmers	4 months to 1 year
Industrialists	Next production target
Politicians	Next election, coup
Foresters	10–100 years
Geneticists	Up to 10 000 years
'Green' ecologists	'Eternity'

rium to another (*multiple stable equilibrium states*), which is virtually a contradiction in terms (Holling 1973; May 1977). But always, the system finds an equilibrium level of some kind, thus becoming 'sustainable'. As Wiens (1984) has somewhat wryly observed: 'belief in equilibrium theory amounts to verification of a paradigm due to faith in that paradigm.' All this assumes, however, that external environmental changes (e.g. precipitation patterns) are such that they will sustain such equilibria, and there are rarely time limits set on the concept of 'sustainability'. One person's 'sustainability' may, in fact, be another person's 'change' (Table 12.1).

Towards disequilibrium

A simplistic view of tropical rainforests, for example, is that they are equilibrium systems in balance with the climate of the humid tropics (the old climatic climax concepts of Clements (1936) and Tansley (1935)), and that they have existed, largely undisturbed, under such a climate for millions of years. For many 'green' ecologists, these 'Orientalist' wonders of biodiversity should unquestionably be kept as such in perpetuity. Large-scale human disturbance is seen as disrupting the age-old equilibrium, thereby reducing sustainability and causing ecological mayhem.

We now know that this view is totally fallacious, and that humid tropical climates, just like those of the high latitudes, have altered dramatically over geological time, and that the tropical rainforests themselves are dynamic systems, with gaps, building and mature phases, and an ever-changing flux of species, each responding individualistically to environmental change. It is somewhat salutary to reflect that the Malay peninsula, a very bastion of the 'green' image of

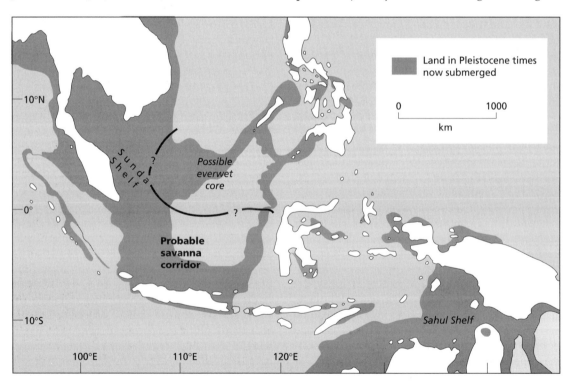

Fig. 12.1 A reconstruction of the ecology of the Malay peninsula some 18 000 years ago. Note the presence of a savanna corridor. After Whitmore (1987).

tropical rainforests, was actually a dry savanna only 18 000 years ago (Fig. 12.1). There was certainly no million-year sustainability or equilibrium there.

Moreover, the fossil pollen record has consistently shown us that each separate species has responded individually to environmental change (Fig. 12.2). Birks (1989) has produced a range of isochrone maps based on 135 radiocarbon-dated pollen diagrams to illustrate the pattern of tree-spreading in the British Isles over the last 10 000 years. Birch (*Betula*), hazel (*Corylus avellana*), elm (*Ulmus*), oak (*Quercus*), ash (*Fraxinus excelsior*) and beech (*Fagus sylvatica*) all exhibit strongly individualistic behaviour in terms of their arrival, direction of spread, timings, rates of spread and attainment of ecological and geographical limits; there is simply no such thing as an 'equilibrium forest' which moves *en masse*, as an entity (see Chapters 4 and 7).

It may well be, therefore, that the very concepts of sustainability and equilibrium are intrinsically flawed, and that the reality is one of disequilibrium systems, which gain their essential resistance and persistence through a constant individualistic response to change (DeAngelis & Waterhouse 1987). Such appears to be increasingly the message from both anthropological and ecological studies being carried out in Africa south of the Sahara, where, in particular, the intrinsic variability of precipitation has long been recognized. One such system is the Ngisonyoka Turkana pastoral system studied recently by Coughenour *et al.* (1985), Ellis & Swift (1988) and Ellis *et al.* (1993). The Ngisonyoka herders buffer the spatial and temporal variability in primary productivity by tracking forage availability using a nomadic mobility strategy; any attempt to 'stabilize' these dry and wet season patterns would be totally counterproductive, and could be damaging to the environment. Likewise, any attempt to impose on Africa the concepts of 'equilibrium' and 'sustainability' might just represent the very worst development scenario of them all.

This should not, perhaps, be surprising, because it is important to remember that these concepts really derive from the now totally outdated vegetation climax concepts developed in northern Europe and North America during the early part of the 20th century, where anything which was not 'forest' tended to be perceived as a 'sub-climax',

Table 12.2 Area of key biomes in Africa south of 22°S. After Rutherford & Westfall (1994).

	Area (km²)	Percentage of land surface
Non-forest biomes		
Savanna	959 067	46.16
Nama-Karoo	541 127	26.05
Grassland	343 216	16.52
Succulent Karoo	111 212	5.35
Fynbos	69 875	3.36
Desert	52 893	2.55
Forested biomes		
Afromontane Sub-tropical lowland Tropical lowland forest	309	0.01

below the 'optimum' (and forest invariably represented the 'optimum'); such sub-climaxes would return to forest when the relevant disturbance factor (such as fire or grazing) was removed. In Africa immediately south of the Sahara, however, there is very little that has ever been forest; in southern Africa, south of 22°S, less than 1% of the land carries any 'natural' forest today (Table 12.2; Fig. 12.3), and it probably carried even less at certain times in the past. These biomes are not 'derived' from forest; they never were forest in the first place. Increasingly, however, they are seen as disequilibrium systems, which persist and adapt to change through their non-equilibriation.

Such simple facts have profound effects on what people regard as 'important' in ecology. For most Northern foresters, for example, fire was long regarded as an 'enemy', causing change and disrupting the forest equilibrium. In North America, the firefighter soon became a national hero, an image perfectly captured in Stephen Pyne's (1989) 'documentary' novel, *Fire on the Rim: a Firefighter's Season in the Grand Canyon*. By contrast, in the non-forested tropics and subtropics, fire is normally a vital 'friend', used to manage the indigenous savanna and tropical grasslands more effectively. Not surprisingly, in *Anthills of the Savanna*, the famous West African novelist, Chinua Achebe, writes totally differently about his subject: 'The trees had become hydra-headed bronze statues so ancient that only blunt residual features remained on their faces, like anthills surviving to tell the new grass of the savannah about last year's brush fires'. The tone of Achebe is also more accepting of change, less

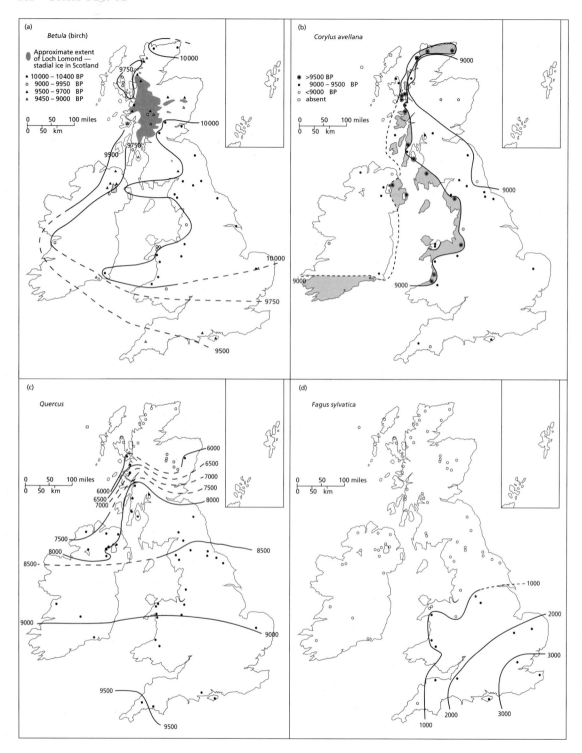

Fig. 12.2 Isochrone maps showing the rational limits of the pollen of four tree species in the British Isles over the last 10 000 years. (a) *Betula* (birch) pollen; (b) *Corylus avellana* (hazel) pollen; (c) *Quercus* (oak) pollen; (d) *Fagus* (beech) pollen. Note the individualistic behaviour of each pollen type. After Birks (1989). Note that the Loch Lomond shadial represents the Younger Dryas.

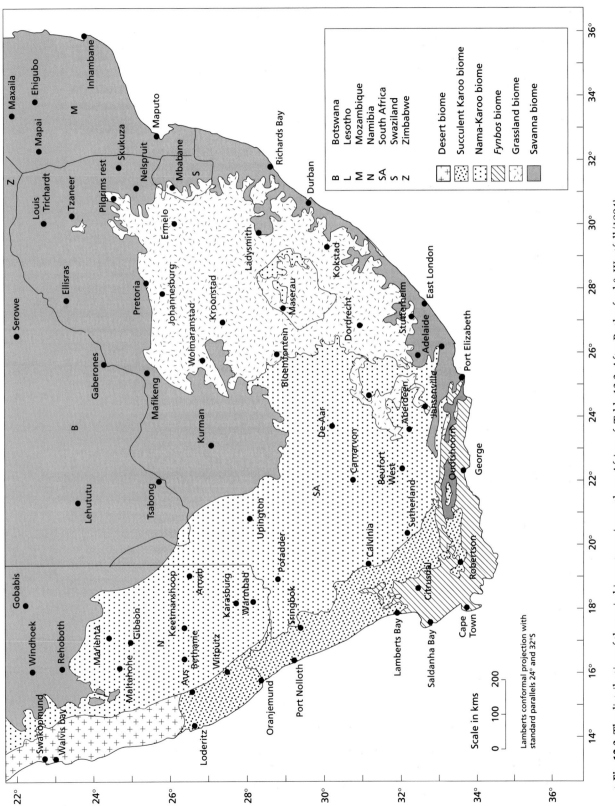

Fig. 12.3 The distribution of the main biome types in southern Africa. cf. Table 12.2. After Rutherford & Westfall (1994).

aggressive. It is interesting to note that savannas, in particular, are quintessential non-equilibrium systems (Young & Solbrig 1992), which are not near thermodynamic equilibrium, and in which people, fire and grazing animals are integral constituents. Savannas oscillate continually and often chaotically due to their own internal dynamics. But they persist, and have great resilience.

These basic conflicts in ecological perception are perhaps most clearly exemplified in many general textbooks on ecology, which tend to carry virtually no references to fire, because their 'ecology' has been derived from the North. The classic text by Begon *et al.* (1996) is a case in point, providing only three references to fire in over 900 pages of text, two references referring inevitably to North America; this text is thus of limited value to sub-tropical and tropical ecologists, its inherent agenda being entirely Northern in character.

The real problem, however, may be far more fundamental, and we may desperately require a paradigm shift from 'Northern' to 'Southern' ecology, from equilibrium to disequilibrium systems, from a forest hegemony to a non-forested world; the programmes of the two are completely different, a fact recognized by South Africa's scholar-politician, General Smuts, a long time ago. And people do not normally live in the forests.

Economics and sustainability: extending the fallacy

We have thus seen that both the concepts of sustainability and equilibration are probably deeply flawed in ecological terms; the adoption of the concept of 'sustainability' by economists is therefore likely to represent an unfortunate extension of these fallacies. Yet, this is a commonplace in the new, so-called, 'green' economics (e.g. Pearce *et al.* 1989), which strives to balance the triumvirate of profit (economic efficiency), equity (present and future), and sustainability (environmental integrity), as illustrated by Young & Solbrig's (1992) 'buzz-word' title: *Savanna Management for Ecological Sustainability, Economic Profit and Social Equity*.

The initial concept of sustainable development was popularized by the Brundtland Commission (World Commission on Environment and Develop-

ment 1987), but the concept has stubbornly resisted easy definition. Most definitions involve at least three elements (Young 1992), namely: (i) the effects of present-day activities on the future; (ii) the importance of maintaining ecological processes; and (iii) the benefits of improving the quality of life now without denying future generations a similar opportunity. The link with intergenerational equity is thus obvious, although the concept of 'equity' is much wider than this, involving all questions of access to and control over natural resources. But these questions, as we shall see, are essentially political in character.

The main problem for the environmentalist economist in confronting global change is the fact that most aspects of such change do not enter a conventional market (Adams 1991). As Adams says: 'Clean air, peace and quiet, habitats of endangered species, views with nostalgic associations, buildings or places considered sacred, life, and global climatic *stability* [our italics] all present unresolved practical and technical cash valuation problems'. Indeed such concepts remain—despite numerous efforts to the contrary—'externalities' for the average economist. It is important to note, however, that, although Adams is undoubtedly correct in putting economics in its place—a much more lowly one than normally allowed—he himself still falls into the traps of wanting 'stability' as the costed value and embracing 'values' which are heavily weighted to the concerns of the North (e.g. 'peace and quiet'). Archaeologists in Buddhist countries like Thailand have a real problem because 'merit' is largely attached to the building of new temples, not to the conservation of old ones, which are often regarded as little more than 'heaps of old stones'. In other words, change and the new is valued over preservation and the old.

Nevertheless, Adams has rightly identified the fundamental problem for the economist, namely that, for environmental goods and concepts that are not naturally traded in markets, there is no way of deducing 'willingness-to-accept' (WTA) values (in contrast to 'willingness-to-pay' (WTP) values) from market behaviour. As Adams concludes: 'Since it takes only one infinity to wreck a whole cost–benefit analysis, taking willingness-to-accept answers at face value renders the cost–benefit method useless; it gives every prospective loser a veto'. Economics, for example, can contribute hardly anything in the case of a wealthy contractor, such as a gold or uranium mining company, wishing to develop a site, such as the 'Dreamland' habitats of Kakadu National Park in the Northern Territories, Australia, which are sacred to the penniless Aborigines who live there. 'Ability-to-pay' (ATP) values are clearly even more meaningless than others in such circumstances. Economics, including 'green' economics, is therefore of little real help in dealing with the central issues of global environmental change. This is especially so, however, where the extension to 'green' economics also embraces ecological concepts, such as equilibration and sustainability, which are themselves suspect.

Economists thus tend to regard 'development' as sustainable over the long term when it maintains the ecological processes and functions that underpin it. Unfortunately, this rather 'motherhood-and-apple pie' viewpoint belies the fact that both the economics and the 'development' ecologies involved remain rooted in the colonial ecological mind-set. And, as we have already argued, this mind-set may be totally inappropriate for both the ecologies and economies of the South, and may actually prevent necessary change, or deflect the traditional directions of change. The planting of exotic tree species in the open veld of South Africa can actually create erosion, not prevent it.

The Northern 'green' paradigm of change is thus drawn up on a different agenda, centred on self-preservation and on a Northern-related set of values. We must therefore address the central issue of global environmental change, namely its relationship to political ecology. As Bryant (1991) states: 'ecological change is imbued with political meaning'.

Political ecology

Political ecology has been defined as 'an inquiry into the political sources, conditions, and ramifications of environmental change' (Bryant 1991), and, in contrast to other ecological literature and approaches, it focuses on the interplay of diverse socio-political forces, and the relationships of these forces to environmental change. Bryant (1992) argues that political ecology addresses at least three distinct, but of necessity interlinked, discourses.

The first of these concerns the contextual sources of environmental change, focusing on the general environmental impacts of the state and its policies, on interstate relations, and on global capitalism. The second examines the location-specific aspects of environmental change, concentrating on conflict over access to natural resources and documenting the resistance of the relatively powerless, such as poor peasants, as they fight to protect the environmental foundations of their traditional livelihoods. The third weighs the political ramifications of environmental change by assessing the effects of such change on socio-economic and political relationships, looking at how change affects socio-economic inequalities, and how unequal exposure to environmental change leads to political confrontation and the formation of environmental movements.

The basic truth is that global environmental change is unequal in both its effects and its causes. Global environmental change is never neutral, in that someone always benefits from it, while others will lose out. The simple idea, for example, that enhanced global warming will be equally disastrous for all is not tenable. States and intrastate regions will inevitably view the prospect of global warming differently. If global warming is predicted to increase and diversify agricultural production in an area, the process will be looked on favourably; if, on the other hand, agriculture is likely to suffer because of the changes, or lowland cities become flooded through a world-wide rise in sea level consequent on polar-ice melting, the attitude will be negative. The crucial point is that global warming will alter the political map of the world, and the paradigm arguments about enhanced greenhouse warming will be more readily accepted by those polities which envisage a decline in political power. Change is good for some, but not for others.

Similarly, the causes of change are not equally distributed across the globe, leading to very vitriolic arguments about who is responsible for 'acid rain', carbon dioxide (CO_2) emissions or oceanic pollution. It is now widely argued that future wars are more likely to arise over such conflicts, and over the management of certain increasingly scarce resources, especially water, than over any other political issues. These tensions are especially well exemplified by the current arguments concerning who is really 'responsible' for the enhanced greenhouse effect, whether or not this effect is actually important, and whether global warming is indeed taking place (see Chapter 5). The North blames the South; the South blames the North, and many, especially in the South, do not believe it is happening at all, or that it matters not one jot if it is. The basic science is weak and unpredictive at any meaningful scale, while the politics are powerful, especially in the drive of the North, and its expensive scientists, to maintain their hegemony over the South.

The arguments of the North are particularly disreputable in this respect. First, if the enhanced greenhouse effect were to prove of any significance, it is clearly because of the emissions which have taken place from the industrialized North over the last 200 years. It is thus vital to recognize that most studies, such as the *World Resources Report* (World Resources Institute 1991), only take into account current emissions, thus neatly side-stepping the selection of a particular time frame, which inevitably leads to a neglect of the cumulative effect of gas emissions. Moreover, the lifetimes of the different greenhouse gases vary greatly, as does the degree to which they affect the energy balance. Basically, the North has not paid any environmental levy for its historic emissions, and many countries in the South, like Malaysia, argue quite cogently that they have no obligation to do anything at all to curb their current contributions until this 200 years of emissions is taken into account. Why should they now stop their push to develop? The arguments certainly will not divert them from industrializing and competing vigorously with the North.

The North then tries to shift the blame onto the South, pointing to the burning of forests and savannas throughout the tropics and sub-tropics, conveniently forgetting that these formations have always burnt, even, periodically, the tropical rainforests of Borneo and the Amazon. In fact, most tropical formations, such as savannas, were burning, through lightning strike and other natural causes of fire, before a single hominid set foot on them. But, unlike the story for Europe and North America, these areas must not be allowed to clear *their* forests in the process of development. And, of course, China and India must certainly not use their coal reserves to develop as the North did using its energy resources.

The particular focus on CO_2 emissions is espe-cially useful in bolstering these dubious arguments, although, of course, it is only one of the many 'greenhouse' gases. Little is heard about the most important greenhouse gas of all, water vapour, for example. Nevertheless, the real global responsibilities become somewhat clearer if more gases than CO_2 are actually taken into the equation. This has been well demonstrated by Neela Mukherjee (1992).

In her fascinating corrective analysis, Mukher-jee compares the relative contributions of differ-ent nations to the emissions of three key greenhouse gases, namely CO_2, methane and chlo-rofluorocarbons (CFCs). She suggests 'measures' of emissions and relative efficiency in economic activities, relating these to the per capita income of each country studied (Table 12.3). Her main conclusions are somewhat salutary:

1 On the basis of the scores, both unweighted and weighted, for greenhouse gas emissions, high-income countries clearly have the highest scores and contribute most to global warming (Fig. 12.4).

2 Low-income countries make such a small con-tribution to global warming that they can safely be excluded from any responsibility for it.

3 Some middle-income countries, however, can be included in the allocation of responsibility.

4 The reduction in emissions must come from the high emitters in relation to the global average and not from those at the bottom.

The responsibility for global warming thus lies

Table 12.3 The political ecology of greenhouse warming. The real 'culprits' and the real 'oppressed': the top and bottom five emitters of greenhouse gases, respectively, (unweighted CO_2, methane and CFCs). After Mukherjee (1992).

	Income per capita (US$)
Top five	
United Arab Emirates	15 830
Canada	15 160
Australia	11 100
Kuwait	14 610
Denmark	14 930
Bottom five	
Zaire	150
Togo	290
Congo P.R.	870
Guatemala	950
Haiti	360

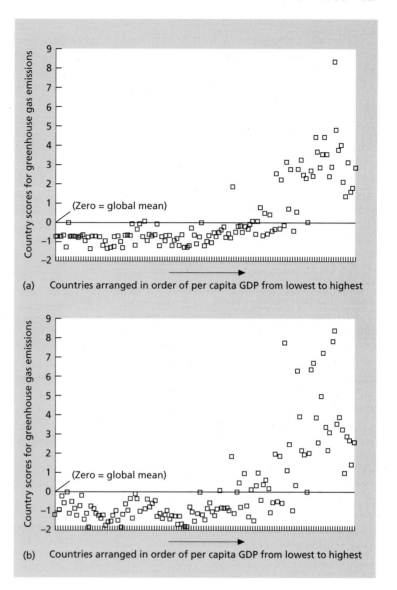

Fig. 12.4 Per capita gross domestic product (GDP) against (a) unweighted greenhouse gas emissions and (b) weighted greenhouse gas emissions. After Mukherjee (1992).

firmly with the high-income countries, and, if one considers Table 12.3, with a range of countries which normally pride themselves on their 'green' credentials, such as Canada, Australia and Denmark, along with other countries which the North does not wish to offend politically, namely the oil-rich states of the Middle East. Fundamentally, as Mukherjee asserts, 'The rich countries have made disproportionate use of the atmospheric sink which the poor countries are now expected to clean up by emitting less, slowing growth and preserving tropical forests'. The poor thus remain poor, and, by slight of hand, the culpable North even tries to shift the blame onto these countries, which have virtually no responsibility for global warming. And, of course, all this assumes that global warming is actually taking place through the emissions of gases, which in reality represent just a minute fraction of the billions of variables which govern the earth–atmosphere system. It is Alice-in-Wonderland science. But we are in the realms of political ecology, not of science. It is all about power and control, not about saving the planet.

Conclusion

In this chapter, therefore, we have begun to question the whole basis of the global environmental change paradigm. We have seen that change is the norm, and that humans have constantly adjusted to change through movement and adaptation. We have also considered the nature of change, and found that equilibrium ideas, and associated concepts such as 'sustainability', are inherently flawed. We may actually be living in a disequilibrium world. We have further argued that 'green' economics is likewise flawed, and that economics is a spent force when faced with the major issues of global environmental change. Finally, we have shown that a key approach to any understanding of the global environmental change paradigm must lie in political ecology, and neither in science nor economics.

In Chapter 13, we will conclude by examining the political ecology of one of the most 'orientalist' of the current global environmental change themes, namely the deforestation of the tropics, and we will contrast this with the history of forest change in a selected area of the high-income, developed world.

Further reading

Atkinson, A. (1991) *Principles of Political Ecology*. London, Belhaven.

Bryant, R.L. (1992) Political ecology: an emerging agenda in Third-world studies. *Political Geography* **11**, 12–36.

Kenny A. (1994) The earth is fine; the problem is the Greens. *The Spectator* 12 March 1994, 9–11.

Mukherjee, N. (1992) Greenhouse gas emissions and the allocation of responsibility. *Environment and Urbanization* **4**, 89–98.

Young, M.D. (1992) Sustainable investment and resource use. Equity environmental integrity and economic efficiency. (*MAB Series* Vol. 9). Paris, UNESCO and Parthenon.

Chapter 13

Changing Values in a Changing World

God's Wonderful World Colouring Book: Rainforest—'Buy this book and help save the rainforest' (Fig. 13.1). 'You can save the rainforest'—'The equatorial rainforests help to support all life on earth—but they're in danger of destruction! Find out what you can do'.

All over the world, but especially in Europe and North America, the developed countries of the

Fig. 13.1 Marketing the 'green' paradigm of tropical rainforests to children: '*God's Wonderful World Colouring Book: Rainforest*. Lion Publishing Inc., Oxford, 1995.

209

North, children can buy rainforest colouring books, just like the one portrayed in Fig. 13.1, many offering competitions, with even prize trips to the 'fascinating rainforest conservatory' at Kew Gardens in London. Some books are well-meaning and carefully thought out; others are poor, shoddy products, simply trading on the global environmental change paradigm we have discussed in Chapter 12. But all illustrate one thing, the fact that rainforests have become, probably more than any other biome or 'green' topic, the subject of a new 'orientalism'. It is now practically axiomatic that rainforests are the exotic 'other'—'the biological wonder of the world'—that rainforest clearance is a 'bad' thing, and that we shall all suffer if the rainforests are destroyed (Park 1992). Simply consider the contrasting values expressed by the following advertising blurbs for one such colouring book: 'Huge areas of land are being cleared for mining or to grow crops or graze animals, while the trees are being cut down for timber'. 'And if the rainforests disappear we are all in danger'.

But where are the indigenous forests of Europe and North America? When the 17th and 18th century Puritan settlers opened up the wooded lands of Massachusetts in New England for mining, farming and forestry, initiating their contribution to the multifarious America we all know today, where were the colouring books then? Was the American continent in danger? Was the world destroyed? And has America ultimately suffered because these forests were cleared? Was John Winthrop's 'City upon a Hill', Boston, doomed to fail as the forests fell to the axe and the plough?

'God's Wonderful World Colouring Book: the city upon a hill'—'Buy this book and help create a new Commonwealth'. 'You can develop New England for God'—'Hunting, farming, mining, felling, and trading for the Commonwealth. Find out what you can do'. Colour in a pumpkin, not a toucan.

Whose values in a changing world?

You may feel that such heavy irony is out of place in a science textbook such as this, but we have already seen that current ecological science is largely a political construct, reflecting the values of the North rather than those of the South. Tropical rainforests have many 'values', which

likewise vary from North to South, country to country, and people to people (Bryant *et al.* 1993). Savage (1984; see also Vayda & Padoch 1989) has analysed the history of Western impressions of the tropical forests of South-East Asia, and what strikes the reader most is the extremely diverse and contradictory views that have existed, which range from a 'cornucopia of nature' to a 'white mans' graveyard', from 'a land of gold and birds of paradise' to 'a pestilential region of poisonous vapours and cannibals'. Today, the diversity of viewpoints is no less. For some, tropical forests are a sign of 'underdevelopment' and a lack of progress; to others they are vital for the very ecology and survival of the whole earth. The question thus arises: whose values dominate decision-making concerning rainforests in our changing world?

In June 1991, the University of London School Examinations' Board set a most revealing question on its General Certificate of Education Examination Paper in Geography for 16–19 year olds. This question showed a tree (Fig. 13.2), a very temperate tree, with a list of 'misuses' of trees on the left-hand side. Students were then asked to provide a contrasting list of '*four* ecological benefits' of trees on the right-hand side, and to expand on their answers in the space below. The so-called 'misuses' included: acid rain, cattle ranching, deforestation, firewood, charcoal burning, and soil disruption from road building. Such a set of values could only have been conflated in the North. The majority of people in the world depend on firewood and charcoal for all their energy requirements, indeed for their very livelihood. Moreover, this firewood and charcoal are often produced by systems which have been maintained for many generations. They are actually 'sustainable' to use the equilibrium term. From the forests to the savannas, multi-purpose tree species sustain both life and land in the tropics.

The 'values' exhibited in this question are those of a comfortable Northern suburbia, long divorced from the forests it seeks to save; a suburbia where lumber is bought from the store, chopped for fun on a Sunday, and used for a picturesque Yuletide fire in winter, although the real warmth, of course, comes from an energy-sapping central heating system. The examination question shows no understanding whatsoever of the values of most of the world; it is frankly a non-

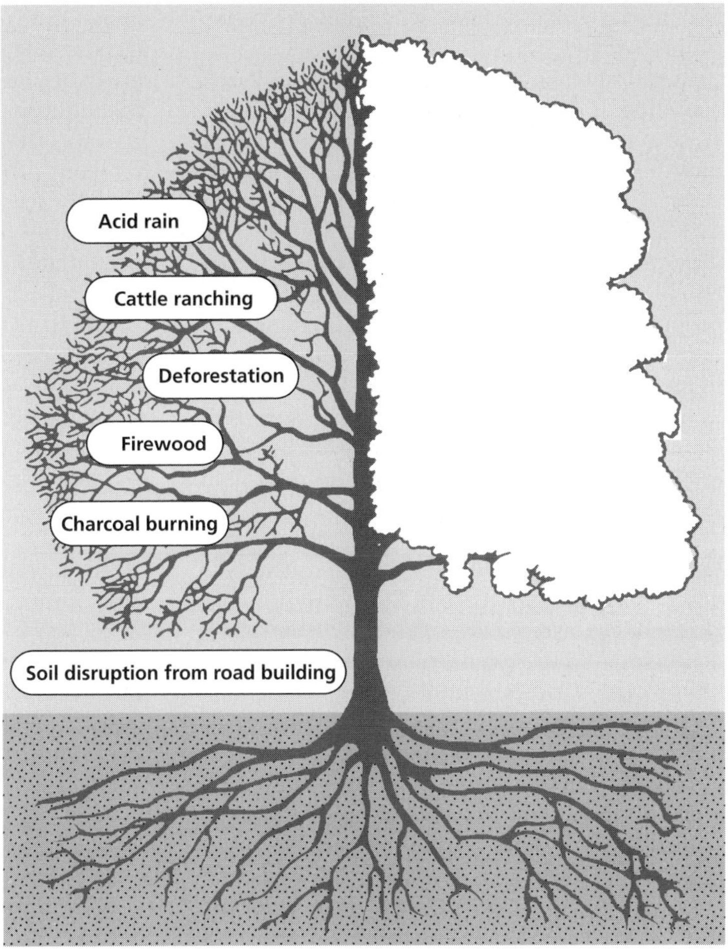

Fig. 13.2 Part of an examination question, Fig. 2 for use with Question 2, set in June 1991 by the University of London School Examinations' Board for its General Certificate of Education Examination Paper in Geography for 16–19-year olds (Section on 'Ecosystems and Human Activity'). The question read as follows:

Study Figure 2, in the separate insert provided, which could be used as a poster to illustrate the uses and misuses of trees.

(a) Annotate the right-hand side to show *four* ecological benefits of trees. Expand your answer in the space below Figure 2.

(b) With reference to *one* of the misuses shown in Figure 2, explain why it has occurred and why it is a cause for concern.

See text (p. 210) for a full explanation.

sense, and was largely unanswerable. Yet, it illustrates all too cogently the immense gulf between those seeking to continue to control the tropics and those who live there.

'Little green lies'

The North has thus adopted the rainforests as its special concern, and imbued them with its own values. Whether the countries of the North have done this out of guilt for the clearing of their own forests, or because they wish to extend the physical control of their colonial past through the mind-set of ecology, it is hard to know. Unquestionably, however, the Northern middle classes see the preservation of 'rainforest' as a key item on their global 'stability' agenda. And having agreed this agenda, they have started a process of

scientific 'myth-making' to bolster their case. These myths are the 'little green lies' of the current paradigm. The fundamental aim behind them is to make us all believe that we really do *need* rainforests, not just want them or like them.

The following statement, which opens a book entitled *Tropical Rainforests* by Chris Park (1992), is a fine example of the standard formula encapsulating some of the myths generated over the last three decades: 'Tropical rainforests are the most complex ecosystems on earth. Rainforests (better known to many people as jungles) have been the dominant form of vegetation in the tropics for literally millions of years, and beneath their high canopy lives a diversity of species which is unrivalled anywhere else on earth'. Not surprisingly, Park's book ends in gloom: 'The time-bomb of ecological, environmental, climatic and human damage caused by deforestation continues to tick...'.

Regrettably, not one of the elements in Park's statement is scientifically tenable when subjected to scrutiny. Tropical rainforests are not the most complex ecosystems on earth, and there are many competing claimants for this accolade, depending, of course, on how you define 'complexity'. The word 'jungle' actually comes from a Hindi word meaning 'scrub forest', and what most people call 'jungle' is in reality the impenetrable margins of the forest which have been subjected to human interference (Whitmore 1990). 'Jungle' is normally a human product! Moreover, it is worth noting that some disturbed habitats and secondary forests are richer in species than their undisturbed counterparts.

Most ludicrous of all, however, is the idea that rainforests are '*literally* [our italics] millions of years' old. We have already, in many previous chapters, emphasized the individualistic response of species to environmental change, and the fossil record clearly indicates that the majority of present-day 'rainforests' are less than 18 000 years old, having been subject to drought, fire and cold during the peak of the last ice age. Finally, the plant diversity beneath the canopy is actually quite low, and again it depends on how you define 'diversity'; if anything, the assertion is more likely to be true of the canopy itself, and is certainly more appropriate for the invertebrates than for any other groups of organisms.

The whole statement is thus a farago of 'little green lies', all, nevertheless, supporting the current political ecological values placed on rainforests by the North. But the statement is not comprehensive, and there are a series of further scientific 'myths' which are also widely employed to bolster the case for needing rainforests. These additional scientific myths can be summarized as follows:

1 The rainforests are the 'lungs of the world'. In fact, because of their decomposition processes, most rainforests tend to use up as much as or even more oxygen than they give out (see Chapter 5).

2 Rainforests are a vital carbon sink, and their cutting and burning will lead to enhanced greenhouse warming. We have already laid this myth to rest in Chapters 5 and 12, but it is also worth noting that when the forest is removed, it is not a case of perfect carbon storage being replaced by no carbon storage. The replacement systems, such as grassland, will also have complex carbon cycles, which need to be taken into account.

3 Rainforests are vital in preventing soil erosion. In some circumstances, this may indeed be true, but many other types of surface vegetation, when properly managed, such as grasslands, can be more effective. In the Himalayas, it is now recognized that the erosion and flooding once thought to have been caused by deforestation probably represents the normal erosion only to be expected through the continual geological uplift of the mountains, which lie over an important subduction zone between the Indian (–Australian) and the Eurasian plates. There is even some evidence that agriculture and terracing have been beneficial in slowing down the erosion (Ives & Messerli 1989).

This cavalcade of 'little green lies' has three main purposes, all relating to the essential political ecology of rainforests. The first is the desire of the North to maintain a controlling interest over the resources of the South. The second is the worry that changes in the South, both ecological and economic, might damage the polities of the North. The third is the desperate drive of scientists to obtain continued funding for research into the questions that the current paradigm places on the agenda.

And the power of a paradigm should never be underestimated. Scientists, especially in Australia, frequently use the term 'rainforest' for formations such as monsoon forests and savanna forests, neither of which has anything in common with the ecology of true tropical rainforests. The reason for this is simple: it is easier to get funding for research on 'rainforests' than on these less well-known, seasonal formations. Yet the tragedy goes deeper; the less well-known formations cover far more of the tropics than do the real rainforests, but the Northern public focus on 'rainforests' has deprived them of their rightful place in the pantheon of world ecology. The North stresses rainforests; the people of the South live in the monsoon and savanna lands. Herein lies an essential tension.

Whose biodiversity?

One final 'little green lie', however, requires a more detailed consideration. The axiom of the 'biodiversity' of the tropical rainforest has become central to tropical rainforest myth-making, either being employed as an argument to keep the rainforests completely undisturbed or supporting the concept of the 'sustainable use' of the genetic basis of rainforest trees, and other organisms (e.g. bushmeat), by local populations in what have become termed 'extractive reserves'. The latter are often deemed to be more economically valuable in the long term than deforestation and its resultant replacement systems, such as agriculture and mining.

Unfortunately, this whole argument is dangerously flawed, and research increasingly shows that the genetic resources in the replacement systems, like trees in fields and fallow, tend to be more prized by local populations. Moreover, as soon as a genetic resource is really 'valued', an industrial replacement is quickly synthesized by the North, often leaving the original producers in a worse position than they were before. Most biodiversity arguments turn out to be cruel deceptions, hindering competitive development in the South, condemning people to extremely hard lives, and deflecting the ordinary processes of change.

Recent research in West Africa illustrates this point particularly well. When clearing their forests for fields, farmers in Benin retain an average of 63 trees per hectare, of which a large proportion are karité trees (Vitellaria paradoxa), oil palms and néré trees (Parkia globosa) (K. Schreckenberg, pers. comm., 1995). These have been traditionally used to produce the vegetable oils employed in cooking and soap-making. However, such resources are much easier to exploit in the fields than in the forests, and, not unsurprisingly, collectors aspire to less arduous and more lucrative methods. Moreover, the traditional sources are now being increasingly replaced by the production of groundnut oil and soybean. The argument that the maintenance of the forests is necessary to sustain the production of non-timber forest products does not hold. Change is taking place, with new resources being exploited, as an easier and more profitable life is sought.

Of course, some of the biodiversity (e.g. sources of medicines) will be of use to humans. But in approaching the topic of biodiversity, we must remember certain salutary facts and, as Dr Johnson recommended, 'clear our minds of cant'.

99.99% of all life that has ever existed on the surface of the earth has become extinct, from the trilobites through dinosaurs to mammoths. Extinction is a normal process of change, which does not result in the destruction of the earth. Extinction is an integral aspect of global environmental change, and, for the species concerned, it is somewhat irrelevant whether the agent of extinction is catastrophic or gradual. The real problem arises when humans put relative 'values' on species. In doing so, we set a new, anthropocentric, agenda, inevitably favouring some species over others. It would indeed be intriguing, for example, to see the smallpox virus replace the giant panda on the collecting cans and badges of the World Wide Fund for Nature.

Humans do not have an absolute interest in 'all' biodiversity, only that biodiversity which is perceived to be of value to Homo sapiens and to the survival of H. sapiens, such as new crops, drug plants, pretty flowers, and cuddly animals. By stark contrast, potential enemies, like viruses, bacteria, and even some insect, bird and mammal pests and disease vectors, should, of course, be eradicated at all costs. There is nothing intrinsically wrong in arguing the case for maintaining certain selected species, so long as we are honest about it: we are doing so because these species are thought to be useful to us or because we like them. They possess an evolutionary benefit for H. sapiens or, at the very least, they do not threaten or harm us. Yet, a moral case can even be made for letting the giant panda—a favourite animal once distributed all round the northern hemisphere, but now naturally on the way out—go extinct with dignity, rather than shipping it ignominiously across the world to breed, subjecting it to artificial insemination, and using it to keep badly-run and outdated zoos afloat financially.

As we have argued previously, the perceived values for biodiversity will vary from one human being to another, between cultures, between nations, and between the North and South. There is no agreed, common agenda. An African elephant has many differing and contradictory values, including a destroyer of crops, a destroyer of woodlands, an opener of lands, a hunting trophy, a source of ivory, a source of tourist revenues, an ecological element of the savannas and forests, and, like the rainforests, an 'orientalist' value as a 'biological wonder'. But, as with the mammoth, the extinction of the African elephant does not threaten the survival of the earth. We do not *need* it. Some may want to keep it, and many would be saddened by its loss; but these are different questions.

Likewise, we do not *need* tropical rainforests, despite the panoply of myths that have been concocted to support the paradigm. It is a false agenda. Some people may *like* rainforests; others may think that they have special values in certain circumstances; but we do not ultimately need them for the earth's survival, any more than we needed the dodo or the temperate forests of Massachusetts. Needing, wanting and liking are very different.

A study in environmental change: the Harvard Forest

The 3000-acre Harvard Forest, near Petersham, Massachusetts, is one of the most richly studied areas of forest in the world, both in terms of its history and its current ecology. The story of Harvard Forest probably tells us more about the nature of global environmental change than anything we have said so far in this book (Foster 1993). In fact, it is a symbol of change, which we should be careful to heed when we try to enforce our current paradigms on the changing forested lands of the tropics. Between 1931 and 1941, the story of Harvard Forest was captured in a series of models designed by R.T. Fisher, the first Director of the Harvard Forest (Harvard University 1975). These models are now housed in the Fisher Museum, where they are employed as a major educational resource. Plates 13.1–13.7 show the historical sequence of the models, which are described below.

The pre-European forest of Central New England consisted of a mixture of coniferous and broad-leaved trees, supporting a few scattered Indian communities (see Fig. 11.13, which illustrates the history of forest clearance in North America). But it was not a stable system, even then. Atlantic winds, storms and hurricanes, coupled with fire, destroyed the forest periodically, creating a patchwork of trees, each patch starting anew after disturbance. Change and disturbance were the norm. Stands of pure white pines often formed a transient stage in the progression to shade-enduring hardwoods, such as red and white oaks, and hemlock (Plate 13.1, facing p.136).

In June 1630, however, a band of around one thousand English men, women and children sailed into what was to become Boston Harbour, the first of many Puritans fleeing the England of Charles I. Initially, they joined the small English settlement of Charlestown, situated across the Charles River estuary from the Shawmut Peninsula. Soon, however, the Reverend William Blaxton (or Blackstone), a somewhat shadowy figure who had come to the Massachusetts Bay area in 1623, invited them over to his side of the river where there was a better water supply and where he had settled around 1627 on land today occupied by the elegant houses of Spruce Street and the famous Louisburg Square, Beacon Hill, Boston. By 1635, the newly named settlement of Boston had grown in to a town with a population of almost 4000, and Blaxton, always somewhat a loner, had left for Rhode Island, where he died in 1675. But the tide of Puritan immigration was underway, soon to spread out west, north and south through the forested lands of Massachusetts.

The town of Petersham, in which the majority of Harvard Forest now lies, is situated north-west of Boston and it was granted to 71 proprietors in 1733. The first settlers came there from the 'frontier' towns of Rutland, Lancaster, Lunenburg and Brookfield in the early spring to begin the clearing of the forest. A sawmill was immediately established. There was little need to conserve timber in a land where the supply of wood seemed inexhaustible. Small fields were cut, surrounded by rough stone walls, and simple dug outs, lean-tos and log cabins were built (Plate 13.2, facing p. 136). Unwanted wood and timber were then burned to fertilize the earth. The inexorable process of opening up the forests of New England was well underway—but there was no 'green' movement back in England predicting ecological doom and gloom should the forests fall to the axe and the plough.

A hundred years on, by 1840, the town of Petersham had grown to a population of 1775, the highest in its history. Over 70% of the land was now under agriculture, being used for tillage, pasture and orchards (Plate 13.3, facing p.136). There was very little undisturbed forest left, the remaining woodlands being employed for timber, firewood, or coverts for hunting. Indeed, wood was now becoming a scarce and valuable commodity. Inevitably, however, change came with the introduction of coal for heating in the cities, and new timber sources were exploited, first transported by sea, and later by railroad, from New York, Pennsylvania, and the Great Lakes States. The pre-European settlement forest had all but vanished; yet ecological disaster had not occurred.

Then, around 1850, change once again came to the landscape of Petersham. The construction of the Erie Canal, the spread of the railroad, gold in California, and the American Civil War all drew the New England farmer away to the richer farm lands of the West. The population declined, and the fields and pastures were quickly abandoned. Soon white pine began to recolonize the pioneer land, sprouting from seed stores buried in the once-ploughed soil, or spreading from the coverts and remaining patches of woodland (Plate 13.4, facing p.136). These even-aged stands of young pines grew quickly into thicker woodlands, and, as they matured, with higher canopies, more light came in and the old hardwoods, red oak, white ash, black cherry, maple and chestnut, all began to reappear. The mixed forest started its return.

By 1890, these 'old field' stands of volunteer white pines (Plate 13.5, facing p.136) took on an economic value, so much so that portable sawmills and wood-using factories began to appear, dotted around the landscape. The factories produced boxes, matchsticks, heels for shoes, toys, pails and other wooden goods for the burgeoning cities; some cuts were even good enough for finished lumber. A hundred acre woodlot of volunteer white pines might yield some $30 000, the only real cost of exploitation being the inescapable taxes. Here you could reap what ecology had sown. It has been estimated that between 1895 and 1925, fifteen billion board feet of second growth pine was harvested in central New England. A new era of ecology and economics had begun. The stands were clear cut, and the logs were dragged on wooden sleds ('scoots') to the portable steam sawmills.

But this very process initiated yet another unexpected change in the landscape (Plate 13.6, facing p.136). Following clear cutting, the land looked devastated, and one could just imagine the outcry of the current 'green' movement if it had existed in Petersham at that time. On closer

inspection, however, you would have seen, among the debris of dead pine poles, litter, and sawdust, a range of scraggy hardwoods. By 1915, these hardwoods had come into their own, the main species returning as sprouts being red and hard maple, red and white oak, white ash, chestnut, black cherry and black birch. A new forest was forming, one that had actually been nurtured underneath the 'old field' forests of white pine. The essential dynamism of ecology was, as always, in tune with change. By 1930, these hardwoods had reached cordwood size (Plate 13.7, facing p.136), and, with good silvicultural practice—identifying the potential crop trees as against the whip trees, weeds, and trainers—they would be easily turned into a new and valuable asset.

Today, Harvard Forest is once more a rich mixture of forest types, of gaps, of new growth, and old growth forest and plantations. But beneath the varied trees, you can still see the rough stone walls, often double, which once surrounded the 75-acre subsistence farm of John and Molly Sanderson, whose wooden clapboard house, built sometime after 1763, still stands by the edge of the forest, a reminder of part of the long and ever-changing story of Harvard Forest, which appears to have come full circle.

Yet, there are other changes too. Now few mature chestnuts grace the forest, chestnut blight killing the saplings before they can become trees. In 1938, a great hurricane swept across the region. Studies at the Pisgah Tract in south-western New Hampshire provide a particularly dramatic record of the impacts of this catastrophic event on the forests (Foster 1988a, b) (Fig. 13.3). In 1957, a forest fire burned 78 acres, stopped only by the arrival of a pumper truck. Since then, the forest has come up again, with multi-stemmed white and grey birches, red oaks and red maples sprouting from the fire-killed trees. But there is also modern scientific research on the very topic of this book, global environmental change.

On 15 October 1988, the Harvard Forest was designated a Long-Term Ecological Research (LTER) site by the National Science Foundation (NSF), one of 17 established at the time. The approved research agenda was inevitably set within the global environmental change paradigm, and includes studies on forest–atmosphere trace gas fluxes, the long-term history of climate, distur-

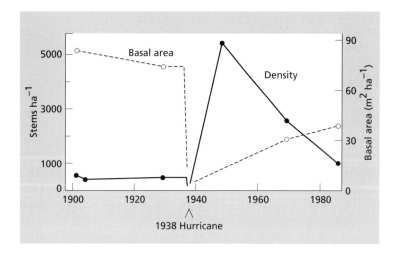

Fig. 13.3 The 1938 hurricane and its effect on the forests in the Pisgah Tract, south-western New Hampshire, USA. Note the progressive trend of natural thinning and increasing basal area following the hurricane in a forest composed of beech, hemlock, maple and birch. After Foster (1988a).

bance and vegetation, and community, population and plant architectural responses to disturbance, several subjects covered or mentioned in the previous chapters of this book. The Harvard Forest, once a forest home of Indians, then a land of pioneer European settlers, an open agricultural countryside, a landscape of abandoned fields, followed by volunteer white pines, clearcut chaos, and a hardwood resurrection, is now a centre for earnest scientists, all working under the global environmental change paradigm. Today, the landscape even has demonstration 'pulldowns', where the forest has been artificially 'blown' over to simulate hurricane damage, like that experienced in 1938 (Foster 1988a). Scientists measuring root mass, trunk angles, light, moisture, and wind conditions are then able to produce computer programmes to tell us what will happen after a hurricane has passed through. Dynamic forests and computer models go well together (Shugart 1984).

Replicating change

This complex history of Harvard Forest is, in reality, not so special, but rather the norm (Cronon 1983; Whitney, 1994). Its particular significance lies in the fact that we know such a lot about the story, because of the excellent records relating to the site and the intensive work of many different scientists and scholars. Nevertheless, the story could be replicated in many places throughout the world (Christensen 1989), like Seton Falls, an urban park in New York City (Loeb 1989) (Fig.

13.4). The message is simple: change is constant, and for each change there will be a new mixture of ecology, economics and politics.

The global environmental change paradigm must thus accept change as the essence of things. We predict, for example, that the tropical rainforests will have their own complex histories, paralleling those at Harvard Forest and Seton Falls, and that one day people will also be able to sit on a stone wall beneath the trees and believe that they are in an undisturbed, forested land. To try to stop the processes of change is both futile and dangerous, and we must have extremely telling arguments indeed if we are to impose our desire for 'stability' on others and the world. Yet, as we have seen, much of the global environmental change paradigm rests on false myths, scientific fabrications, and flawed arguments.

Planet Earth is intrinsically dynamic, and humans, for their very survival, must be dynamic too, moving, changing and evolving. Stability is never an option. As Emily Dickinson worried, way back in 1859, at the family Homestead on Main Street, Amherst, just a little distance from Harvard Forest:

Perhaps the 'Kingdom of Heaven's' changed—
I hope the 'Children' there
Won't be 'new fashioned' when I come—
And laugh at me—and stare—
                    The Complete Poems
                    1970, Poem 70

By the year 2400 AD, our late-20th-century environmental concerns will seem as 'old fash-

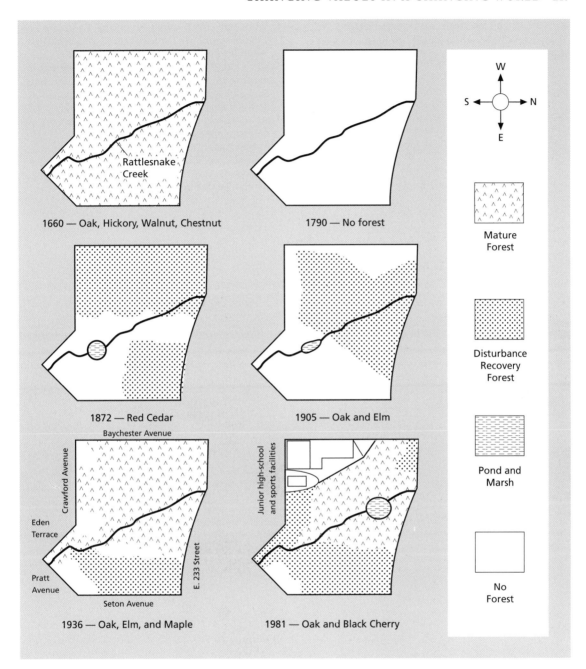

Fig. 13.4 Seton Falls Park, New York City: maps showing historical changes in the dominant species and forest structure. Compiled by Loeb (1989).

ioned' as the Puritan aims for Boston, John Winthrop's 'City upon a Hill'. People will surely laugh and stare.

Further reading

Bryant, R.L., Rigg, J. & Stott, P. (1993) The political ecology of southeast Asian forests: transdisciplinary discourses. *Global Ecology and Biogeography Letters* (Special Edition) **3** (4–6), 101–296.

Christensen, N.L. (1989) Landscape history and ecological change. *Journal of Forest History* **33** (3), 116–125.

Cronon, W. (1983) *Changes in the Land. Indians, Colonists and the Ecology of New England*. Hill & Wang, New York.

Foster, D.R. (1993) Land-use history and forest transformations in central New England. In: M. McDonald & S. Pickett (eds) *Humans as Components of Ecosystems*. pp. 91–110 Springer, New York.

Stott, P. (1991) Harvard Forest. *Global Ecology and Biogeography Letters* **1** (4), 99–101.

Whitney, G.G. (1994) *From Coastal Wilderness to Fruited Plain: a History of Environmental Change in Temperate North America from 1500 to the Present*. Cambridge University Press, Cambridge.

Glossary of Terms

Acid rain the acidification of rainwater as a result of aerial pollution with oxides of sulphur and nitrogen as a result of industrial activity and the use of internal combustion engines.

Aerobic of organisms or metabolic processes requiring oxygen.

Allerød a warm episode in the closing stages of the Pleistocene approximately 10 800 to 11 900 radiocarbon years ago (see Fig. 4.1).

Amino acid the nitrogen-containing organic molecules that form the building blocks of proteins.

Anaerobic of organisms or metabolic processes functioning only in the absence of oxygen, or intolerant of oxygen (see Table 2.2).

Angiosperms the flowering plants (= Magnoliophyta = Anthophyta). Plants in which the ovules, which are to become seeds at maturity, are enclosed within an ovary (c.f. gymnosperms in which the ovules are not so enclosed).

Aridity index the mean annual precipitation of an area divided by the mean annual potential evaporation.

Arthropods the largest phylum of invertebrates and the first to colonize the land. Their bodies are segmented and have an external, more or less rigid, skeleton. Includes the insects, spiders, scorpions, lobsters and crabs.

Asthenosphere the uppermost layer of the earth's mantle, immediately underlying the lithosphere. It is more fluid than the lithosphere and can respond to high levels of stress by plastic flow (see Fig. 2.6).

Autotrophic of an organism, making its own foodstuff from simple substances either by photosynthesis (using light as the energy source) or chemosynthesis (using chemical energy) (see Table 2.2).

Biodiversity the totality of species present in an ecosystem or in an area (including plants, animals and microbes). Some would extend the use of the term even to the genetic diversity found within species.

Biogeochemical cycle the global cycling of an element through atmosphere, hydrosphere, geosphere and biosphere.

Biosphere the totality of the earth's living organisms.

'Business as usual' a scenario (often associated with a particular model) in which it is assumed that currently prevailing conditions will continue into the future.

C3 plant a plant in which the initial stage of photosynthesis involves the addition of a molecule of carbon dioxide to a five-carbon sugar molecule, resulting in the production of two 3-C molecules.

C4 plant a plant in which the initial stage of photosynthesis involves the addition of a molecule of carbon dioxide to a three-carbon molecule to form a 4-C unit. This is temporary and the process is later reversed in specialized vascular bundle cells and the normal C3 system comes into operation. Its advantages are that it can operate more efficiently at low carbon dioxide concentrations, at high light intensities, at high temperatures and in conditions of water stress.

CAM plant similar to a C4 plant in its biochemistry, but it lacks specialized vascular bundle cells. It operates a C4 initial carbon dioxide fixation system at night, stores the acidic products (hence 'Crassulacean Acid Metabolism') and then indulges in the conventional system during daytime while its stomata are closed. It is particularly efficient under arid conditions and where carbon dioxide concentrations are low (as occasionally in aquatic systems).

Carbon neutral a process which absorbs as much carbon dioxide as it produces, such as the continued growth of mature forest.

Carrying capacity the upper limit of population that can be sustained under any given set of conditions.

Catastrophism an alternative to uniformitarianism. The present-day geological setting is the outcome of occasional, perhaps never to be repeated, crises in history. The term was first used of the idea of a Biblical Flood as an influential geological event.

CFCs chlorofluorocarbons. Compounds of carbon, hydrogen, chlorine and fluorine. Industrially produced as propellants for aerosols, as refrigerants and fillers for plastic foams. They dissociate under UVR in the stratosphere and form active radicals which speed up the breakdown of ozone.

Coccolithophorids microscopic planktonic organisms that secrete calcium carbonate cases around their cells.

Convection a process driving the flow of warmer water (or gas) to rise relative to adjacent cooler (denser) water as a result of the downward movement of the cooler water. The same process operates within fluid rock in the earth's mantle, where local heating may cause a plume of hotter, lighter rock to rise through cooler, denser magma.

Crisis phase in paradigms, a period characterized by the re-assessment of former observational data, new theoretical thinking and free speculation. May lead to either the establishment of a new paradigm or the strengthening of the original paradigm.

Cyanobacteria bacteria containing chlorophyll a and able to carry out photosynthesis (= 'blue–green algae'). Probably the first photosynthetic organisms on earth.

Dendrochronology the use of annual tree-rings in the dating of woods.

Denitrification the process by which certain microbes under conditions of low oxygen availability (such as waterlogged conditions) use the nitrate ion as a means of oxidizing organic materials in their respiration. Nitrogen gas is thus generated and the element becomes less available to plants.

Dobson's stations sites where atmospheric ozone has been logged over a number of years using Dobson spectrophotometers, which use the characteristic absorption spectrum of that gas as a means of measuring it.

El Niño Southern Oscillation (ENSO) a periodic disturbance of the Pacific Ocean/climate system, affecting movements of winds, ocean currents and rainfall across the whole Pacific region. It also affects fisheries and hence bird and terrestrial life on the Pacific margin. See Figs 6.2 to 6.4.

Energy transfer efficiency the efficiency with which energy is transferred from one feeding (trophic) level to another. In aquatic ecosystems, for example, each transfer of energy involves a loss of about 90% of the energy available.

Epicontinental of seas which are shallow and occupy a partially land-enclosed part of the continental shelf (e.g. the North Sea) or may extend well into the interior of a continent (e.g. Hudson Bay).

Erosion the process by which inorganic substrates, such as soils, are transported from one area to another. Wind, water and gravity may be involved in such transport.

Eustatic a term applied to changing sea levels caused by the melting of ice-caps during warm conditions and the expansion of ice masses during cold conditions.

Eutrophication the flushing of excess plant nutrients, particularly nitrogen and phosphorus, into an ecosystem. Often used of the enrichment of water bodies by excessive inputs of nutrient elements as a result of human activity, such as fertilizer application or the ploughing of grassland.

Fixation used of the process whereby gaseous substances (e.g. carbon dioxide, nitrogen) are converted from that form into a solid (fixed) form, e.g. as in photosynthesis where carbon dioxide is fixed into a carbohydrate form.

Flux the flow of some component of the earth system, usually in the context of its movement in a cycle. May be quantified as mass per unit time (e.g. gigatonnes per year). If the site of the flow is to be emphasized, the flux may be referred to as a pathway.

Food web (or chain) the relationships between organisms in terms of food supply (e.g. grass–grasshopper–lark–hawk). Links between species are often complex and branching (food webs), but may in simple systems be relatively linear (food chains).

Foraminifera microscopic, planktonic organisms that construct tiny shells (tests) around themselves made of calcium carbonate.

Fossil fuel organic materials (coal, oil, gas, etc.) formed by past photosynthesis and now preserved within geological sediments from which they can be extracted and used as a source of energy.

Fynbos literally meaning 'fine-leaved bush'; a highly distinctive biome of open to closed, dwarf shrubby, shrub-woodland, found in the southwestern and southern Cape Province of South Africa. The *fynbos* biome approximates to the Cape Floristic Kingdom, where diversities are probably higher than anywhere else in the world. Noted for many endemic plant species.

General circulation model (GCM) a computer model of the earth's atmospheric and oceanic circulation systems on the basis of which predictions can be made concerning the impact of anticipated changes in conditions, e.g. a rise in the level of atmospheric carbon dioxide.

Geosphere the rock component of the earth in its totality, excluding all the water (the hydrosphere) and all living material (the biosphere) and the atmospheric component (the atmosphere).

Geothermal energy heat generated within the earth's interior.

Gigatonne (Gt) a measure of mass, being equivalent to 10^9 (a billion) tonnes or 10^{15} g.

Glacial stage a period of time within which cold conditions prevail that are of sufficiently low temperature and which last sufficiently long to permit the expansion of glaciers in high-latitude and high-altitude areas. In the Pleistocene they often lasted for tens of thousands of years.

Greenhouse effect the process by which the earth's atmosphere retains some of the long-wave radiation reflected from the earth's surface.

Greenhouse gas any gas, such as water vapour, carbon dioxide, methane, etc. that absorbs strongly in the infra-red region of the spectrum and which contributes to the greenhouse effect.

Gymnosperms the naked-seeded plants; a group comprising all seed-bearing plants which are not angiosperms (in which the ovule is enclosed in an ovary at the time of pollination). Comprises the conifers, cycads and ginkgos amongst living forms.

Heinrich events cold episodes in the last glacial stage during which masses of icebergs were released into the North Atlantic.

Hemispheric metabolism the cycling of elements such as carbon within one of the earth's (northern or southern) hemispheres. The different extents of land masses, human industrialization (see Fig. 5.6) and the degree of isolation of the atmospheric circulation patterns can lead to differences between the conditions of the two hemispheres.

Heterotrophic of an organism which consumes foodstuffs derived from other organisms rather than producing its own (as with autotrophs). It may eat plants (herbivores), digest organic matter externally (many fungi and bacteria) or consume animals (carnivores).

Holocene the current interglacial, approximately the last 10 000 years.

Hominid the evolutionary branch of the great apes that includes the human species.

Hydrosphere all the water in the oceans, in rivers, lakes, the atmosphere and as ground water within the crust. The totality of global water (c.f. atmosphere, geosphere).

Ice core a column of ice obtained by drilling into an icecap.

Inter-tropical convergence zone the region of convergence of tropical air masses resulting from convectional movements in the neighbourhood of the equator where solar energy input to the earth is greatest. The location of the convergence zone varies with season as the angle of incident light alters. See Fig. 11.8.

Interglacial stage a period of time within which warm conditions prevail such as to cause the retreat of glaciers and the development of temperate vegetation in their stead. Pleistocene interglacials characteristically lasted about 10 000–20 000 years.

Interstadial an episode in which climatic warming occurs but is of limited duration, often allowing only the development of boreal vegetation.

Invertebrate animals lacking backbones (contrast vertebrates). Comprising a number of phyla, not particularly closely related but all sharing this negative character.

Ion a charged atom or molecule resulting from its dissociation on becoming dissolved in water.

Isostatic a term applied to changes in sea level relative to land surfaces resulting from warping of the earth's crust, as for example under the load of a glacial ice mass.

Isotherm a line joining all geographical points of the same temperature at any given time.

Keystone species a species of animal, plant or microbe that occupies a significant position in an ecosystem, such that its loss would cause extensive repercussions, possibly a cascade of losses of other dependent organisms.

Leaching the process in which the percolation of water (often acid in reaction from dissolved carbon dioxide and from the presence of organic acids) removes ions from the soil and leaves it more acid in its pH and lower in its fertility.

Leaf area index the area of plant leaf area per unit of ground. It is an index of the complexity of structure or architecture of plant canopies and may also provide some indication of the efficiency with which solar radiation is being tapped by plants.

Lithosphere the outermost part of the solid earth, both the land surface and the layers beneath the oceans, comprising the rigid rock of the crust and the outer part of the mantle which forms the plates (which have moved in the process of plate tectonics) (see Fig. 2.6).

Little Ice Age an episode of prevailing cold climatic conditions during historic times dating approximately from the 14th to the 19th century AD.

Loess wind-blown sandy material characteristic of very cold dry conditions.

Migration the predictable (often seasonal) movements of animals from one part of the earth to another.

Milankovitch cycles changes in the orbit of the earth around the sun and in the position of the earth's axis that operate in a predictable manner in time and may form the basis of climate cycles, such as glacials and interglacials (see Fig. 3.20).

Mineralization the conversion of an element from an organic form to an inorganic one, e.g. the conversion of an amino acid to ammonia and carbon dioxide.

Mire a peat-forming ecosystem, including swamps, fens and bogs.

Missing sink an unaccounted for loss of an element from a reservoir, such as the unaccountable loss of some carbon dioxide from the atmosphere (see Fig. 5.5).

Mycorrhiza a close association between a plant root and a fungus in which the fungus benefits from the losses of organic compounds within and around the root tips while the decomposing and mineral accumulation activities of the fungus enable the plant root to take up required elements more efficiently.

Nama-Karoo a distinctive grassy dwarf-shrubland largely confined to the summer rainfall areas of southern Africa.

Nitrification the process by which certain microbes obtain energy by the oxidation of ammonium ions to nitrite and then nitrate ions. The process is of particular importance to those plants that take up most of their nitrogen from the soil in the form of nitrate ions.

Nitrogen fixation the activities of certain microbes (some free-living and others associated with particular plants) in which they are able to convert atmospheric nitrogen into ammonia and hence into amino acids and proteins.

Nuclear winter cold conditions which might be anticipated in the event of global nuclear war. Atmospheric dust, in particular, would cause an effect similar to massive volcanic eruption and would reduce incoming energy and lower temperatures.

Oceanic conveyor belt the global circulation pattern of the oceans that depends on the salinity and temperature of the ocean waters and that has a strong influence on global climates (see Fig. 3.22).

Organochlorine compounds organic compounds containing chlorine that have often been used as pesticides (such as DDT). Some are particularly persistent and others degenerate to give rise to chlorine in the atmosphere where they may react with ozone.

Orientalism a complex discourse on power, domination and hegemony developed by Edward Said (1993) concerning the relationships between the West (Occident) and the Rest (Orient).

Oxygen isotope stages changes in the ratio of different oxygen isotopes have been used to divide the history of the Pleistocene into a series of climatic stages numbered from the top of the column.

Oxygen isotopes forms of oxygen that differ in their atomic weight. Their differential movements under different climatic regimes has permitted their use as indicators of past climates when extracted from ice-caps and marine sediments (see Fig. 3.9).

Ozone a reactive form of oxygen in which there are three atoms of oxygen per molecule, rather than the two characteristic of oxygen gas. Formed naturally by the action of UVR on oxygen in the stratosphere, and in the troposphere by the action of various anthropogenic pollutant gases in the presence of sunlight.

Paradigms 'universally recognised scientific achievements that for a time provide model problems and solutions to a community of practitioners' (Kuhn 1962). An approach to the history of science.

Paradigm shift a significant change in the dominant framework of thought and research (e.g. from creationism to evolution).

Periglacial climatic conditions that are not sufficiently cold to result in glacier ice development but in which frost causes permanently frozen subsoils and vegetation is of a tundra character.

Permafrost permanently frozen subsoil. In summer the upper layers of the soil may thaw, but the lower layers remain frozen.

pH a logarithmic scale of acidity (the concentration of hydrogen ions). See Fig. 9.12.

Photochemical smog a high density of charged particles and water droplets in the atmosphere resulting from the accumulation of various pollutants under stable atmospheric conditions and high light intensities. Vehicle exhausts are the most frequent causes of such pollution.

Photorespiration the loss of photosynthetic efficiency in C3 plants because of competition between carbon dioxide and oxygen for locations on the photosynthetic enzyme rubisco. Thus photosynthesis is more efficient in an atmosphere of nitrogen (together with CO_2). C4 plants, because of the separation of C3 metabolism into specialized cells, have overcome this problem.

Photosynthetic relating to photosynthesis, the process by which plants and those bacteria containing chlorophyll use light energy to take up atmospheric carbon dioxide to produce carbohydrate and in the process release oxygen (see Table 2.2).

Photovoltaic cell an implement that is able to convert the radiant energy of light into electrical energy.

Phytoplankton the microscopic algae living in the upper, illuminated part of a lake or the ocean (photic zone). The photosynthesizing phytoplankton take carbon dioxide from the surface water and represent the primary producers in the open water environment.

Polar front the boundary between cold polar waters and the warmer southern waters in the North Atlantic (see Fig. 4.2).

Polarity reversals periodic changes in the polarity of the earth's magnetic field. Such changes are recorded in the rocks and can be used as a basis of dating geological materials.

Political ecology 'an inquiry into the political sources, conditions and ramifications of environmental change' (Bryant 1991).

Precambrian the span of geological time from the origin of the earth, 4600 million years ago, until the start of the Cambrian Period, approximately 570 million years ago (see Fig. 2.9).

Primary productivity the rate at which energy from the sun is incorporated into organic materials by the process of photosynthesis in plants. It is expressed as dry weight or energy accumulated per unit area per unit time.

Radiocarbon years estimates of age based on the radiocarbon system of dating. This method assumes a constant production of ^{14}C in the atmosphere, which has not proved to be correct, but it is possible to translate radiocarbon 'dates' into true solar dates by the application of a calibration curve (see Fig. 3.11).

Radionuclide an unstable compound that decomposes to release radioactive radiation.

Ruminant animal animals such as cows that have a 'second stomach', or rumen, in which associated bacteria digest cellulose from herbage and thus generate methane.

Savannas tropical and sub-tropical woody-grass formations (mostly with C4 grasses) developed under conditions of seasonal water availability. They occupy over one-third of the world's land surface.

Silviculture the management process whereby a forest is tended, removed and replaced by a new forest crop.

Sink a reservoir of an element into which it is being absorbed. For example, the oceans absorb carbon dioxide and can therefore be regarded as a sink for carbon.

Soil respiration the process by which carbon dioxide is generated by the soil as a result of the activities of animal detritivores and fungal and bacterial metabolism as they consume organic materials.

Spread the extension of the range of distribution of an organism into new areas (see migration).

Stadial an intensely cold episode of limited duration during which glacial expansion often occurs.

Stomata pores in the surfaces of plants, particularly in leaves, through which gaseous exchange for photosynthesis and respiration take place. When open they also lose water in the process of transpiration, which may be a disadvantage to a plant in dry conditions, but which does serve the purpose of reducing leaf temperature.

Stratosphere that part of the atmosphere from about 15

to 50 km altitude. It lies above the troposphere and merges above into the mesosphere (see Fig. 8.1).

Succulent-Karoo a vegetation noted for its succulent plants, characterizing the winter rainfall areas of southern Africa. Famous for its mass flowering displays of annuals and ephemerals (e.g. Asteraceae), especially in spring.

Tectonic relating to the deformation and movement of the earth's rigid crustal plates.

Thermocline the temperature gradient from the (normally) warmer upper layers of the ocean or a lake down to the cooler water below. The term is applied particularly to the steepest part of that gradient, where the change from warm to cool water is most abrupt. It is then used as though referring to a surface, the interface between warm and cool water bodies.

Thermohaline applied to the circulation of ocean water, usually in the vertical dimension, when water temperature and differing salinity combine to give differences of density to the water which drive the flow (c.f. convection).

Till unsorted rock detritus deposited by ice from a glacier.

Timberline the upper altitudinal limit of tree cover.

Transfer functions numerical constructs by means of which it is possible to translate the composition of an assemblage of fossils into a set of climatic conditions.

Uniformitarianism the belief that nature behaves in a regular and predictable manner so that the geological processes operating today also operated in the past. Current landforms are the product of the long-term and gradual operation of current processes.

UVR, UV-A, UV-B and UV-C ultraviolet radiation (UVR) is part of the radiation received from the sun and is composed of three components representing different wave-bands within the UV spectrum, from shorter wavelength (UV-C) to longer (UV-A). See Fig. 8.3.

Varves bands in lake sediments caused by a seasonal cycle of freeze/thaw in the soils of the catchment. Being annual in deposition, they can be used as a method of dating lake sediments.

Vertebrates animals with a backbone (vertebral column) including the fish, amphibians, reptiles, birds and mammals.

Whip (in forestry) a tall tree with a small narrow crown which sways in the wind, thus 'whipping' or injuring adjacent timber.

Younger Dryas a cold (stadial) episode at the close of the Pleistocene, about 10 000 to 10 800 radiocarbon years ago (see Fig. 4.1).

References

Achebe, C. (1987) *Anthills of the Savannah*. Heinemann, London.

Ackerman, S. (1989) European prehistory gets even older. *Science* **246**, 28–30.

Adam, D.P. (1985) Quaternary pollen records from California. In: Bryant, V.M. & Holloway, R.G. (eds) *Pollen Records of Late-Quaternary North American Sediments*. American Association of Stratigraphic Palynologists Foundation, Dallas, Texas, pp. 125–140.

Adams, J.G.U. (1991) On being economical with the environment. *Global Ecology and Biogeography Letters* **1**, 161–163.

Adams, W.M. (1990) *Green Development: Environment and Sustainability in the Third World*. Routledge, London.

Addiscott, T. (1988) Farmers, fertilisers and the nitrate flood. *New Scientist* 18 October, 50–54.

Addiscott, T. & Powlson, D. (1989) Laying the ground rules for nitrate. *New Scientist* 23 April, 28–29.

Alderton, D.H.M. (1985) Sediments. In: *Historical Monitoring*. Technical Report of the Monitoring and Assessment Research Centre, University of London, 1–95.

Allan, T. & Warren, A. (1993) *Deserts: The Encroaching Wilderness*. Mitchell Beazley, London.

Alley, R.B., Meese, D.A., Shumuan, C.A. *et al.* (1993) Abrupt increase in Greenland snow accumulation at the end of the Younger Dryas event. *Nature* **362**, 527–529.

Alm, T. & Birks, H.H. (1991) Late Weichselian flora and vegetation of Andoya, Northern Norway—macrofossil (seed and fruit) evidence from Nedre Aerasvatn. *Nordic Journal of Botany* **11**, 465–476.

Amman, B. & Lotter, A.F. (1989) Late-glacial radiocarbon- and palynostratigraphy on the Swiss Plateau. *Boreas* **18**, 109–126.

Anderson, I. (1994) Sunny days for solar power. *New Scientist* **143**, 21–25.

Andreae, M.O. & Raemdonck, H. (1983) Dimethyl sulfide in the surface ocean and the marine atmosphere: a global view. *Science* **221**, 744–747.

Andrews, P. (1992) Evolution and environment in the Hominoidea. *Nature* **360**, 641–646.

Arensburg, B., Tillier, A.M., Vandermeersch, B., Duday, H., Schepartz, L.A. & Rak, Y. (1989) A Middle Palaeolithic human hyoid bone. *Nature* **338**, 758–760.

Ascher, W. (1995) *Communities and Sustainable Forestry in Developing Countries*. Institute for Contemporary Studies, San Francisco.

Asfaw, B., Beyene, Y., Suwa, G. *et al.* (1992) The earliest Acheulean from Konso-Gardula. *Nature* **360**, 732–735.

Ashmore, M. (1990) The greenhouse gases. *Trends in Ecology and Evolution* **5**, 296–297.

Ashmore, M.R. & Bell, J.N.B. (1991) The role of ozone in global change. *Annals of Botany* **67** (Suppl. 1), 39–48.

Atkinson, A. (1991) *Principles of Political Ecology*. Belhaven, London.

Ayres, R., Schlesinger, W. & Socolow, R. (1994) Human impacts on the carbon and nitrogen cycles. In: *Industrial Ecology and Global Change* (eds R. Socolow, C. Andrews, F. Berkhout & V. Thomas). Cambridge University Press, Cambridge, pp. 121–172.

Baker, R.G., Maher, L.J., Chumbley, C.A. & Van Zant, K.A. (1992) Patterns of Holocene environmental change in the Midwest. *Quaternary Research* **37**, 379–389.

Barbier, E.B., Burgess, J.C. & Folke, C. (1994) *Paradise Lost: the Ecological Economics of Biodiversity*. Earthscan, London.

Barghoorn, E.S. & Tyler, S.A. (1965) Microorganisms from the Gunflint chert. *Science* **147**, 424–27.

Barinaga, M. (1990) Fish, money and science in Puget Sound. *Science* **247**, 631.

Battarbee, R.W., Flower, R.J., Stevenson, A.C., Jones, V.J., Harriman, R. & Appleby, P.G. (1988) Diatom and chemical evidence for reversibility of acidification of Scottish lochs. *Nature* **332**, 530–532.

Becker, B., Kromer, B. & Trimborn, P. (1991) A stable-isotope tree-ring timescale of the Late Glacial/Holocene boundary. *Nature* **353**, 647–649.

Beddington, J. (1995) Fisheries: the primary requirements. *Nature* **374**, 213–214.

Beerling, D.J. & Chaloner, W.G. (1994) Atmospheric CO_2 changes since the last glacial maximum; evidence from the stomatal density of fossil leaves. *Review of Palaeobotany and Palynology* **81**, 11–17.

Begon, M., Harper, J.L. & Townsend, C.R. (1996) *Ecology: Individuals, Populations and Communities*, 3rd edn. Blackwell Science, Oxford.

Benjamin, M.M. & Honeyman, B.D. (1992) Trace metals. In: *Global Biogeochemical Cycles* (eds S.S. Butcher, R.J. Charlson, G.H. Orians & G.V. Wolfe). Academic Press, London, pp. 317–352.

Bennett, K.D. (1983) Postglacial population expansion of forest trees in Norfolk, UK. *Nature* **303**, 164–167.

Berkhout, F. (1994) Nuclear power: an industrial ecology that failed? In: *Industrial Ecology and Global Change* (eds R. Socolow, C. Andrews, F. Berkhout & V. Thomas). Cambridge University Press, Cambridge, pp. 319–327.

Berner, R.A. (1994) 3Geocarb II: A revised model of atmospheric CO_2 over Phanerozoic time. *American Journal of Science* **291**, 56–91.

Beyea, J. (1991) Energy policy and global warming. In: *Global Climate Change and Life on Earth* (ed. R.L. Wyman). Chapman & Hall, London, pp. 224–242.

Birks, H.J.B. (1989) Holocene isochrone maps and patterns of tree-spreading in the British Isles. *Journal of Biogeography* **16**, 503–540.

Bjorkman (1975) Environmental and biological control of photosynthesis. In: *Environmental and Biological Control of Photosynthesis* (ed. R. Marcelle). Junk, The Hague, pp. 1–52.

Blackford, J.J. & Chambers, F.M. (1991) Proxy records of climate from blanket mires: evidence for a Dark Age (1400 BP) climatic deterioration in the British Isles. *The Holocene* **1**, 63–67.

Blank, L.W. (1985) A new type of forest decline in Germany. *Nature* **314**, 311–314.

Blumenthal, C., Barlow, S. & Wrigley, C. (1990) Global warming and wheat. *Nature* **347**, 235.

Böhlmann, D. (1991) *Ökologie von Umweltbelastungen in Boden und Nahrung*. Gustav Fischer, Stuttgart.

Bonnefille, R., Roeland, J.C. & Guiot, J. (1990) Temperature and rainfall estimates for the past 40 000 years in equatorial Africa. *Nature* **346**, 347–349.

Briffa, K.R., Jones, P.D., Schweingruber, F.H., Shiyatov, S.G. & Cook, E.R. (1995) Unusual twentieth-century summer warmth in a 1000-year temperature record from Siberia. *Nature* **376**, 156–159.

Broecker, W.S. (1993) Defining the boundaries of the late-glacial isotope episodes. *Quaternary Research* **38**, 135–138.

Broecker, W.S. (1994) Massive iceberg discharges as triggers for global climate change. *Nature* **372**, 421–424.

Broecker, W.S. & Denton, G.H. (1990) What drives glacial cycles? *Scientific American* **262** (1), 43–51.

Broeker, W.S., Kennett, J.P., Flower, B.P. *et al.* (1989) Routing of meltwater from the Laurentide Ice Sheet during the Younger Dryas cold episode. *Nature* **341**, 318–321.

Brown, A. (1992) (ed.) *The U.K. Environment*. HMSO, London.

Brown, H.S., Kasperson, R.E. & Raymond, S. (1990) Trace pollutants. In: *The Earth as Transformed by Human Action* (ed. B.L. Turner). Cambridge University Press, Cambridge, pp. 437–454.

Brown, J.H. (1993) Assessing the effects of global change on animals in western North America. In: *Earth System Responses to Global Change*, (eds H.A. Mooney, E.F. Fuentes & B.I. Kronberg). Academic Press, San Diego, pp. 267–284.

Brown, L.R. (1992) Grainland shrinks. In: *Vital Signs: The Trends that are Shaping Our Future 1992–1993* (eds L.R. Brown, C. Flavin & H. Kane). Earthscan, London.

Brown, L.R. (1993) Fertilizer use falls. In: *Vital Signs: The Trends that are Shaping Our Future* (eds L.R. Brown, C. Flavin & H. Kane). Earthscan Publications, London, pp. 40–41.

Bryant, R.L. (1991) Putting politics first: the political ecology of sustainable development. *Global Ecology and Biogeography Letters* **1**, 164–166.

Bryant, R.L. (1992) Political ecology: an emerging research agenda in Third-World studies. *Political Geography* **11**, 12–36.

Bryant, R.L., Rigg, J. & Stott, P. (1993) The political ecology of southeast Asian forests: transdisciplinary discourses. *Global Ecology and Biogeography Letters* (Special Edition) **3** (4–6), 101–296.

Bryant, J.P., Swihart, R.K. Reichardt, P.B. & Newton, L. (1994) Biogeography of woody plant chemical defense against snowshoe have browsing: comparison of Alaska and eastern North America *Oikos* **70**, 385–95.

Burga, C.A. (1988) Swiss vegetation history during the last 18 000 years. *New Phytologist* **110**, 581–602.

Butcher, S.S., Charlson, R.J., Orians, G.H. & Wolfe, G.V. (eds) (1992) *Global Biogeochemical Cycles*. Academic Press, London.

Cahoon, D.R., Stocks, B.J., Levine, J.S., Cofer, W.R. & O'Neill, K.P. (1992) Seasonal distribution of African savanna fires. *Nature* **359**, 812–815.

Caldwell, M.M., Teramura, A.H. & Tevini, M. (1989) The changing solar ultraviolet climate and the ecological consequences for higher plants. *Trends in Ecology and Evolution* **4** (12), 363–367.

Campbell-Platt, G. (1988) The food we eat. *Inside Science (New Scientist)* No. 10, 1–4.

Cane, M.A., Eshel, G. & Buckland, R.W. (1994) Forecasting Zimbabwean maize yield using eastern equatorial Pacific sea surface temperature. *Nature* **370**, 204–205.

Chappellaz, J., Barnola, J.M., Raynaud, D., Korotkevich, Y.S. & Lorius, C. (1990) Ice-core record of atmospheric methane over the past 160 000 years. *Nature* **345**, 127–131.

Charlson, R.J., Anderson, T.L. & McDuff, R.E. (1992) The sulfur cycle. In: *Global Biogeochemical Cycles* (eds S.S. Butcher, R.J. Charlson, G.H. Orians & G.V. Wolfe). Academic Press, London, pp. 285–300.

Christensen, N.L. (1989) Landscape history and ecological change. *Journal of Forest History* **33** (3), 116–125.

Clements, F.E. (1936) Nature and structure of the climax. *Journal of Ecology* **24**, 252–284.

Cofer III, W.R., Levine, J.S., Winstead, E.L. & Stocks, B.J. (1991) New estimates of nitrous oxide emissions from biomass burning. *Nature* **349**, 689–691.

COHMAP Members (1988) Climatic changes of the last 18 000 years: observations and model simulations.

Science **241**, 1043–1052.

Cope, M.J. & Chaloner, W.G. (1985) Wildfire: an interaction of biological and physical processes. In: *Geological Factors and the Evolution of Plants* (ed. B.H. Tiffney). Yale University Press, New Haven, Conn., USA.

Coughenour, M.B., Ellis, J.E., Swift, D.M. *et al.* (1985) Energy extraction and use in a nomadic pastoral ecosystem. *Science* **230**, 619–624.

Cox, C.B. & Moore, P.D. (1993) *Biogeography: An Ecological and Evolutionary Approach*, 5th edn. Blackwell Science, Oxford.

Crawford, R.M.M. & Abbott, R.J. (1994) Pre-adaptation of arctic plants to climate change. *Botanica Acta* **107**, 271–278.

Creber, G.T. & Chaloner, W.G. (1985) Tree growth in the Mesozoic and Early Tertiary and the reconstruction of palaeoclimates. *Palaeogeography, Palaeoclimatology, Palaeoecology*, **52**, 35–60.

Cronon, W. (1983) *Changes in the Land. Indians, Colonists and the Ecology of New England*. Hill & Wang, New York.

Crowley, T.J. & North, G.R. (1991) *Paleoclimatology*. Oxford University Press, New York.

Culbard, E.B., Thornton, I., Watt, J., Wheatley, M., Moorcroft, S. & Thompson, M. (1988) Metal contamination in British urban dusts and soils. *Journal of Environmental Quality* **17**, 226–234.

D'Arrigo, R.D., Jacoby, G.C. & Fung, I.Y. (1987) Boreal forests and atmosphere–biosphere exchange of carbon dioxide. *Nature* **329**, 321–323.

Dansgaard, W., Johnsen, S.J., Reeh, N., Gundestrup, N., Clausen, G.B. & Hammer, C.U. (1975) Climate changes. Norsemen and modern man. *Nature* **225**, 24–28.

Dansgaard, W., White, J.W.C. & Johnsen, S.J. (1989) The abrupt termination of the Younger Dryas climate event. *Nature* **339**, 532–534.

Darwin, C. (1859) *The Origin of Species by Means of Natural Selection*. Murray, London.

Darwin, C. (1868) *The Variation of Animals and Plants under Domestication*. Murray, London.

Davis, M.B. (1983) Quaternary history of deciduous forests of North America and Europe. *Annals of the Missouri Botanical Garden* **70**, 550–563.

DeAngelis, D.L. & Waterhouse, J.C. (1987) Equilibrium and nonequilibrium concepts in ecological models. *Ecological Monographs* **57**, 1–21.

Denning, A.S., Fung, I.Y. & Randall, D. (1995) Latitudinal gradient of atmospheric CO_2 due to seasonal exchange with land biota. *Nature* **376**, 240–243.

Diamond, J.M. (1987) Human use of world resources. *Nature* **328**, 479–480.

Diamond, J. (1989) Were Neanderthals the first humans to bury their dead? *Nature* **340**, 344.

Dickinson, E. (1990) *The Complete Poems* (ed. Thomas H. Johnson). Faber & Faber, London & Boston.

Dittmann, S., Marencic, H., Roy, M. & Vogt, B. (1994) *Okosystem Forschung, Wattenmeer Niedersachsen*. National Park Niedersächsisches Wattenmeer, Wilhelmshaven.

Doll, R., Evans, H.J. & Darby, S.C. (1994) Paternal exposure not to blame. *Nature* **367**, 678–679.

Driscoll, C.M.H. (1993) Coordinated measurement of solar radiation. *Radiological Protection Bulletin* **143**, 10–14.

Dunbar, R. (1984) Scapegoat for a thousand deserts. *New Scientist* **104**, 30–33.

Ehleringer, J.R. (1978) Implications of quantum yield differences on the distribution of C_3 and C_4 grasses. *Oecologia* **31**, 255–267.

Ellenberg, H. (1988) *Vegetation Ecology of Central Europe*, 4th edn. Cambridge University Press, Cambridge.

Ellis, J.E. & Swift, D.M. (1988) Stability of African pastoral ecosystems: alternative paradigms and implications for development. *Journal of Range Management* **41**, 450–459.

Ellis, J.E., Coughenour, M.B. & Swift, D.W. (1993) Climate variability, ecosystem stability, and the implications for range and livestock development. In: *Range Ecology at Disequilibrium: New Models of Natural Variability and Pastoral Adaptation in African Savannas* (eds R.H. Behnke, I. Scoones & C. Kerven). ODI & IIED, London, pp. 31–41.

Elwin, H.V.H. (1964) *The Tribal World of Verrier Elwin: An Autobiography*. Oxford University Press, Oxford.

Emiliani, C. (1972) Quaternary paleotemperatures and the duration of high-temperature intervals. *Science* **178**, 398–401.

Engstrom, D.R., Hansen, B.C.S. & Wright, H.E. Jr (1990) A possible Younger Dryas record in Alaska. *Science* **250**, 1383–1385.

Fairbanks, R.G. (1989) A 17 000-year glacio-eustatic sea level record: influence of glacial melting rates on the Younger Dryas event and deep-ocean circulation. *Nature* **342**, 637–642.

Fells, I. (1992) Global environmental implications for future energy supply and use. In: *Managing the Human Impact on the Natural Environment* (ed. M. Newson). Belhaven Press, London, pp. 232–241.

Flavin, C. (1992a) Oil production falls. In: *Vital Signs* (eds L.R. Brown, C. Flavin & H. Kane). Earthscan, London, pp. 44–45.

Flavin, C. (1992b) Natural gas production climbs. In: *Vital Signs* (eds L.R. Brown, C. Flavin & H. Kane). Earthscan, London, pp. 46–47.

Flavin, C. & Lenssen, N. (1992) Nuclear power at a standstill. In: *Vital Signs* (eds. L.R. Brown, C. Flavin & H. Kane). Earthscan, London, pp. 48–49.

Flower, R.J. & Battarbee, R.W. (1983) Diatom evidence for recent acidification of two Scottish lochs. *Nature* **305**, 130–133.

Foster, D.R. (1988a) Disturbance history, community organisation and vegetation dynamics of the old-growth Pisgah Forest, southwestern New Hampshire,

U.S.A. *Journal of Ecology* **76**, 105–134.

Foster, D.R. (1988b) Species and stand response to catastrophic wind in central New England. *Journal of Ecology* **76**, 135–151.

Foster, D.R. (1993) Land-use history and forest transformations in central New England. In: *Humans as Components of Ecosystems* (eds M. McDonald & S. Pickett). Springer, New York, pp. 91–110.

Foucault, A. & Stanley, D.J. (1989) Late Quaternary palaeoclimatic oscillations in East Africa recorded by heavy minerals in the Nile delta. *Nature* **339**, 44–46.

Ganeshram, R.S., Pedersen, T.F., Calvert, S.E. & Murray, J.W. (1995) Large changes in oceanic nutrient inventories from glacial to interglacial periods. *Nature* **376**, 755–758.

Gasse, F., Arnold, M., Fontes, J.C. *et al.* (1991) A 13 000-year climate record from western Tibet. *Nature* **353**, 742–745.

Gates, D.M. (1993) *Climate Change and its Biological Consequences*. Sinauer, Sunderland, Massachusetts.

Gerlach, T. (1991) Etna's greenhouse pump. *Nature* **351**, 352.

Glantz, M.H. (ed.) (1994) *Drought Follows the Plough*. Cambridge University Press, Cambridge.

Gleason, H.A. (1926) The individualistic concept of the plant association. *Bulletin of the Torrey Botanical Club* **53**, 7–26.

Goldberg, E.D. (1993) Uses and abuses of ocean space. In: *Surviving with the Biosphere* (eds N. Polunin & J. Burnett). Edinburgh University Press, Edinburgh, pp. 112–119.

Gorham, E. (1990) Biotic impoverishment of the northern peatlands. In: *The Earth in Transition; Patterns and Processes of Biotic Impoverishment* (ed. G.M. Woodwell). Cambridge University Press, Cambridge, pp. 65–98.

Gould, S.J. (1989) *Wonderful Life*. Hutchinson.

Green, D.H., Eggins, S.M. & Yaxley, G. (1993) The other carbon cycle. *Nature* **365**, 210–211.

Gribbin, J. (1990) *Hothouse Earth: The Greenhouse Effect and Gaia*. Black Swan, London.

Grimm, E.C., Jacobson, G.L., Watts, W.A., Hansen, B.C.S. & Maasch, K.A. (1993) A 50 000-year record of climate oscillations from Florida and its temporal correlation with the Heinrich Events. *Science* **261**, 198–200.

GRIP (Greenland Ice Core Project) Members (1993) Climate instability during the last interglacial period recorded in the GIRP ice core. *Nature* **364**, 203–207.

Grun, R., Beaumont, P.B. & Stringer, C.B. (1990) ESR dating evidence for early modern humans at Border Cave in South Africa. *Nature* **344**, 537–539.

Gulland, J.A. (1993) Uses and abuses of the sea: fishing. In: *Surviving with the Biosphere* (eds N. Polunin & J. Burnett). Edinburgh University Press, Edinburgh, pp. 120–140.

Gunn, J.M. & Keller, W. (1990) Biological recovery of an acid lake after reductions in industrial emissions of sulphur. *Nature* **345**, 431–433.

Guzman, R. & Iltis, H.H. (1991) Biosphere reserve established in Mexico to protect rare maize relative. *Diversity* **7**, 82–84.

Hall, C.A.S., Ekdahl, C.A. & Wartenberg, D.E. (1975) A fifteen-year record of biotic metabolism in the Northern Hemisphere. *Nature* **255**, 136–138.

Harvard University (1975) *The Harvard Forest Models*. Harvard University Printing Office, Cambridge, Mass.

Hawksworth, D.L. (1994) Biodiversity in microorganisms and its role in ecosystem function. In: *Biodiversity and Global Change* (eds O.T. Solbrig, H.M. van Emden & P.G.W.J. van Oordt). CAB International, Wallingford, pp. 85–95.

Helas, G. (1995) Emissions of atmospheric trace gases from vegetation burning. *Philosophical Transactions of the Royal Society of London A* **351**, 297–312.

Herndl, G.J., Muller-Niklas, G. & Frick, J. (1993) Major role of ultraviolet-B in controlling bacterioplankton in the surface layer of the ocean. *Nature* **361**, 717–719.

Heusser, L.E. & Morley, J.J. (1990) Climatic change at the end of the last glaciation in Japan inferred from pollen` in three cores from the northwest Pacific ocean. *Quaternary Research* **34**, 101–110.

Hill, A., Ward, S., Deino, A., Curtis, G. & Drake, R. (1992) Earliest *Homo. Nature* **355**, 719–722.

Hodell, D.A., Curtis, J.H. & Brenner, M. (1995) Possible role of climate in the collapse of Classic Maya civilization. *Nature* **375**, 391–394.

Hogan, K.B., Hoffman, J.S. & Thompson, A.M. (1991) Methane on the greenhouse agenda. *Nature* **354**, 181–182.

Hoganson, J.W. & Ashworth, A.C. (1992) Fossil beetle evidence for climatic change 18 000–10 000 years BP in south-central Chile. *Quaternary Research* **37**, 101–116.

Holling, C.S. (1973) Resilience and stability of ecological systems. *Annual Review of Ecology and Systematics* **4**, 1–23.

Hooghiemstra, H. (1984) *Vegetation and Climatic History of the High Plain of Bogota, Columbia: a Continuous Record of the Last 3.5 Million Years*. Cramer, Vaduz, Leichtenstein.

Houghton, J.T., Jenkins, G.J. & Ephraums, J.J. (eds) (1990) *Climate Change—the IPCC Scientific Assessment*. Cambridge University Press, Cambridge.

Houghton, J.T., Callander, B.A. & Varney, S.K. (eds) (1992) Climate Change 1992. *The Supplementary Report to the IPCC Scientific Assessment*. Cambridge University Press, Cambridge.

Houghton, R.A. (1991) The role of forests in affecting the greenhouse gas composition of the atmosphere. In: *Global Climate Change and Life on Earth* (ed. R.L. Wyman). Routledge, Chapman & Hall, New York, pp. 43–55.

Hubbard, J.M. (1989) Photovoltaics today and tomorrow. *Science* **244**, 297–304.

Hughes, M.K., Kelly, P.M., Pilcher, J.R. & LaMarche, V.C. (eds) (1982) *Climate from Tree Rings*. Cambridge University Press, Cambridge.

Huntley, B. (1991) How plants respond to climate change: Migration rates, individualism and the consequences for plant communities. *Annals of Botany* **67** (Suppl. 1), 15–22.

Huntley, B. & Birks, H.J.B. (1983) *An Atlas of Past and Present Pollen Maps for Europe: 0–13 000 Years Ago*. Cambridge University Press, Cambridge.

Huttunen, A., Huttunen, R.-L., Vasari, Y., Panovska, H. & Bozilova, E. (1992) Late-glacial and Holocene history of flora and vegetation in the Western Rhodopes Mountains, Bulgaria. *Acta Botanica Fennica* **144**, 63–80.

Imbrie, J., Van Donk, J. & Kipp, N.G. (1973) Paleoclimatic investigation of a late Pleistocene Caribbean deep-sea core: comparison of isotopic and faunal methods. *Quaternary Research* **3**, 10–38.

Intergovernmental Panel on Climate Change (IPCC) (1992) *Climate Change 1992. The Supplementary Report to the IPCC Scientific Assessment*. Cambridge University Press, Cambridge.

IPCC (1990) *Climate Change: The IPCC Scientific Assessment*. Cambridge University Press, Cambridge.

Ives, J.D. & Messerli, B. (1989) *The Himalayan Dilemma: Reconciling Development and Conservation*. Routledge/UNU, London & New York.

Jacobson, G.L., Webb, T. III & Grimm, E.C. (1987) Patterns and rates of vegetation change during the deglaciation of eastern North America. In: *North America and Adjacent Oceans During the Last Deglaciation* (eds W.F. Ruddiman & H.E. Wright Jr.). Geological Society of America, The Geology of North America, v. K-3, pp. 277–288.

Jaffe, D.A. (1992) The nitrogen cycle. In: *Global Biogeochemical Cycles* (eds S.S. Butcher, R.J. Charlson, G.H. Orians & G.V. Wolfe). Academic Press, London, pp. 263–284.

Jahnke, R.A. (1992) The phosphorus cycle. In: *Global Biogeochemical Cycles* (eds S.S. Butcher, R.J. Charlson, G.H. Orians & G.V. Wolfe). Academic Press, London, pp. 301–315.

Jansen, E. & Sjoholm, J. (1991) Reconstruction of glaciation over the past 6 Myr from ice-borne deposits in the Norwegian Sea. *Nature* **349**, 600–603.

Jenkins, D.G. (1987) Was the Pliocene-Pleistocene boundary placed at the wrong stratigraphic level? *Quaternary Science Reviews* **6**, 41–42.

Jessen, K. & Farrington, A. (1938) The bogs at Ballybetagh, near Dublin, with remarks on Late-glacial conditions in Ireland. *Proceedings of the Royal Irish Academy* **44** B10, 205–260.

Jewitt, S. (1995) Europe's 'Others'? Forestry policy and practices in colonial and postcolonial India. *Environment and Planning D: Society and Space* **13**, 67–90.

Johnsen, S.J., Clausen, H.B., Dansgaard, W. *et al.* (1992) Irregular glacial interstadials recorded in a new Greenland ice core. *Nature* **359**, 311–313.

Johnson, B. (1984) *The Great Fire of Borneo. Report of a visit to Kilimantan-Timur a year later, May 1984*. World Wildlife Fund.

Jones, P.D. (1990) The climate of the past 1000 years. *Endeavour* **14**, 129–136.

Jones, P.D. & Wigley, T.M.L. (1990) Global warming trends. *Scientific American* **263** (2), 66–73.

Jouzel, J., Barkov, N.I., Barnola, J.M. *et al.* (1993) Extending the Vostok ice-core record of palaeoclimate to the penultimate glacial period. *Nature* **364**, 407–412.

Jowett, B. (ed.) (1892) *The Dialogues of Plato*. Oxford University Press, London.

Kane, H. (1992) Fish catch falls. In: *Vital Signs 1992–1993* (eds L.R. Brown, C. Flavin & H. Kane), pp. 30–31.

Kauppi, P.E., Mielikainen, K. & Kuusela, K. (1992) Biomass and carbon budget of European forests, 1971 to 1990. *Science* **256**, 70–74.

Keeley, S.C. & Mooney, H.A. (1993) Vegetation in Western North America, Past and Future. In: *Earth System Responses to Global Change* (eds H.A. Mooney, E.F. Fuentes & B.I. Kronberg). Academic Press, San Diego, pp. 209–237.

Keeling, C.D., Whorf, T.P., Wahlen, M. & van der Plicht, J. (1995) Interannual extremes in the rate of rise of atmospheric carbon dioxide since 1980. *Nature* **375**, 666–670.

Keeling, R.F., Piper, S.C. & Heimann, M. (1996) Global and hemispheric CO_2 sinks deduced from changes in atmospheric O_2 concentrations.

Kelche, W.R., Stone, C.R., Laws, S.C., Gray, L.E., Kemppainen, J.A. & Wilson, E.M. (1995) Persistent DDT metabolite *p,p'*-DDE is a potent androgen receptor antagonist. *Nature* **375**, 581–585.

Keller, M., Veldkamp, E., Weitz, A.M. & Reiners, W.A. (1993) Effect of pasture age on soil trace-gas emissions from a deforested area of Costa Rica. *Nature* **365**, 244–246.

Kelly, P.M. & Wigley, T.M.L. (1992) Solar cycle length, greenhouse forcing and global climate. *Nature* **360**, 328–330.

Kenny A. (1994) The earth is fine; the problem is the Greens. *The Spectator*, 12 March 1994, 9–11.

Kerr, J.B. & McElroy, C.T. (1993) Evidence for large upward trends of ultraviolet-B radiation linked to ozone depletion. *Science* **262**, 1032–1034.

Korhola, A. (1992) Mire induction, ecosystem dynamics and lateral extension on raised bogs in the southern coastal area of Finland. *Fennia* **170**, 25–94.

Kudrass, H.R., Erlenkeuser, H., Vollbrecht, R. & Weiss, W. (1991) Global nature of the Younger Dryas cooling event inferred from oxygen isotope data from Sulu Sea cores. *Nature* **349**, 406–409.

Kuhlbusch, T.A., Lobert, J.M., Crutzen, P.J. & Warneck, P. (1991) Molecular nitrogen emissions from denitrification during biomass burning. *Nature* **351**, 135–137.

Kuhn, T.S. (1962, 1970) *The Structure of Scientific Revolutions*. University of Chicago Press, Chicago.

Kukla, G. (1987) Loess stratigraphy in Central China and correlation with an extended oxygen isotope stage scale. *Quaternary Science Reviews* **6**, 191–219.

Kuo, C., Lindberg, C. & Thomson, D.J. (1990) Coherence established between atmospheric carbon dioxide and global temperature. *Nature* **343**, 709–714.

Kutzbach, J.E. (1981) Monsoon climate of the early Holocene: climate experiment with the earth's orbital parameters for 9000 years ago. *Science* **214**, 59–61.

Lamb, H.H. (1972) *Climate Present, Past and Future*. Methuen, London.

Lambeck, K. & Nakada, M. (1992) Constraints on the age and duration of the last interglacial period and on sea-level variations. *Nature* **357**, 125–128.

Landsberg, H.E. (1970) Man-made climatic changes. *Science* **170**, 1265–1274.

Langner, J., Rodhe, H., Crutzen, P.J. & Zimmermann, P. (1992) Anthropogenic distribution of tropospheric sulphate aerosol. *Nature* **359**, 712–716.

Lara, A. & Villalba, R. (1993) A 3620-year temperature record from Fitzroya cupressoides tree rings in southern South America. *Science* **260**, 1104–06.

Lazcano, A. (1992) Origins of life: the historical development of recent theories. In: *Environmental Evolution* (eds L. Margulis & L. Olendzenski). M.I.T. Press, Cambridge, Mass.

Lean, J. & Warrolow, D.A. (1989) Simulation of the regional climatic impact of Amazon deforestation. *Nature* **342**, 411–413.

Learmonth, A. (1988) *Disease Ecology*. Basil Blackwell, Oxford.

Ledru, M.-P. (1993) Late Quaternary environmental and climatic changes in Central Brazil. *Quaternary Research* **39**, 90–98.

Lester, R.N. (1989) Evolution under domestication involving disturbance of genic balance. *Euphytica* **44**, 125–132.

Levesque, A.J., Mayle, F.E., Walker, I.R. & Cwynar, L.C. (1993) A previously unrecognized Late-glacial cold event in eastern North America. *Nature* **361**, 623–626.

Levine, J.S. (1991) *Global Biomass Burning: Atmospheric, Climatic and Biospheric Implications*. MIT Press, Cambridge M.A.

Lewin, R. (1984) *Human Evolution: An Illustrated Introduction*. Blackwell Scientific Publications, Oxford.

Lewin, R. (1988) A new tool maker in the hominid record. *Science* **240**, 724–725.

Lezine, A.M. (1989) Late Quaternary vegetation and climate in the Sahel. *Quaternary Research* **32**, 317–334.

Likens, G.E. & Bormann, F.H. (1995) *Biogeochemistry of a Forested Ecosystem*. 2nd edition. Springer, New York.

Lobert, J.M., Scharffe, D.H., Hao, W.M. & Crutzen, P.J.

(1990) Importance of biomass burning in the atmospheric budgets of nitrogen-containing gases. *Nature* **346**, 552–554.

Loeb, R.E. (1989) The ecological history of an urban park. *Journal of Forest History* **33** (3), 134–143.

Long, S.P. & Hall, D.O. (1987) Nitrogen cycles in perspective. *Nature* **329**, 584–585.

Lorius, C., Jouzel, J., Raynaud, D., Hansen, J. & Le Treut, H. (1990) The ice-core record: climate sensitivity and future greenhouse warming. *Nature* **347**, 139–145.

Lotter, A.F. (1991) Absolute dating of the late-glacial period in Switzerland using annually laminated sediments. *Quaternary Research* **35**, 321–330.

Lovelock, J. (1988) *The Ages of Gaia*. Oxford University Press, Oxford.

Lovelock, J. (1995) *Gaia: A New Look at Life on Earth*. 3rd edition. Oxford University Press, Oxford.

Lowe, J.J. (1992) Lateglacial and early Holocene lake sediments from the northern Apennines, Italy—pollen stratigraphy and radiocarbon dating. *Boreas* **21**, 193–208.

Lowe, J.J. & Walker, M.J.C. (1984) *Reconstructing Quaternary Environments*. Longman, London.

Lynch, J.F. (1992) Distribution of overwintering Nearctic migrants in the Yucatan Peninsula II. Use of native and human-modified vegetation. In: *Ecology and Conservation of Neotropical Migrant Landbirds* (eds J.M. Hagan & D.W. Johnston). Smithsonian Institute, Washington, pp. 178–195.

MacCracken, M.C. (1995) The evidence mounts up. *Nature* **376**, 645–646.

MacDonald, G.M., Edwards, T.W.D., Moser, K.A., Pienitz, R. & Smol, J.P. (1993) Rapid response of treeline vegetation and lakes to past climate warming. *Nature* **361**, 243–246.

MacDonald, R.B. & Hall, F.G. (1980) Global crop forecasting. *Science* **208**, 670–678.

MacKenzie, D. (1991) The West pays up for Third World seeds. *New Scientist* 11 May, 18–19.

MacKenzie, D. (1994) How safe is safe? *New Scientist* **144**, 12–13 (27 August 1994).

Mackie, R.M. (1993) Ultraviolet radiation and the skin. *Radiological Protection Bulletin* **143**, 5–9.

Mangerud, J., Sonstegaard, E. & Sejrup, H.-P. (1979) Correlation of the Eemian (interglacial) Stage and the deep-sea oxygen-isotope stratigraphy. *Nature* **277**, 189–192.

Martinson, D.G., Pisias, N.G., Hays, J.D., Imbrie, J., Moore, T.C. & Shackleton, N.J. (1987) Age dating and the orbital theory of ice ages: development of a high-resolution 0 to 300 000-year chronostratigraphy. *Quaternary Research* **27**, 1–29.

Maslin, M. (1993) Waiting for the polar meltdown. *New Scientist* **139** (1889), 36–41.

Matthews, M.A. (1959) The earth's carbon cycle. *New Scientist* 8 October, 644–646.

May, R.M. (1976) In: *Theoretical Ecology: Principles and Applications* (ed. R.M. May). Blackwell Scientific

Publications, Oxford, pp. 142–162.

May, R.M. (1977) Thresholds and breakpoints in ecosystems with a multiplicity of stable states. *Nature* **269**, 471–477.

Mayewski, P.A., Lyons, W.B., Spencer, M.J., Twickier, M.S., Buck, C.F. & Whitlow, S. (1990) An ice-core record of atmospheric response to anthropogenic sulphate and nitrate. *Nature* **346**, 554–556.

Mayewski, P.A., Meeker, L.D., Whitlow, S. *et al.* (1993) The atmosphere during the Younger Dryas. *Science* **261**, 195–97.

McClure, H.A. (1976) Radiocarbon chronology of late-Quaternary lakes in the Arabian Desert. *Nature* **263**, 755–756.

McManus, J.F., Bond, G.C., Broeker, W.S., Johnsen, S., Labeyrie, L. & Higgins, S. (1994) High-resolution climate records from the North Atlantic during the last interglacial. *Nature* **371**, 326–29.

McElroy, M. (1992) Comparison of planetary atmospheres: Mars Venus and Earth. In: *Environmental Evolution* (eds L. Margulis & L. Olendzenski). M.I.T. Press, Cambridge, Mass.

Mellanby, K. (1967) *Pesticides and Pollution*. Collins New Naturalist, London.

Mercer, J.H. (1969) The Allerød oscillation: a European climate anomaly? *Arctic and Alpine Research* **1**, 227–234.

Mercier, N., Valladas, H., Joron, J.-L., Reyss, J.-L., Leveque, F. & Vandermeersch, B. (1991) Thermoluminescence dating of the late Neanderthal remains from Saint-Cesaire. *Nature* **351**, 737–739.

Miller, G.H., Hollin, J.T. & Andrews, J.T. (1979) Aminostratigraphy of U.K. Pleistocene deposits. *Nature* **281**, 539–543.

Mohnen, V.A. (1988) The challenge of acid rain. *Scientific American* **259** (2), 14–22.

Moore, P.D. (1981) The varied ways plants tap the sun. *New Scientist* **81**, 394–397.

Moore, P.D. (1994) Does plankton hold key to carbon budgets? *Science Watch* **5** (6), 7–8.

Moore, P.D., Webb, J.A. & Collinson, M.E. (1991) *Pollen Analysis*, 2nd edn. Blackwell Science, Oxford.

Morel, F.M.M. & Chisholm, S.W. (1991) *Limnology and Oceanography*; preface, **36** (8).

Mukherjee, N. (1992) Greenhouse gas emissions and the allocation of responsibility. *Environment and Urbanization* **4**, 89–98.

Myers, N. (1994) *The Gaia Atlas of Planet Management*, 2nd edn. Gaia Books, London.

Nandy, A. (1983) *The Intimate Enemy: Loss and Recovery of Self under Colonialism*. Oxford University Press, New Delhi.

NASA. (1988), *Earth Systems Science*. National Aeronautics and Space Administration, Washington D.C.

NERC (1989) *Our Future World*. 2nd edn. NERC, Swinden, UK.

Nevitt, G.A., Veit, R.R. & Kareiva, P. (1995) Dimethyl sulphide as a foraging cue for Antarctic Procellariiform seabirds. *Nature* **376**, 680–682.

Newell, R.E., Reichle, H.G. & Seiler, W. (1989) Carbon monoxide and the burning earth. *Scientific American* **261** (4), 58–64.

Nisbet, E.G. (1991) *Living Earth—A Short History of Life and its Home*. Harper Collins, London.

Nisbet, E.G. (1991) *Leaving Eden: To Protect and Manage the Earth*. Cambridge University Press, Cambridge.

Nisbet, E.G. & Fowler, C.M.R. (1995) Is metal disposal toxic to deep oceans? *Nature* **375**, 715.

Northrup, R.R., Zengshou, Y., Dahlgren, R.A. & Vogt, K.A. (1995) Polyphenol control of nitrogen release from pine litter. *Nature* **377**, 227–229.

Oechel, W.C., Hastings, S.J., Vourlitis, G., Jenkins, M., Riechers, G. & Grulke, N. (1993) Recent change of Arctic tundra ecosystems from a net carbon dioxide sink to a source. *Nature* **361**, 520–523.

Ozenda, P. & Borel J.-L. (1995) Possible responses of mountain vegetation to a global climatic change: the case of the Western Alps. In: *Potential Ecological Impacts of Climate Change in the Alps and Fennoscandian Mountains* (eds A. Guisan, J.I. Holten, R. Spichiger & L. Tessier). Botanical Garden of the City of Geneva, Switzerland, pp. 137–144.

Pachur H.J. & Kropelin S. (1987) Wadi Howar; paleoclimatic evidence from an extinct river system in the southeastern Sahara. *Science* **237**, 298–300.

Park, C.C. (1992) *Tropical Rainforests*. Routledge, London & New York.

Parry, M. (1990) *Climate Change and World Agriculture*. Earthscan, London.

Parry, M.L. & Swaminathan, M.S. (1992) Effects of climate change on food production. In: *Confronting Climate Change* (ed. I.M. Mintzer). Cambridge University Press, Cambridge, pp. 113–126.

Pauly, D. & Christensen, V. (1995) Primary production required to sustain global fisheries. *Nature* **374**, 255–257.

Payette, S., Filion, L., Delwaide, A. & Begin, C. (1989) Reconstruction of tree-line vegetation response to long-term climate change. *Nature* **341**, 429–432.

Pearce, D., Markandya, A. & Barbier, E. (1989) *Blueprint for a Green Economy*. Earthscan, London.

Pearce, F. (1987) Acid rain. *New Scientist (Inside Science)*, 5 November 1987, pp. 1–4.

Pearman, G.I. & Fraser, P.J. (1988) Sources of increased methane. *Nature* **332**, 489–90.

Pielou, E.C. (1991) *After the Ice Age: The Return of Life to Glaciated North America*. University of Chicago Press, Chicago.

Penck, A. & Bruckner, E. (1909) *Die Alpen im Eiszeitalter*. Tauchnitz, Leipzig.

Pigott, C.D. & Huntley, J.P. (1981) Factors controlling the distribution of *Tilia cordata* at the northern limits of its geographical range. III. Nature and causes of seed sterility. *New Phytologist* **87**, 817–839.

Pimentel, D., Harvey, C., Resosudarmo, P., Sinclair, K., Kurz, D., McNair, M., Crist, S., Shpritz, L., Fitton , L., Saffouri, R. & Blair, R. (1995) Environmental and economic costs of soil erosion and conservation benefits. *Science* **267**, 1117–1123.

Pine, S.J. (1989) *Fire on the Rim: a Firefighter's Season at the Grand Canyon*. Ballantine Books, New York.

Pittock, A.B., Whetton, P. & Wang, Y. (1994) Climate and food supply. *Nature* **371**, 25.

Ponamperuma, C. (1992) Cosmochemical evolution and the origins of life. In: *Environmental Evolution* (eds L. Margulis & L. Olendzenski). M.I.T. Press, Cambridge, Mass.

Porter, S.C. (1986) Pattern and forcing of northern hemisphere glacier variations during the last millennium. *Quaternary Research* **26**, 27–48.

Rackham, O. (1986) *The History of the Countryside*. M. Dent, London.

Ramaswamy, V. (1992) Explosive start to the last ice age. *Nature* **359**, 14.

Rasmussen, R.A. & Khalil, M.A.K. (1983) Global production of methane by termites. *Nature* **301**, 700–702.

Rampino, M.R. & Self, S. (1992) Volcanic winter and accelerated glaciation following the Toba super-eruption. *Nature* **359**, 50–52.

Raynaud, D. & Barnola, J.M. (1985) An Antarctic ice core reveals atmospheric CO_2 variations over the past few centuries. *Nature* **315**, 309–311.

Reille, M., de Beaulieu, J.L. (1995) Long Pleistocene pollen records from the Praclaux Crater, south-central France. *Quaternary Research* **44**, 205–215.

Reilly, J. (1994) Crops and climate change. *Nature* **367**, 118–119.

Rind, D., Rosenzweig, C. & Goldberg, R. (1992) Modelling the hydrological cycle in assessments of climate change. *Nature* **358**, 119–122.

Ritchie, J.C. (1986) Climate change and vegetation response. *Vegetation* **67**, 65–74.

Robbins, C.S., Sauer, J.R., Greenberg, R.S. & Droege, S. (1989) Population declines in North American birds that migrate to the Neotropics. *Proceedings of the National Academy of Science* **86**, 7658–7662.

Roberts, L. (1989) How fast can trees migrate? *Science* **243**, 735–737.

Roberts, N. (1990) Ups and downs of African lakes. *Nature* **346**, 107.

Roberts, R.G., Jones, R. & Smith, M.A. (1990) Thermoluminescence dating of a 50 000-year-old human occupation site in northern Australia. *Nature* **345**, 153–156.

Rogers, J.W. (1993) *A History of the Earth*. Cambridge University Press, Cambridge.

Rogers, R.A., Martin, L.D. & Nicklas, T.D. (1990) Ice-age geography and the distribution of native North American languages. *Journal of Biogeography* **17**, 131–143.

Rosenberg, R., Elmgren, R., Fleischer, S., Jonsson, P., Persson, G. & Dahlin, H. (1990) Marine eutrophication case study in Sweden. *Ambio* **19** (3), 102–108.

Rosenzweig, C. (1985) Potential CO_2-induced climate effects on North American wheat-production regions. *Climate Change* **7**, 367–389.

Rosenzweig, C. (1994) Maize suffers a sea-change. *Nature* **370**, 175–176.

Rosenzweig, C. & Parry, M.L. (1994) Potential impact of climate change on world food supply. *Nature* **367**, 133.

Roset, J.P. (1984) The prehistoric rock paintings of the Sahara. *Endeavour* **8**, 75–84.

Rossignol-Strick, M., Nesteroff, W., Olive, P. & Vergnaud-Grazzini, C. (1982) After the deluge; Mediterranean stagnation and sapropel formation. *Nature* **295**, 105–110.

Rowlson, D.S., Goulding, K.W.T. (1995) Agriculture, the nitrogen cycle and nitrate. In: *Nitrate Control Policy, Agriculture and Land Use* (ed. J. North). University of Cambridge, Cambridge, pp. 5–25.

Ruddiman, W.F. & McIntyre, A. (1981) The North Atlantic during the last deglaciation. *Palaeogeography, Palaeoclimatology, Palaeoecology*, **35**, 145–214.

Rutherford, M.C. & Westfall, R.H. (1994) *Biomes of Southern Africa: an Objective Categorization*. National Botanical Institute, Pretoria.

Sabloff, J.A. (1995) Drought and decline. *Nature* **375**, 357.

Sage, R.F. (1995) Was low atmospheric CO_2 during the Pleistocene a limiting factor for the origin of agriculture? *Global Change Biology* **1**, 93–106.

Sahagian, D.L., Schwartz, F.W. & Jacobs, D.K. (1994) Direct anthropogenic contributions to sea level rise in the twentieth century. *Nature* **367**, 54–57.

Said, E.W. (1993) *Culture and Imperialism*. Chatto & Windus, London.

Savage, V. (1984) *Western Impressions of Nature and Landscape in South-East Asia*. Singapore University Press, Singapore.

Schell, D.M. (1983) Carbon-13 and carbon-14 abundances in Alachan aquatic organisms: delayed production from peat in Arctic food webs. *Science* **219**, 1068–1071.

Schimel, D.S. (1995) Terrestrial ecosystems and the carbon cycle. *Global Change Biology* **1**, 77–91.

Schindler, D.W. (1988) Effects of acid rain on freshwater ecosystems. *Science* **239**, 149–157.

Schlesinger, W.H. (1990) Vegetation an unlikely answer. *Nature* **348**, 679.

Schmidt, K. (1994) Scientists count a rising tide of whales in the seas. *Science* **263**, 25–26.

Schneider, S.H. (1989) The changing climate. *Scientific American* **261** (3), 38–47.

Schneider, S.H. & Thompson, S.L. (1988) Simulating the climatic effects of nuclear war. *Nature* **333**, 221–227.

Schulze, E.-D. (1989) Air pollution and forest decline in

a spruce (*Picea abies*) forest. *Science* **244**, 776–783.

Schwarzback, S.E. (1995) CFC alternatives under a cloud. *Nature* **376**, 297–298.

Scorer, R.S. (1993) Recording fires by satellite. *Nature* **365**, 215.

Seddon, B. (1971) *Introduction to Biogeography*. Duckworth, London.

Serebryanny, L.R. & Tishkov, A.A. (1996) Quaternary environmental changes and ecosystems of the European Arctic. In Gilmanov, T., Holten, J., Maxwell, B. Oechal, W.C. & Sveinbjornsson, B. (eds), *Global Change and Arctic Terrestrial Ecosystems*, Norwegian Institute for Nature Research, Trondheim.

Sharpe, R.M. (1995) Reproductive biology: another DDT connection. *Nature* **375**, 538–539.

Shennan, (1989) *Journal of Quaternary Science* **4**, 77–89.

Shugart, Jr., H.H. (1984) *A Theory of Forest Dynamics: The Ecological Implications of Forest Succession Models*. Springer, New York.

Siegenthaler, U. & Sarmiento, J.L. (1993) Atmospheric carbon dioxide and the ocean. *Nature* **365**, 119–125.

Sifrin, G. (1983) Power scheme will raise the Dead Sea. *New Scientist* **97**, 311–313.

Simons, E.L. (1989) Human origins. *Science* **245**, 1343–1350.

Simonich, S.L. & Hites, R.A. (1995) Global distribution of persistent organochlorine compounds. *Science* **269**, 1851–1854.

Singh, G., Joshi, R.D., Chopra, S.K. & Singh, A.B. (1974) Late-Quaternary history of vegetation and climate of the Rajasthan Desert, India. *Philosophic Transactions of the Royal Society London B* **267**, 467–501.

Skye, E. (1989) Changes to climate and Flora of Hopen Island during the last 110 years. *Artic* **42**, 323–332.

Smil, V. (1990) Nitrogen and phosphorus. In: *The Earth as Transformed by Human Action* (ed. B.L. Turner). Cambridge University Press, Cambridge, pp. 423–436.

Smith, T.M. & Shugart, H.H. (1993) The transient response of terrestrial carbon storage to a perturbed climate. *Nature* **361**, 523–25.

Spicer, R.A., Rees, P. McA & Chapman, J.L. (1993) Cretaceous phytogeography and climate signals. *Philosophical Transactions of the Royal Society B* **341**, 277–286.

Steele, L.P., Dlugokencky, E.J., Land, P.M., Tans, P.P., Martin, R.C. & Masarie, K.A. (1992) Slowing down of the global accumulation of atmospheric methane during the 1980s. *Nature* **358**, 313–316.

Stewart, W.N. & Rothwell, G.W. (1993) *Paleobotany and the Evolution of Plants*, 2nd edn. Cambridge University Press, Cambridge.

Stott, P. (1991) Harvard Forest. *Global Ecology and Biodiversity Letters*, **1**, 99–101

Street-Perrott, F.A. & Harrison, S.P. (1985) Late-levels and climate reconstruction. In: *Paleoclimate Analysis and Modeling* (ed. A.D. Hecht). John Wiley, New York, pp. 291–340.

Street-Perrott, F.A. & Perrott, R.A. (1990) Abrupt climatic fluctuations in the tropics—the influence of Atlantic Ocean circulation. *Nature* **343**, 607–612.

Stringer, C.B. & Andrews, P. (1988) Genetic and fossil evidence for the origin of modern humans. *Science* **239**, 1263–1268.

Stringer, C.B. & Grun, R. (1991) Time for the last Neanderthals. *Nature* **351**, 701–702.

Stringer, C.B., Grun, R., Schwarcz, H.P. & Goldberg, P. (1989) ESR dates for the hominid burial site of Es Skhul in Israel. *Nature* **338**, 756–758.

Sundquist, E.T. & Miller, G.A. (1980) Oil shales and carbon dioxide. *Science* **208**, 740–741.

Swiss Federal Office of Environment, Forests and Landscape (FOEFL) (1994) *Global Warming and Switzerland: Foundations for a National Strategy*. Berne, Switzerland.

Tallis, J.H. (1991) *Plant Community History*. Chapman & Hall, London.

Tansley, A.G. (1935) The use and abuse of vegetational concepts and terms. *Ecology* **16**, 284–307.

Teeri, J.A. & Stowe, L.G. (1976) Climatic patterns and the distribution of C_4 grasses in North America. *Oecologia* **23**, 1–12.

Thery, I., Gril, J., Vernet, J.L., Meignen, L. & Maury, J. (1995) First use of coal. *Nature* **373**, 480–481.

Thomas, D.S.G. & Middleton, N.J. (1994) *Desertification, Exploding the Myth*. Wiley, Chichester.

Thomas, V. & Spiro, T. (1994) Emissions and exposure to metals: cadmium and lead. In: *Industrial Ecology and Global Change* (eds R. Socolow, C. Andrews, F. Berkhout & V. Thomas). Cambridge University Press, Cambridge, pp. 297–318.

Thompson, L.G., Mosley-Thompson, E., Davis, M.E. *et al.* (1989) Holocene—Late Pleistocene climatic ice core records from Qinghai—Tibetan Plateau. *Science* **246**, 474–477.

Tivy, J. (1990) *Agricultural Ecology*. Longman, London.

Tobias, P.V. (1993) Earliest *Homo* not proven. *Nature* **361**, 307.

Tolba, M.K. (1992) *Saving Our Planet: Challenges and Hopes*. Chapman and Hall, London.

Tolba, M.K. & El-Kholy, O.A. (eds) (1992) *The World Environment 1972–1992*. Chapman & Hall, London.

Torr, M., (1984) A new image of the atmosphere. *New Scientist*.

Turco, R.P., Toon, O.B., Ackerman, T.P., Pollack, J.B. & Sagan, C. (1990) Climate and smoke: an appraisal of nuclear winter. *Science* **247**, 166–176.

Turner, C. (1970) The middle Pleistocene deposits at Marks Tey, Essex. *Philosophic Transactions of the Royal Society, London B* **257**, 373–440.

UNEP (1987a) *The Greenhouse Gases*. UNEP/GEMS Environmental Library No. 1. UNEP. Nairobi, Kenya.

UNEP (1987b) *The Ozone Layer*. UNEP/GEMS Environment Library **2**, 1–36.

UNEP (1991) *Environmental Data Report*. 3rd edition 1991–2. Blackwell Scientific Publications, Oxford.

UNEP (1992) *World Atlas of Desertification*. Edward Arnold, Sevenoaks.

UNEP (1993) *Environmental Data Report 1993–94*. Blackwell Scientific Publications, Oxford.

Vayda, A.P. & Padoch, C. (1989) Review of Shugart, Jr., H.H. 1984. A Theory of Forest Dynamics: The Ecological Implications of Forest Succession Models. Springer, New York. *Journal of Forest History* **33** (3), 151–152.

Vitousek, P.M., Ehrlich, P.R., Ehrlich A.H. & Matson, P.A. (1986) Human appropriation of the products of photosynthesis. *BioScience* **36**, 368–373.

Walker & Low (1994).

Wallace, J.M. & Vogel, S. (1994) El Nino and Climate Prediction. *Reports to the Nation on our Changing Planet* **3**, 1–24. Office of Global Programs, National Oceanic and Atmospheric Administration: US Department of Commerce, Washington DC, USA.

Warrick, R.A. & Barrow, E.M. (1990) Climate and sea level change: A perspective. *Outlook on Agriculture* (CAB International) **19** (1), 5–8.

Warrick, R.A. & Oerlemans, H. (1990) Sea level rise. In: *Climate Change—the IPCC Scientific Assessment* (eds J.T. Houghton, G.J. Jenkins & J.J. Ephraums). Cambridge University Press, Cambridge.

Watson, A.J. *et al.* (1994) Minimal effect of iron fertilization on sea-surface carbon dioxide concentrations. *Nature* **371**, 143–145.

Watson, H. (1971) Size, structure and activity of the productive system of crops. In: *Potential Crop Production* (eds P.F. Wareing & J.P. Cooper). Heinemann, London, pp. 76–88.

Watts, W.A. (1980) Regional variation in the response of vegetation to late-glacial climatic events in Europe. In: *Studies in the Late-glacial of North-west Europe* (eds. J.J. Lowe, J.M. Gray & J.E. Robinson). Pergamon Press, Oxford, pp. 1–21.

Webb, J. (1994) Can we learn to love the wind? *New Scientist* **143**, 12–14 (16 July 1994).

West, R.G. (1977) *Pleistocene Geology and Biology* 2nd edn. Longman, London.

West, R.G. (1991) *Pleistocene Palaeoecology of Central Norfolk*. Cambridge University Press, Cambridge.

Wetherald, R.T. (1991) Changes of temperature and hydrology caused by an increase of atmospheric carbon dioxide as predicted by general circulation models. In: *Global Climate Change and Life on Earth* (ed. R.L. Whyman). Routledge, Chapman & Hall, New York, pp. 1–17.

White, R.M. (1990) The great climate debate. *Scientific American*, **263** (1), 18–25.

Whiting, G.J. & Chanton, J.P. (1993) Primary production control of methane emission from wetlands. *Nature* **364**, 794–795.

Whitmore, T.C. (ed.) (1987) *The Biogeographical Evolution of the Malay Archipelago*. Clarendon Press, Oxford.

Whitmore, T.C. (1990) *An Introduction to Tropical Rain Forests*. Clarendon Press, Oxford.

Whitney, G.G. (1994) *From Coastal Wilderness to Fruited Plain: A History of Environmental Change in Temperate North America from 1500 to the Present*. Cambridge University Press, Cambridge.

Wiens, J. (1984) On understanding a non-equilibrium world: myth and reality in community patterns and processes. In: *Ecological Communities: Conceptual Issues and the Evidence* (eds D.R. Strong, D. Simberloff, L.G. Abele & A.B. Thistle). Princeton University Press, Princeton, N.J., pp. 439–457.

Wigley, T.M.L. (1989) Measurement and prediction of global warming. In: *Ozone Depletion: Health and Environmental Consequences* (eds R.R. Jones & T. Wigley). John Wiley, Chichester, pp. 85–97.

Wigley, T.M.L. & Raper, S.C.B. (1990) Natural variability of the climate system and detection of the greenhouse effect. *Nature* **344**, 324–27.

Wigley, T.M.L., Ingram, M.J. & Farmer, G. (eds) (1981) *Climate and History: Studies in Past Climates and Their Impact on Man*. Cambridge University Press, Cambridge.

Winker, K., Warner, D.W. & Weisbrod, A.R. (1992) The Northern Waterthrush and Swainson's Thrush as transients at a temperate inland stopover site. In: *Ecology and Conservation of Neotropical Migrant Landbirds* (eds J.M. Hagan & D.W. Johnston). Smithsonian Institute, Washington, pp. 384–402.

Winstanley, D. (1973) Rainfall patterns and general atmospheric circulation. *Nature* **245**, 190–194.

Wittwer, S.H. (1985) Carbon Dioxide levels in the Biosphere – Effects on Plant Productivity. *Critical Reviews in Plant Science* **2**, 171–198.

WOCE. (1993) *WOCE Report*. WOCE Canada Secretariat, Department of Oceanography, University of British Columbia, Vancouver, Canada.

Wolff, C.G. (1988) *Emily Dickinson*. (Radcliffe Biography Series). Addison-Wesley, Reading Mass.

Wood, B. (1992) Origin and evolution of the genus *Homo*. *Nature* **355**, 783–90.

Wood, B. (1994) The oldest hominid yet. *Nature* **371**, 280–281.

Woodward, F.I. (1987a) Stomatal numbers are sensitive to increases in CO_2 from pre-industrial levels. *Nature* **327**, 617–618.

Woodward, F.I. (1987b) *Climate and Plant Distribution*. Cambridge University Press, Cambridge.

Woodward, F.I. & Bazzaz, F.A. (1988) The responses of stomatal density to CO_2 partial pressure. *Journal of Experimental Botany* **39**, 1771–1781.

World Commission on Environment and Development (1987) *Our Common Future*. Oxford University Press, Oxford.

World Resources Institute (1991). *World Resources 1990–91*. World Resources Institute, Washington.

Wright, H.E. Jr. (1992) Patterns of Holocene climatic change in the Midwestern United States. *Quaternary Research* **38**, 129–134.

Wright, H.E. Jr (1989) The amphi-Atlantic distribution of the Younger Dryas. *Quaternary Science Reviews* **8**, 295–306.

Yechieli, Y., Magaritz, M., Levy, Y. *et al.* (1993) Late Quaternary geological history of the Dead Sea area, Israel. *Quaternary Research* **39**, 59–67.

Young, M.D. (1992) *Sustainable Investment and Resource Use. Equity, Environmental Integrity and Economic Efficiency.* (MAB Series Vol. 9). UNESCO and Parthenon, Paris.

Young, M.D. & Solbrig, O.T. (1992) *Savanna, Management for Ecological Sustainability, Economic Profit and Social Equity.* (MAB Digest 13). UNESCO, Paris.

Zhisheng, A., Porter, S.C., Weijian, Z. *et al.* (1993) Episode of strengthened summer monsoon climate of Younger Dryas age on the loess plateau, Central China. *Quaternary Research* **39**, 45–54.

Index